T0296781

Geoinformatics for Geosciences

Advanced Geospatial Analysis using RS, GIS, and Soft Computing

Earth Observation Series

Geoinformatics for Geosciences

Advanced Geospatial Analysis using RS, GIS, and Soft Computing

Edited by

Nikolaos Stathopoulos
Operational Unit "BEYOND Centre for Earth Observation Research and Satellite Remote Sensing", Institute for Astronomy, Astrophysics, Space Applications and Remote Sensing, National Observatory of Athens, Greece

Andreas Tsatsaris
Department of Surveying and Geoinformatics Engineering, University of West Attica (UniWA), Athens, Greece

Kleomenis Kalogeropoulos
Department of Surveying and Geoinformatics Engineering, University of West Attica (UniWA), Athens, Greece

Series Editor

George Petropoulos

Elsevier
Radarweg 29, PO Box 211, 1000 AE Amsterdam, Netherlands
The Boulevard, Langford Lane, Kidlington, Oxford OX5 1GB, United Kingdom
50 Hampshire Street, 5th Floor, Cambridge, MA 02139, United States

ISBN: 978-0-323-98983-1

For Information on all Elsevier publications
visit our website at https://www.elsevier.com/books-and-journals

Publisher: Candice Janco
Acquisitions Editor: Peter Llewellyn
Editorial Project Manager: Ali Afzal-Khan
Production Project Manager: Kumar Anbazhagan
Cover Designer: Mark Rogers

Typeset by MPS Limited, Chennai, India

Contents

Section 4
Geospatial modeling and analysis

List of contributors

Nadia Abbaszadeh Tehrani Aerospace Research Institute, Ministry of Science, Research and Technology, Tehran, Iran

Mohamed A.E. AbdelRahman Division of Environmental Studies and Land Use, National Authority for Remote Sensing and Space Sciences (NARSS), Cairo, Egypt

Abdulla Al-Rawabdeh Department of Earth and Environmental Sciences, Laboratory of Applied Geoinformatics, Yarmouk University, Irbid, Jordan

Sylvia Antoniadi Operational Unit "BEYOND Centre for Earth Observation Research and Satellite Remote Sensing", Institute for Astronomy, Astrophysics, Space Applications and Remote Sensing, National Observatory of Athens, Athens, Greece

Vyron Antoniou Hellenic Army Academy, Athens, Greece

Muheeb Awawdeh Department of Earth and Environmental Sciences, Laboratory of Applied Geoinformatics, Yarmouk University, Irbid, Jordan

Christos Chalkias Department of Geography, Harokopio University of Athens, Athens, Greece

Isidora Isis Demertzi Department of Geography, Harokopio University of Athens, Athens, Greece; School of Rural and Surveying Engineering, National Technical University of Athens, Zografou Campus, Athens, Greece

Spyridon E. Detsikas Department of Geography, Harokopio University of Athens, Athens, Greece

Antigoni Faka Department of Geography, Harokopio University of Athens, Athens, Greece

Habes Ghrefat Department of Earth and Environmental Sciences, Laboratory of Applied Geoinformatics, Yarmouk University, Irbid, Jordan

Emilie Hendrickx Avia-GIS NV, Zoersel, Belgium

Guy Hendrickx Avia-GIS NV, Zoersel, Belgium

Fares Howari College of Arts and Sciences, Fort Valley State University, Fort Valley, GA, United Arab Emirates

Xiao Huang Department of Geoscience, University of Arkansas, Fayetteville, AR, United States

Emmanouela Ieronimidi Operational Unit "BEYOND Centre for Earth Observation Research and Satellite Remote Sensing", Institute for Astronomy, Astrophysics, Space Applications and Remote Sensing, National Observatory of Athens, Athens, Greece

Milad Janalipour Aerospace Research Institute, Ministry of Science, Research and Technology, Tehran, Iran

Bin Jiang Urban Governance and Design Thrust, Society Hub, The Hong Kong University of Science and Technology (Guangzhou), Guangzhou, P.R. China

Kleomenis Kalogeropoulos Department of Surveying and Geoinformatics Engineering, University of West Attica (UniWA), Athens, Greece

Katerina Karagiannopoulou Operational Unit "BEYOND Centre for Earth Observation Research and Satellite Remote Sensing", Institute for Astronomy, Astrophysics, Space Applications and Remote Sensing, National Observatory of Athens, Athens, Greece

Efthimios Karymbalis Department of Geography, Harokopio University of Athens, Athens, Greece

Charalampos Kontoes Operational Unit "BEYOND Centre for Earth Observation Research and Satellite Remote Sensing", Institute for Astronomy, Astrophysics, Space Applications and Remote Sensing, National Observatory of Athens, Athens, Greece

Penelope Kourkouli ICEYE Oy, Espoo, Finland

Demetris Koutsoyiannis School of Civil engineering, Department of Water Resources and Environmental Engineering, Laboratory of Hydrology and Water Resources, National Technical University of Athens, Athens, Greece

Xiao Li Texas A&M Transportation Institute, Texas A&M University, College Station, TX, United States

Panagiota Louka Department of Natural Resources Development and Agricultural Engineering, Agricultural University of Athens, Athens, Greece; Department of EU Projects, NEUROPUBLIC S.A., Piraeus, Greece

Nikos Mamasis School of Civil engineering, Department of Water Resources and Environmental Engineering, Laboratory of Hydrology and Water Resources, National Technical University of Athens, Athens, Greece

Cedric Marsboom Avia-GIS NV, Zoersel, Belgium

Gaëlle Nicolas Avia-GIS NV, Zoersel, Belgium

Emmanouil Oikonomou Department of Surveying & Geoinformatics Engineering, University of West Attica, Athens, Greece

Mina Petrić Avia-GIS NV, Zoersel, Belgium

George P. Petropoulos Department of Geography, Harokopio University of Athens, Athens, Greece

Evangelos Pissias Department of Surveying and Geoinformatics Engineering, University of West Attica (UniWA), Athens, Greece

Christos Polykretis Department of Natural Resources Management and Agricultural Engineering, Agricultural University of Athens, Athens, Greece

Athanasios Psarogiannis Department of Surveying and Geoinformatics Engineering, University of West Attica (UniWA), Athens, Greece

Farzaneh Shami Department of Surveying Engineering, Islamic Azad University, South Tehran Branch, Tehran, Iran

Nikolaos Stathopoulos Operational Unit “BEYOND Centre for Earth Observation Research and Satellite Remote Sensing”, Institute for Astronomy, Astrophysics, Space Applications and Remote Sensing, National Observatory of Athens, Athens, Greece

Aqil Tariq Department of Wildlife, Fisheries and Aquaculture, Mississippi State University, MS, United States; State Key Laboratory of Information Engineering in Surveying, Mapping and Remote Sensing (LIESMARS), Wuhan University, Wuhan, P.R. China

Andreas Tsatsaris Department of Surveying and Geoinformatics Engineering, University of West Attica (UniWA), Athens, Greece

Ioanna Tselka Department of Geography, Harokopio University of Athens, Athens, Greece; School of Rural and Surveying Engineering, National Technical University of Athens, Zografou Campus, Athens, Greece

Demetrios E. Tsesmelis Laboratory of Technology and Policy of Energy and Environment, School of Applied Arts and Sustainable Design, Hellenic Open University, Patras, Greece

Alexia Tsouni Operational Unit “BEYOND Centre for Earth Observation Research and Satellite Remote Sensing”, Institute for Astronomy, Astrophysics, Space Applications and Remote Sensing, National Observatory of Athens, Athens, Greece

Roger Venail Avia-GIS NV, Zoersel, Belgium

Di Yang Wyoming Geographic Information Science Center, University of Wyoming, Laramie, WY, United States

Melpomeni Zoka Operational Unit “BEYOND Centre for Earth Observation Research and Satellite Remote Sensing”, Institute for Astronomy, Astrophysics, Space Applications and Remote Sensing, National Observatory of Athens, Athens, Greece

Lei Zou Department of Geography, Texas A&M University, College Station, TX, United States

About the editors

Dr. Nikolaos Stathopoulos is a senior researcher—geoscientist and scientific/technical project manager at the Operational Unit "BEYOND Centre for Earth Observation Research and Satellite Remote Sensing", of the Institute for Astronomy & Astrophysics, Space Applications and Remote Sensing, of the National Observatory of Athens (NOA), Greece. He holds a PhD in analysis and modeling of geoenvironmental hazards and water resources via GIS and remote sensing, and two MSc diplomas in science and technology of water resources and in geoinformatics. His Bachelor's degree was in mining engineering specializing in Earth sciences—geotechnology. His research interests mainly focus on natural disasters (hazard modeling, vulnerability and risk assessment, management, etc.), land degradation and soil erosion, water resources, and climate change impacts on natural processes.

Prof. Andreas Tsatsaris is a rural and surveying engineer (PhD, MSc, and BSc) and head of the Surveying and Geoinformatics Engineering Department at the University of West Attica (UniWA), Greece, since 2019. He is also repeatedly elected as the director of the Research Laboratory "GAEA" since 2013. He teaches GIS, thematic cartography, spatial epidemiology, and medical geography, in both undergraduate and postgraduate study programs of the Faculty of Engineering at UniWA, as well as of the Faculties of Medicine at the Universities of Crete, Aristotle University of Thessaloniki, and National and Kapodistrian University of Athens.

Dr. Kleomenis Kalogeropoulos is a surveying engineer (PhD, MSc, and BSc) and an adjunct lecturer teaching GIS and thematic cartography at the Department of Surveying and Geoinformatics Engineering, University of West Attica, Greece. His research interests are focused on spatial data infrastructures, spatial analysis, natural disaster modeling, geoarchaeology, digital cartography, and geography. He is a certified evaluator of the General Secretariat in research and technology and an expert in Earth sciences in the use of GIS and remote sensing applications for the Institute of Educational Policy of the Greek Ministry of Education.

Introduction

The 21st century is characterized by significant social, economic, and environmental changes but also by rapid progress and technological development. A key benchmark of the century is climate change and its impacts, a global phenomenon that directly and indirectly affects human life and the environment. The most important effects of this phenomenon are the change in climate in many regions, the looming rise in the average temperature of the earth, the rise in average sea level locally, the melting of the ice caps, etc., as well as the effects of these changes, such as the steadily increasing frequency of natural hazards and disasters, changes in the biodiversity of regions, and changes in natural processes. An important aid in the proper management and response to this reality is the modern and very effective tools that have been created, developed, tested, and applied in recent years, through the technological and scientific explosion that has taken place in parallel with climate change.

To be more specific, human activities and climate change are the modern catalysts for natural processes and the geo-environmental risks that accompany them. Globally, natural disasters (floods, fires, landslides, etc.) and other natural phenomena (soil erosion, desertification, etc.), exacerbated by anthropogenic factors, are rapidly increasing in intensity, frequency, spatial density, and significant spread of their occurrence areas. The impact of these processes is devastating in terms of human life, private property, infrastructure, etc., and on a larger scale it has a very negative impact on the social, environmental, and economic status of the affected area.

Research, management, and response to natural disasters and processes have been the focus of interest and engagement of the scientific community and decision-makers in the last decade. The modern environmental, social, and economic challenge, among others, due to climate change, is the prediction of natural disasters and the shielding of vulnerable areas against imminent natural hazards and ever-increasing natural processes with negative impacts. A typical example in this direction is Directive 2007/60 of the European Union (EU), which aims to inform and improve flood prevention, management, and restoration methods in the Member States. More generally, this directive attempts to set environmental objectives for Member States' water management policy, with a particular focus on groundwater, which is also threatened by natural factors, combined with human overexploitation, with a deterioration in its quantitative and qualitative characteristics. Similar initiatives have been taken worldwide with very significant results.

In parallel, a wide range of ongoing or rising challenges, that affect human well-being, need to be addressed and coped with. Health crisis is at the top of the list, with the pandemic of COVID-19 being the protagonist followed by vector-borne diseases and other climate-induced, transmitted diseases, which show a rising trend in the last decade. At the same time, the energy crisis adds a lot of pressure on households and national economies as well, and innovative solutions need to be adopted and put into action.

Last but not least, food security, environmental and ecosystem preservation, land restoration, organic agriculture, sustainable management of water resources, and all other EU and global goals never stopped being at the top of the priorities list. In specific, all the above are core targets of various international (EU and global) initiatives and directives, such as the UN's Agenda 2030 with the 17 sustainable development goals, WHO's One Health approach, and the European Green Deal and its derivatives (e.g., Adaptation to Climate Change strategy strategy and soil mission).

All these needs and priorities can be countered by, or rather greatly supported by, building upon the technological development and scientific progress, that took place at the end of the 20th and beginning of the 21st century and as it seems is set to continue at an exponential rate in the years to come. Under the wider umbrella of geoinformatics, remote sensing/Earth observation, geographic information systems, and data analytics (artificial intelligence, machine learning, etc.) are able to offer important techniques, tools, models, databases, etc., for addressing all the aforementioned topics. These technologies and scientific practices already occupy a considerable part of the world's literature and the field of academic research.

In this light, this book attempts to showcase integrated and applied methods based on the wider range of geoinformatics and dig deep into the literature for state-of-the-art research articles treating all the aforementioned challenges. It approaches interdisciplinary topics across various fields, like environment and climate, geomorphology and hazards, natural resources, health, education, socioeconomic life, crowdsourcing, etc. It also attempts to highlight various data sources, databases, and specialized software (open or commercial). Furthermore, the presented methods aim to reduce the time needed to produce usable results, to overcome major obstacles that arise both in office or/and field research, and above all to create targeted guidelines for its application where it is needed. All presented articles are developed under specific and common principles, that is, to be scalable and transferable.

The aim for both the research and science as well as the practical and applied contribution of the presented methods is to:

- have very low costs for their development and implementation (use of free data and tools where possible),

- be relatively easy to implement for researchers,
- produce fast, useful, and reliable results,
- overcome problems that for field research would be a deterrent (e.g., accessibility in difficult or inaccessible areas, weather conditions, etc.),
- be able to assist and facilitate large parts of the field survey (thus reducing the cost and time of the survey),
- direct field research toward targeted actions (further reducing both costs and time of research),
- can be used for policy-making (forecasting, protection, exploitation, management, etc.), and
- have prospects for the future development and improvement, in line with new knowledge and technological progress.

Therefore this book intends to become a benchmark in the literature, on applied geoinformatics in the wide field of geosciences. It focuses on showcasing applied methodologies on advanced geospatial analysis via RS, GIS, and soft computing and highlights important works, in these multidisciplinary fields, through literature research. Hopefully, this ambitious book project will find the place it deserves in scientific literature and above all become a useful guide and tool for students, researchers, stakeholders, and everyone else involved in the fields of geoinformatics and geosciences.

Nikolaos Stathopoulos
Andreas Tsatsaris
Kleomenis Kalogeropoulos

Section 1

Geospatial human environment

Chapter 1

Geoinformatics, spatial epidemiology, and public health

Andreas Tsatsaris[1], Kleomenis Kalogeropoulos[1] and Nikolaos Stathopoulos[2]

[1]Department of Surveying and Geoinformatics Engineering, University of West Attica (UniWA), Athens, Greece, [2]Operational Unit "BEYOND Centre for Earth Observation Research and Satellite Remote Sensing", Institute for Astronomy, Astrophysics, Space Applications and Remote Sensing, National Observatory of Athens, Athens, Greece

1.1 Introduction

The concepts of interdisciplinarity and multidisciplinarity are necessary in understanding the role geography played, is still playing, and will be playing in the future. Geographical interdisciplinarity is defined as the study of geographical phenomena from the viewpoint of the interactive relation between space and society by using analytical tools of geography, as well as tools of those sciences that cope with registering, explaining, and interpreting social, economic, political, cultural, and historical phenomena (Leontidou & Sklias, 2001). On the other hand, Geographical multidisciplinarity is defined as the cooperation of various sciences for the resolution of practical problems in the partially overlapping domain of a certain discipline. This resolution does not affect, neither epistemologically nor methodologically speaking, none of the sciences involved; it does not even advance their superficial communication mode to a more substantial level (Leontidou, 1992).

Unlike multidisciplinarity, which consists of nothing more than conveying concepts from one science to another, interdisciplinarity requires the shift of the frontiers of each science that describes, analyses, and interprets the phenomena individually, so as to make possible its cooperation with the other ones.

Geography, with the exception of certain periods, has always been an interdisciplinary science. Its scientific characteristic is intrinsic interdisciplinarity, imposed by its own scope. On the contrary, multidisciplinarity, which was geography's dominant characteristic, for example, during the 1960s, was

Geoinformatics for Geosciences. DOI: https://doi.org/10.1016/B978-0-323-98983-1.00002-8

extrinsic, imposed for that matter by the Positivistic Paradigm, which limited the multidimensionality of the geographical perspective, thus fragmenting it (Leontidou, 1992).

Interdisciplinarity ensures the multidimensional analysis of the phenomena, by restoring the bi-directional relation between humans and nature, and the historical dimension of the interpreting geographical schema, by approaching space through both its own geographical tools and tools of traditionally relative sciences such as geometry, biology, and geology, together with tools from other sciences, including history, economics, politics, sociology, ethnology, psychology, linguistics, and philosophy, or even statistics, demography, possibility theory, game theory, informatics and medicine, and artificial intelligence, which are by definition so closely related that any vertical segregation would seem inconceivable today.

The chapter at hand falls under the heading of Medical Geography, which constitutes a tangible example of interdisciplinarity. Medical Geography copes with the tensions between natural and social science, in the sense that it deals with both natural and human elements, which, in conjunction, can affect the physical and psychical health of a population (Eyles, 1993). Because of the traditional focus on the relationship between biomedical phenomena and the role of the environment (Eyles & Woods, 1983; Mayer & Meade, 1994), Medical Geography is closely related to Epidemiology, as far as both research subject and approach are concerned. However, there are two distinct characteristics in Medical Geography: disease geography and health care geography. The former includes both descriptive research—which entails investigating the frequency of a disease incidence, who is ailing and who is not, when and where the disease occurs—and analytical research—in the sense that its aim is to test ideas concerning what determines whether an individual or a population will ail or not. Descriptively and analytically, disease geography centers on the relations between diseases and their distribution, and potential and existing environmental factors.

The other distinct characteristic of Medical Geography, that is, health care geography, has basically concentrated on the location, accessibility, and utilization of the facilities. Healthcare geography uses modeling techniques that result from location theory and/or designing of qualitative sample research for the investigation of healthcare facilities' location (Joseph & Bantock, 1982), as well as from users'/patients' behavioral patterns (Joseph & Phillips, 1984; Ross et al., 1994). Therefore healthcare geography constitutes a typical example of spatial analysis, in which geographical tools (e.g., correlation analyses and location-allocation modeling) are used. Nevertheless, the present dissertation does not restrict itself to the isolation of healthcare geography patterns; on the contrary, it goes further than that by examining the relations between a distribution and the characteristics of the areas under discussion, which are often mentioned as ecological analyses.

1.2 Geographical epidemiology

1.2.1 General aspect

The subject of geographical epidemiology is the analysis of the spatial (geographical) distribution of a disease incidence. In its simplest form, this subject concerns the usage and interpretation of location maps for certain disease cases. In essence, although, these two different aspects of the issue under discussion exert their influence on the methodology which has been developed to examine matters that arise in this scientific field.

The present chapter examines the geographical behavior (i.e., distribution and spreading of the disease in space) of leishmaniasis. The examination of the nature of leishmaniasis spatial distribution was based on hypotheses that can be classified into four categories (Lawson, 2001):

1. The way the population and its subgroups that manifest the disease in space and time are affected by the disease distribution (descriptive epidemic analysis).
2. The way the relationship between the spatial distribution of the disease incidence and quantitative factors that interpret the disease at certain levels of spatial clusters of the phenomenon occurs is analyzed (ecological analysis).
3. The way the disease spatial distribution appears—that is, the mathematical models that describe the general distribution of the disease in a geographical map (disease mapping).
4. The way unusual spatial concentrations of the disease are analyzed (disease clustering).

Epidemiology studies examine the disease distribution in relation to the individual, space, and time, or determine the causes of the disease spreading. Because of their differences in designing techniques according to their goal, epidemiology studies are divided into descriptive and analytical.

One could possibly classify the present dissertation as a descriptive epidemiology study, on the condition that s/he would relegate its special emphasis on its content which focuses mainly on the geographical dimension of the problem. However, this dissertation would have been impossible to be realized, had there not existed leishmaniasis epidemiology data, which, to become usable, had to undergo processing imposed by descriptive studies' methodology. As it will become evident in the corresponding chapter of the dissertation's special part, the factors examined concern:

1. Disease distribution in space and time.
2. The populations and subgroups that manifest the disease.
3. Geographical locations where the spreading of the disease is more or less probable.

4. Frequency of occurrence of the disease during a given period of time.
5. Probable seasonal patterns of the disease spreading.

These are the factors pleaded by methods of the development of descriptive epidemiology studies. However, these methods lead to the stipulation of hypotheses regarding the parameters that might affect a disease, so that this disease can be further tested through analytical epidemiology studies. Nevertheless, in the case of the present dissertation, what mostly matters is in essence what is described in the third and fourth categories of the aforementioned hypotheses.

We could also be considered to be following the methodological principles of an ecological study, as on the one hand, it is an exploratory study since there is no source of exposition to a substance, and, on the other hand, it examines the following factors:

- Disease clustering
- Verification of quantitative factors that interpret the disease

1.2.2 Space–time interaction

The chapter deals with measurements that do not directly affect individuals as units (global measures) but refer to group properties (ecological level variables). It takes municipalities' administrative boundaries as the spatial analysis unit and refers to municipalities' populations. In this sense, the specialization of the hypotheses can be achieved by taking into consideration the ecological impact on the groups. However, these groups are not represented by their total population number, but by a designated ratio showing the disease change rate on the population at the municipality's level (ecological effects on group rates).

A disadvantage of this approach is the fact that the hypotheses might not absolutely fit the analysis level. On the other hand, the advantages include, first, the fact that simple statistical analysis works, and, second, greater differences between the parameters critical in disease spreading are analyzed. The latter advantage is owed to the fact that the factors affecting the disease movement are more diversified at the municipality level than at the individual level.

Generally speaking, the statistical methods applied to the geographical distribution analysis of the disease incidence depend on the following factors:

The size of the area under discussion (scale)

- Data availability
- Data geographical location (georeference)
- Data geographical discontinuity

- Data source (taking into consideration that because of their nature—i.e., being epidemiology data—, they incorporate potential properties that refer to geographical distributions of the disease change risk in relation to the population)

More specifically, as for the use of spatial statistical analysis methods, the starting point focuses on the role, methods, and consequences of sampling necessary for the study data. Furthermore, in studies on geographical surveillance, it is customary to detect geographical units in which the disease under discussion shows high incidence (Elliott et al., 1997). Moreover, the data type, the methods of data collection, and the data entry in systems that favor their geographical analysis (Geographic Information Systems—hereafter GIS) conduce to public health care services control and preventive intervention, focusing the interest of high-risk areas.

The use of geographical analysis software systems presupposes precision in determining data, qualitative control filters, and accuracy in designing the data bank, which here is analyzed at natural and logical levels.

The quest for methods and analysis techniques regarding geographical phenomena aimed at making the study and classification of their impact on human health possible was the general idea behind the approach of an issue that is particular in the context of Medical Geography and more specifically of Environmental Epidemiology.

The present chapter's research scope is the formulation of methodologies concerning the production of a model eligible for the study and interpretation of geographical parameters which affect the mechanisms under which diseases, mainly infectious ones, generate spreading and distribution patterns in Greece, so as to establish the presuppositions for their continuous geographical surveillance.

The basic analysis tools for the correlation of geographical space to diseases, with the ulterior aim to synthesize multiparameter models and visualize results, are software packages developed recently for use by earth and space scientists (e.g., GIS, Geographical Analysis and Statistics tools, Remote Sensing) and new developments in medical technology (e.g., Biomolecular Methods for Sample Analysis, Biostatistics tools). Especially GIS technology offers possibilities for the synthesis of geographical and descriptive data, and the development of software applications that allow the informed user to have alternative options in case the geographical distribution of the phenomena is part of the problem. The use of these systems allows the understanding of a disease's spatial distribution, at least to some extent. However, GIS, because of its contribution to analysis and decision-making, can be established as the modern way of managing information that correlates space with human health.

Since the 1970s, in international literature, there has been a plethora of studies directed toward the understanding of the mechanisms governing the correlation between human health and space.

As a matter of fact, relevant bibliographical references first appeared as far back as the late 17th century. Needless to say, these references had to do more with plotting great epidemics on contemporary maps rather than analyzing geographical parameters affecting the occurrence of the epidemic and/or spreading—with the notable exception of John Snow's renowned diagrams (1854) for the causes of cholera spreading in London.

GIS dynamic development brought about a boom in the production of relevant studies, mainly after the mid-1990s, when it was understood that geographical analysis methods should be adjusted in the given systems taking the form of specialized software applications.

There are several specialized international journals, both online and in the paper (e.g., *Health and Place*, *Cadernos de Saude Publica*, *Environmental Health Perspectives*, *Social Science and Medicine*, *International Journal of Health Geographics*, *Journal of Spatial and Spatiotemporal Epidemiology*, *American Journal of Epidemiology*, etc.), but about 60% of the specialized papers are published in journals of other specializations (e.g., *International Journal for Geographic Information*, *Computer & Geosciences*, *International Journal of Environmental Research and Public Health*, *Tropical Medicine and Infectious Disease*, etc.).

Moreover, the Internet constitutes an interesting source of bibliographical references, as today there are several dozens of relevant sites: some of them are centered on Medical Geography and its applications (e.g., https://connect.agu.org/geohealth/home, https://malariaatlas.org/, etc.) and others are official sites of great national and international organizations which push forward GIS and Remote Sensing applications concerning health (e.g., https://www.cdc.gov/gis/, https://www.who.int/data/gho/map-gallery, etc.). These sites are regularly updated, and some offer links to online journals and access to online libraries, specialized software, and free-of-charge data. These organizations coorganize or cofund, together with universities and research centers, at least five annual international conferences on specialized issues of Medical Geography.

In Greece, there is not any notable scientific activity in this specialized field. Paper production is extremely meager, and the scientists involved are few and far between.

For the time being, in Greece, the procedures applied to the surveillance of infectious diseases (e.g., influenza, hepatitis, AIDS, etc.), which concern a large part of the general population, do not take into consideration spatial statistical analysis methods. Due to this fact, the data collected cannot meet the requirements of a study whose main directives are founded on the content of geographical analysis. However, there are infectious diseases that are dangerous not because of their incidence in great population numbers, but because of the difficulty in treating them during the stage they are overtly detectable in humans, and this results in their being fatal. One category of this disease is zoonoses, which have been standardized as geographical diseases. A further strong reason for us to cope with this type of disease is that Greece has been appointed the center for the surveillance of zoonoses in the Mediterranean by the World Health Organization (hereafter WHO).

The present chapter's aim is to cope with one of these diseases, namely leishmaniasis. The quantitative and qualitative development of methodologies for studying the geographical spreading and distribution of leishmaniasis as a spatial phenomenon and the diagnosis of its incidence in humans and their activities are founded basically on the intensity of the phenomenon in certain areas of the country (Papadopoulos & Tselentis, 1994), as well as on the difficulty in its diagnosis. Pinpointing the high-risk areas, and consequently, the potentially affected population aims at proposing the presuppositions of disease surveillance and dealing with the problems the disease causes.

The geographical space that was selected is Attica prefecture. The selection was based on the following hierarchically given factors:

- The intensity of the phenomenon. (Health Department statistics for the years 1961–2004 have shown that about 58% of all leishmaniasis cases detected in humans were reported in Attica prefecture.)
- Availability of relatively recently registered cases—comparability with new data. (Laboratory of Clinical Microbiology and Microbial Pathogenesis, Unit of Zoonoses of Medicine School of the University of Crete collected data for the disease's vectors, reservoirs, and hosts—respectively, as far as leishmaniasis goes, sandflies, stray dogs, and humans—for the period 1992–93. These data were kindly available to us, as part of our collaboration with the Laboratory in the context of the present research.)

Data are grouped according to their qualitative characteristics into three thematic units and can be digital or nondigital:

1. Data designating cartographically administrative, physiographical, and environmental components of the reported geographical space.
2. Data derived from human activities and habits.
3. Specialized data concerning the disease under discussion.

These data were either retrieved from records and literature research or collected from field research. The weight of participation of each unit in the logical and natural designing of the geographical database supports the research analytical and synthetic procedures.

The preliminary process and analysis of the relatively recent original registrations of the disease hosts, in conjunction with the distribution of patients and the recorded and cartographic data, determined the probable high-risk areas for the period under discussion. This made possible the direct focus of on-the-spot field research for the acquisition of new data. Through the production of spatial statistical models of distribution and spreading, the high-risk areas were determined in the present time and the disease surveillance methods were decided upon.

The main process concerning recently collected data had to do with laboratory findings analysis, epidemiology data and services registration records, their correlations with geographical space, their spatial statistical analysis, their distribution and spreading models' verification, and their improvement and completion (feedback).

The present chapter's aims are stipulated epigrammatically as follows:

- Formulation of methodological directives for handling the geographical behavior of the disease under discussion.
- Production of a geographical database to act as the core of developing applications for handling the geographical and epidemiological behavior of the disease under discussion.
- Production of spatial statistical models for spotting high-risk areas in space and time.
- Definite and comprehensible to nonexperts' visual representation in maps of the disease under discussion and of the spatial high risk for the populations in the endemic areas.

As it is stated in several points of the text that follows, on the one hand, such research endeavors are not customary in Greece and, on the other hand, there are international scientific networks that cope with the problem under discussion in various places around the globe. Furthermore, there seem to be great or considerable similarities between the anthropogenous and environmental conditions in Greece and those in the countries adjacent to Greece; therefore the geographical parameters affecting the geographical behavior for the disease under discussion in space are similar throughout the wider region, and the countries adjacent to Greece have any reason to wish the notification of the results, as well as of the research methods. It would naturally follow from what has been said that generalizing, the present dissertation's aims have a clear orientation: the notification at a national and international level of the research methodology and results. Subsequently, this study is conducive to:

- the consolidation of interdisciplinary cooperation;
- the development of scientific research in a new scientific field for Greece;
- added value to the experience of the international scientific community;
- the consolidation of methodology and new technologies used by the services which handle health surveillance; and
- the direction of improving the conditions and services concerned with health.

1.3 Methodological principles

GIS are technological tools that lead to a continuously better understanding of the real world. As far as their importance is concerned, they have been compared to fundamental scientific and technological developments such as the invention of the microscope, the telescope, and the electronic microscope.

In the 17th century, Dutch Anton van Leeuwenhoek, who is regarded as the father of microbiology, improved the microscope to the extent that he was the first to see images of bacteria (Dobell, 1960). His invention eventually led to the revolution of understanding and treating infectious diseases. For environmental epidemiologists, the GIS promise is the increased understanding of the correlation (and even of the causality relations) between human exposure to a disease, due to environmental changes, and human health. This leads to the inevitable question: is this a dream a logical and feasible task? The answer to this question should be considered after putting GIS into a context of a systematic approach to producing and testing hypotheses.

To achieve this goal, we will use Popper's hypothetico-deductive method and its extension by Plat (1964). This approach entails hypotheses that are tested experimentally so as to eliminate the theory that supports them in case, they are proven false. Falsifiability of theories constitutes a systematic way of accumulation of scientific knowledge, with new findings being based on older ones. In practice though, science does not always advance in this fashion. However, this method is useful because it encourages researchers to test their theories objectively.

Some authors pose reasonable questions concerning health's added value. Marbury states that most of the time developments in Environmental Epidemiology require carefully planned studies that would give strictly defined results derived from good measurements at an individual level (Marbury, 1996). Moreover, he concludes, it would be unproductive if one's attention was diverted from this effort just because of the availability of a new tool that adds nothing to what we already know.

This objection reflects an expectation regarding GIS, which goes beyond its usage in estimating an individual's exposure to a disease, in disease mapping, and in public health care surveillance. Indeed, the aforementioned analogy of the microscope implies that GIS as a technological application will lead to fundamental developments in understanding the relationships between health and the environment. At this point, a key question arises: Can we use GIS to formulate and test hypotheses in Environmental Epidemiology? And if this is possible, how could Environmental Epidemiology as a scientific field?

Philosopher Karl Popper described a systematic approach concerning the accumulation of scientific knowledge, which could place the following question on an interesting basis (O' Hear, 1996) (Fig. 1.1). First, a theory or a hypothesis is produced to explain the correlations between the patterns of observed data. Then, testable predictions are deduced from this theory and are tested through experiments, so as to prove that they are either badly formulated or theoretically groundless. If thc prediction gets dismissed, the corresponding theory gets rejected; if the prediction is verified, new predictions are produced and tested for further verification of the theory under discussion.

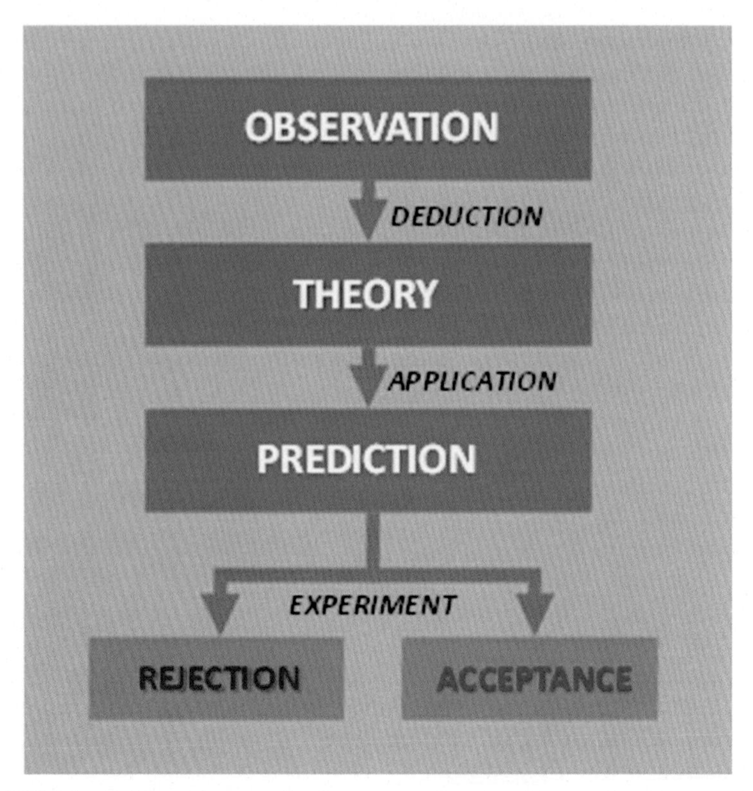

FIGURE 1.1 Karl Popper's suggestion for a systematic approach concerning the accumulation of scientific knowledge.

This approach's important teachings are as follows:

1. Predictions can be proved groundless through experiments (falsifiability). Even if they are verified, they cannot acquire natural law status.
2. Theories can be rejected but cannot be verified beyond any reasonable shadow of a doubt.
3. For the predictions to be useful, they must be proven groundless through experiments and other testing mechanisms.
4. Predictions that cannot be proved groundless are useless.
5. Data from experiments that have been designed for the estimation of predictions must be independent of the original observations used for the theory formulation.
6. Prediction testing for data that have led to the formulation of these predictions is not valid.

John Platt in his paper Strong inference (Platt, 1964) elaborates on the Popperian notion of falsifiability versus verification and extends the hypothetico-deductive method (Fig. 1.2). Platt recognizes that if the

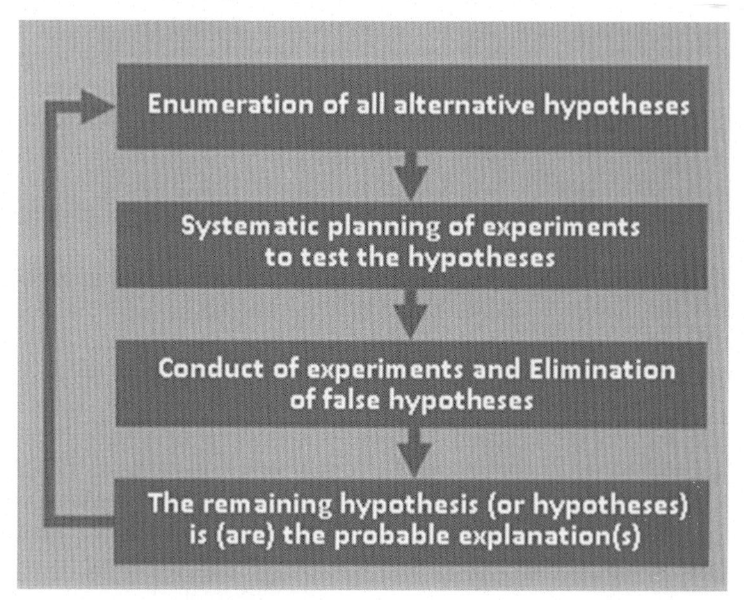

FIGURE 1.2 Method of strong inference.

Popperian approach to proving a prediction groundless is to become useful, it must be used in a systematic way in a context of a world of plausible alternative hypotheses. This approach entails the thorough development of probable explanations, the rejection of those that have been proved groundless, and the acceptance that the remaining explanation, however unbelievable, is the solution to the problem. The Method of Strong Inference is represented in the following diagram.

It is evident that the alternative hypotheses group might change as a result of knowledge acquired during experimenting procedures. Platt's method is a systematic approach to drawing the correct inference.

There are researchers (e.g., Koopman, 1996) who disapprove of this simplifying inference procedure, which is based on individual causality hypotheses. They claim that the hypotheses supporting all theories and patterns are to some extent false, and therefore the causative inference might be impossible. It goes without saying that science, as a knowledge system, advances in many ways which include intuitive leaps and arbitrary circumstances, as well as the systematic estimation of hypotheses. Feyerabend (1981) enumerated several ways for estimating scientific knowledge and recognized criticism, propagation, and realism as the fundamental ideas that continuously play a significant role in the history of science.

The term criticism in this context means that we do not simply accept the phenomena (either observations or greater knowledge systems) as they are; on the contrary, we examine them one by one carefully and cautiously. The term propagation describes the tendency to work on a glut of ideas rather than on isolated ideas. This means, for example, that we can recognize simultaneously and equally on the one hand the contribution of a systematic approach and on the other hand the contribution of arbitrary circumstances (which constitute a long scientific tradition, the discovery of penicillin being the most celebrated). Finally, the term realism recognizes that the contexts of scientific knowledge acquisition are defined by the real world and they are not mere schemata for the mechanical process of events.

In this perspective, the arguments as to which approach is more or less the dominant Paradigm are hardly valid since the use of such paradigms is intrinsically selective.

Popper's hypothetico-deductive method and its extension proposed by Platt in his Method of Strong Inference are nothing more than one way of scientific knowledge acquisition, which leads to useful conclusions concerning GIS, health, and scientific knowledge acquisition. An important contribution of the Method of Strong Inference is that, as it requires predictions that can be proved false, it prevents the use of ambiguous theories that are intrinsically nontestable.

Environmental Epidemiology seeks to justify the causes and correlations of diseases in the context of population morbidity as this manifests itself in the everyday environment. The examination of spatial data through GIS often tries to correlate spatial variables. In practice, the hypothetico-deductive method is not followed strictly; one can mention many studies that examine only the correlation or verification and do not move on to estimating predictions that can be proved false.

GIS is a strong tool for managing georeferenced spatial data, which, among other things, describe events concerning human morbidity results. After producing the spatial database, one of the first steps toward seeking correlations between health and the environment is producing thematic maps.

Thematic Cartography is used for the representation of results arising from spatial analysis procedures between the correlation of geographical and epidemiological data that have populated the geographical database in the form of spatial information layers and tables. Through these procedures, non-expected spatial patterns may appear, which are often the main criterion determining the map to which the attention of further analysis is to be paid. Careful examination of such a map—that is, a map selected because of the nonexpected patterns it represents—often prompts the researcher to formulate explanatory hypotheses.

The use of spatial statistics and/or techniques for recognizing correlated spatial patterns defining as to whether the nonexpected mapped spatial pattern is valid in reality. GIS is increasingly used by professionals in health care who are not familiar with intricate statistical techniques and have no access to the required specialized software for the spatial statistical analysis of mapped data. In such cases, the thematic maps derived from the use of

GIS usually are not analyzed and so it cannot be determined whether the nonexpected spatial patterns are statistically significant.

This is to a great extent attributed to two causes:

- Most GIS software packages do not have the ability to perform spatial statistics, and, conversely, many statistical software packages do not support spatial statistical procedures.
- The "added value" of digital geographical data's spatial statistical analyses is not widely known to healthcare professionals, and therefore they lack the motivation to use spatial statistics.

The great advantage of digital geographical data's spatial statistical analyses is that their use prevents deductions arising from mere observation of nonexpected spatial patterns.

Research based on nonexpected spatial patterns does not produce valid results because of the following three reasons:

1. The basic criterion for selecting thematic maps is the visual impact that is produced from nonexpected spatial patterns, and therefore maps have the tendency to represent patterns where they do not exist.
2. Maps do not undergo statistical tests so as to determine whether an apparent pattern is statistically significant, something that would add value to the examination.
3. The produced hypotheses are of little value because usually they ate not falsifiable.

This brings us back to the aforementioned question: Can we use GIS to formulate and test hypotheses in Environmental Epidemiology? To answer it through the Popperian approach, we take for granted that the production of a GIS requires three steps: spatial data, map production, and examination of whether the represented phenomena give unusual spatial patterns. In other words, we use GIS and spatial statistics as a lens through which we observe the world. Spatial analysis in GIS is used for the production of thematic maps, and spatial statistics determine whether the patterns on these maps are somehow unusual. Useful statistics toward this goal include disease clustering methods, point pattern analysis methods, proximity statistics for determining adjacent areas according to a certain classification, testing of overlapping boundaries, and statistics for estimating the correlation between two statistical variables. The ability of GIS to manage thorough georeferenced data derived from various variables simultaneously supports statistical methods that work with contiguous and noncontiguous digital data, which means in practice various formats of vector and raster files.

Once it is considered feasible for a map to undergo a statistical process, we are eligible to deduce a theory or a hypothesis explaining spatial relations. In this case, we formulate a testable—that is, falsifiable—prediction and design an experiment to test this prediction. It should be noted once again that this approach can prove a theory false, not true.

There are at least three types of experiments that can prove a theory false. We can design an epidemiology study to test the predictions describing the disease incidence in populations. Moreover, we can design a laboratory study when the prediction describes the prediction of disease at a biological level. Finally, we can design a GIS study to estimate epidemiology predictions that include a spatial dimension.

In practice, GIS studies are not very useful in estimating spatial predictions for the following reasons:

1. Managing spatial systems is difficult; therefore designed experiments are either impossible or are conducted with great difficulty. Experimenting on human populations is not acceptable; therefore amendment of population data so as to define the point where morbidity has acute or chronic consequences is not possible. In some cases, although, natural experiments that occur unintentionally (e.g., pathogenic causes arising from a natural disaster) can be proved useful and should be exploited.
2. GIS data used for theory formulation must not be used for testing predictions derived from the same theory, as this would bias the researcher to verify the suggested spatial pattern. This means that undoubtedly one cannot use hypotheses derived from a GIS study, using the very same GIS study.
3. An additional problem is the intrinsic spatial uncertainty of most epidemiological data. This uncertainty has several sources. First of all, it occurs when, instead of patient address points, wider areas are used (e.g., blocks, census zones, etc.). Furthermore, it occurs when raster format data are used; in this case, instead of a precise address, the nearest raster cells are used. Conceptually speaking, positions are almost always uncertain, as people move about; thus the health events that determine the morbidity causal factors can arise anywhere within the space an individual acts. Precise positions do not represent the mobility of modern societies.

1.4 A review

1.4.1 A little bit of history

The term "Medical Geography" has the potential to be considered in different ways depending on where it is addressed. According to some views, the study of the geography of disease is narrowly formulated as a field of medicine or geography (medical geography or geographic medicine).

Medical geography is a multidimensional discipline that approaches understanding the mechanisms through which human health problems are linked to space (Pyle, 1979). Different degrees of economic development, as well as institutional differences in terms of socio-political dimension, bring about similar orientations in scientific practice, resulting in different approaches to understanding the subject of Medical Geography. Furthermore, past and new spheres of influence at the economic and political levels worldwide influence the direction of different orientations in the research of Medical Geography.

Within the narrower boundaries of the science of geography, national educational systems have influenced what might be called types of national or perhaps transnational efforts in the field of Medical Geography.

In Britain and some former Commonwealth countries, several identifiable trends have emerged since the Second World War. Andrew Learmonth's now classic studies on the geographic factors contributing to the ecology of malaria in India (Learmonth, 1957, 1972) as well as his more recent contribution dealing with the geography of hunger, serve as examples of British trends. Melvin Howe's monumental contributions to the field of disease mapping have shown how some simple or more complex variations of cartographic techniques can lead to variable interpretations of health problems (Howe, 1970). Neil McGlashan and J. Harrison have further brought disease mapping closer to the fine arts by refining traditional probabilistic mapping methods with simultaneous epidemiological correlations (McGlashan & Harrington, 1976). In the field of disease spread, R. P. Mansell and Arthur Brownlea, respectively, have shown that malaria has spread in parallel with migration and that hepatitis can be seen to follow patterns of human colonization (Mansell, 1965; Brownlea, 1967).

Efforts in the field of Medical Geography in Germany, Japan, and Belgium show specific trends parallel to those in Britain, but with varying degrees of uniqueness. In Germany, Helmut Jusatz, a pioneering researcher in Medical Geography and one of the major authors of the Welt-Senchen Atlas, continues to set the guidelines in the fields of medical mapping and geographic pathology. Momiyama has shown how climate and the subsequent effects of modernization in Japan have influenced the seasonality of mortality for several decades. Studies of the international geography of cancer, led by Belgian Yola Verhasselt, have gained international recognition.

In France, the foundations of medical geography for several decades were based on the many studies of Max Sorre. Sorre undertook a variety of local studies and is internationally best known for his ideas on the human ecology of human-inhabited space. His major contributions to the field of medical geography included studies of climate and humans, human nutrition, and the human body struggling against the natural environment. In contrast, Henri Picheral has taken a new approach to Medical Geography in France through local and to some extent cartographic analyses of the socioeconomic effects of health problems in the Mediterranean.

Geographical variation in health has long been studied interdisciplinarily and through various approaches, all of which use methodologies that have mainly to do with the health–disease–geographical space relationship (geographical pathology, medical ecology, medical topography, geographical epidemiology, environmental epidemiology, geomedicine, etc.). The methodology of geography has been exploring medical issues for over 50 years. The emerging interest in a new systematic way of medical geography is considered to have been articulated in the first report of the Committee on "Medical Geosystems" within the International Union of Geography in 1952. However, it is only after 15 years that the work of pioneering researchers and teachers covering dozens of regions of the earth has been able to be captured under the term "Medical Geography" and recognized as a distinct scientific field by the international community.

Since biological characteristics, lifestyle, and the nature of medical treatment vary from place to place, it follows that the same must be true of the likelihood of developing a particular disease or even dying from it. Indeed, the incidence of a disease, or the probability of a particular cause of death, varies geographically, on scales ranging from international to local. Take schizophrenia, a disease that strikes with a clear environmental predilection, as an example. Its highest levels are found in Ireland, Scandinavia (particularly some parts of northern Sweden), and Eastern Europe (particularly Croatia). In parts of Western Ireland, for example, 1 in 25 people will be affected in their lifetime. Moderate rates occur in England, Germany, the United States, and Japan. In the United States, about 1 in 100 people eventually develop schizophrenia. Southern Europe has significantly lower rates.

Remarkable variations occur worldwide for a wide range of other diseases. Japan, for example, has the world's highest incidence of stomach cancer. In Nagasaki in particular, the annual rate of this generally fatal disease is 100.2 for men and 51.0 for women per 100,000 inhabitants. These figures contrast with those of Dakar, Senegal, which, with an annual incidence of 3.7 for men and 2.0 for women per 100,000 population, has the world's lowest rate of deaths from stomach cancer.

The wide variations in the risks associated with the disease are not limited to an international scale. The state of Georgia is of particular interest to those involved in the study of cardiovascular disease because the probability of a white male dying from this cause is about twice as high in South and Central Georgia as in the North (Shacklette et al., 1972).

Variations in the risks associated with the disease also occur on much more local scales. Howe (1979), for example, mapped death rates from a variety of diseases in Greater London. The diseases included cancer of the trachea, lung and bronchus, ischemic heart disease, and chronic bronchitis. In general, mortality declined from the city center outwards. Indeed, it is possible that some parameters that control the probability of illness, or death, from a particular disease, are related to residence, or even to a single room. In medieval Europe, certain houses were considered to be prone to cancer, a conclusion that has recently been recognized as being linked to the geology of the substrate and the radon gas appeal, and which is worthy of careful reconsideration.

1.4.2 Disease mapping

The use of maps to examine the spatial incidence of the disease has a long history in the field of geography. Prominent at the beginning of this history is the yellow fever distribution maps made in 1798 (Robinson, 1982). But these maps were followed by other maps with subjects ranging from maps of hospital capacity and the distribution of first aid stations on a battlefield to maps of infected swamps and other hostile health environments. From the beginning of the 19th century, the number of maps began to increase sharply. By the 1820s, the spread of cholera from India to Eurasia and North America led to the creation of a plethora of maps showing the paths of spread, as well as dates and areas of outbreak. Some, such as Rothenburg's Hamburg map of 1836, showed variations in the intensity of cholera in a more complex way and were reprinted by the British General Board of Health when cholera appeared in Britain in 1848–49.

But, as far as the main epidemiological reports were concerned, such maps were usually of secondary importance and incidental rather than central. The big step in disease mapping came in the mid-19th century with the map of cholera produced by Dr. John Snow (1854) to accompany the second edition of his prize-winning On the Mode of Communication of Cholera. What made Snow's work stand out was not the cartography (dotted maps, which were a well-documented tool showing the geographical distribution of individual deaths from cholera), but the deductive logical justification that was the result of studying the map. By showing what he called the "*topography of the epidemic*," Snow was able to conclude the main source of disease transmission. Hirsch's great Handbook of Historical and Geographical Pathology (1883) continued Snow's tradition, giving the worldwide distribution of a wide range of infectious diseases and drawing conclusions about their modes of spread.

The advent of GIS has made the mapping of disease distributions through spatial and point imaging a relatively straightforward matter. But serious work in this area is subject to two approaches for which GIS may ultimately prove critical: first, as an aid to the development of spatial epidemiological models for interpretation and prediction and, second, as a means of testing hypotheses around mapped distributions. A central issue raised by the growing number of disease maps (especially those generated through digital data manipulations) is the extent to which they are effective in improving our understanding of disease evolution. Because maps are often used because of their visual depiction of phenomena (and thus easy understanding of their dispersion in space), the fact that they can obscure evidence and show spurious concentrations of phenomena should not be overlooked. But the careful use of GIS provides the opportunity to very quickly discern visual correlations between disease distributions and to combine specific evidence with the production of statistical and mathematical models. This sharpens the ability to estimate the value of disease maps and reduces interpretation errors.

An important role that modern high-speed computers play in disease mapping is to solve the complex models required for future disease incidence prediction maps. In the biomathematical literature, extensively reviewed by Dietz (1988), Anderson and May (1991), and (Mollison & Levin, 1995), these models almost always refer to a single region. At any time, it is assuming that the total population in a region can be divided into three classes: an at-risk or susceptible population of size St, an infected population of size It, and a recovered population of size Rt. The recovered population consists of people who have had the disease but can no longer transmit it to others because of recovery, isolation following the appearance of obvious symptoms, or death. It is this threefold division of the population that gives such formulations the generic name of SIR (Susceptible, Infective, and Recovered) models.

The simplest use of GIS is to produce maps showing outbreaks as points. Sergeant et al. (1997) described a low-cost solution for mapping disease outbreaks in Laos using a combination of EPIINFO and EPIMAP software. This solution is based on recording the geographic coordinates of villages so that data referring to the villages can be visualized in conjunction with their different levels of administrative boundaries. More advanced use of GIS involves the production of maps showing measures resulting from the combination of multiple thematic layers.

The EPIMAN-FMD project (Sanson et al., 1999) was initiated in late 1988 to create an integrated direct spatial response system for disease epidemiological purposes. The spatial analysis capabilities developed as part of this project include, but are not limited to:

- Immediate printing of maps depicting the locations of properties at risk.
- Printing large-scale maps depicting individually at-risk properties so that veterinarians can directly access them and note the locations of infected animal groups in the event of disease.
- Modeling and mapping of airborne sources of FMD virus to find high-risk areas (WINDSPREAD model).
- The automatic finding of farms located within a user-defined distance radius, and printing of area visit reports.
- Checking the location of farms from which permission to move was requested and whether or not they belong to the monitored areas.
- The creation of a comprehensive spatial model of FMD would allow future hypotheses that answer the question: what if? (INTERSPREAD MODEL).

1.4.3 Main stages toward maturing of Medical Geography and main post-Millennium works

The spatial analysis came to the fore during the "Quantitative Revolution" in the 1960s and 1970s. However, work that takes into account the close relationship between health/disease and GIS/Remote Sensing emerged with a small scattering internationally in the 1980s. It could be argued that the typical geographic dispersion pattern of papers resulting from research applications in Medical Geography using GIS and Remote Sensing technologies provided a means of conveying the concept of the new-for-that-time idea for the development and evolution of these applications (Albert et al., 2000). According to this model, we can distinguish three stages that describe the dispersion of relevant publications internationally and refer to specific periods.

Stage 1: 1980s. This decade identifies a small number of publications internationally that use GIS technology toward understanding the mechanisms through which human health problems are linked to geographical space. The publications in the form of books reporting on applications in medical geography using spatial analysis through GIS/Remote Sensing technologies are less than 5.

Stage 2: 1991–97. From 1991 to 1994, about 20 articles per year are published, mainly in English and fewer in French. The spatial distribution of articles is small (central and northern Europe, the United States, and Canada). The linear publication/year relationship continues to hold between 1995 and 1997, the only difference being that the quantity of articles produced has increased to about 30 per year. The dynamic development of GIS, especially after the mid-1990s, brought about an explosion in the production of this type of research, when it was understood that it was necessary to adapt geographic analysis methods to these systems in the form of specialized software applications. Publications in the form of books on applications in medical geography using spatial analysis through GIS/Remote Sensing technologies number around 50.

Stage 3: 1998 to date. In this stage, an exponential relationship between publications/year is observed. The spatial spread widens considerably, covering other continents (Asia and Africa) and spreading toward South America and Western and South-Western Europe. GIS and Remote Sensing technologies are now well established as classical tools for the study and analysis of geographical data, and it has been consciously inculcated in researchers that investigating the relationship between health/disease and geographical space necessarily requires the assistance of these technologies. Nowadays, especially, there is intense international activity around the issues of Medical Geography, so on the one hand, the production of publications is large and has a significant dispersion and on the other hand, the dissemination of the results of research work and studies with the help of web technology is widespread. The international literature alone (journal articles), both theoretical and specialized articles, mainly referring to case studies, numbers more than 5000 articles in English, French, German, Spanish, Portuguese, and Italian, not counting articles presented at medical geography conferences. There are specialized international electronic and print journals (e.g., Health and Place, Cadernos de Saúde Pública, Environmental Health Perspectives, etc.); however, about 60%

of the specialized articles are hosted in journals of other specialties (e.g., *International Journal for Geographic Information, Computers & Geosciences, Social Science and Medicine, American Journal of Epidemiology*, etc.). The production of this specialized knowledge is indeed enormous, considering what has happened in the period 1999–2006. The number of publications in the form of books and textbooks on applications in medical geography using spatial analysis through GIS/Remote Sensing technologies now exceeds 350 in more than one language worldwide. However, the English language remains dominant.

Works presented case studies of various parasitic diseases, including leishmaniasis, and geospatial technologies (remote sensing, GPS, and GIS), have been widely used. The methods and technologies in use are analyzed here to investigate the influence of environmental factors on the population dynamics of the intermediate hosts of the diseases under investigation and thus to reliably determine the distribution of pathogens (Thomson & Connor, 2000).

A study conducted by Bucheton et al. (2002) assessed the main risk factors for leishmaniasis during an epidemic in a village in eastern Sudan. This study was carried out after the epidemic had struck this area, namely from 1996 to 1999. The risk of developing leishmaniasis was analyzed in relation to environmental, economic, ethnic, and family factors.

Another study from the same year brings the ecological factor into the concerns of researchers investigating the transmission of infectious diseases in the field, asking whether it can be a powerful tool for validating models of infectious disease transmission risk. According to this study, the creation of ecological zone maps is considered an important tool for determining the biological mechanisms that influence the distribution of diseases in space. As the researchers argue, the key factor that could be used to validate these models is the determination of the spatial extent of the disease, which is made possible through the creation of ecological zone maps (Brooker et al., 2002).

Elnaiem et al. (2003) studied the association of some phlebotomine species such as *P. orientalis* and *P. papatasi* with the presence of certain types of vegetation cover in eastern Sudan. *P. orientalis* was found to be abundant in Acacia Seyal forests, while *P. papatasi* was found in the urban fabric of rural areas. Finding the correlation between the main hosts of parasitic diseases such as leishmaniasis with vegetation cover types can lead to correct conclusions on ways to control the disease.

The environmental variables depicted in another study of 190 villages in Gedaref State included rainfall, vegetation condition, soil type, distance from the existing river, the topography of the area, moisture indices, and mean rainfall estimates (Gebgre et al., 2004).

This publication highlights the research hypotheses, and the empirical evidence on the geographical distribution and dispersion of leishmaniasis in Greece and particularly in the wider area of the Athens basin, through spatial analysis in a GIS environment (Tsatsaris et al., 2005, 2016), and (Tsatsaris et al., 2018).

The spatial epidemiology of some diseases often renders surveillance-based methods problematic for estimating the population at risk of infection (Murray et al., 2004; Gething et al., 2006), while the diseases are transmitted (Riley, 2007; Kubiak et al., 2010). By approaching diseases cartographically and using spatial mathematical models, it is possible to better approximate them (Brooker et al., 2002; Tsatsaris et al., 2005; Ferguson et al., 2005; Hay et al., 2010).

More than 1400 species of infectious agents that can cause disease in humans have been identified in the international literature (Cleaveland, et al., 2001; Taylor et al., 2001; Woolhouse & Gowtage-Sequeria, 2005). For this reason, mapping an infectious disease becomes important. This mapping makes it possible to visualize the magnitude of each disease spatially and can help in decision-making (Hay, 2000; Rogers et al., 2002; Cromley, 2002; Rogers & Randolph, 2003; Riley, 2007; Tsatsaris et al., 2005, 2016; Papadimitriou et al., 2006; Antoniou, Christodoulou, et al., 2010; Antoniou, Psaroulaki, et al., 2010; Mazeris et al., 2010; Iliopoulou et al., 2007, 2009, 2018; Pergantas et al., 2012; Sandalakis et al., 2012; Papa et al., 2013, 2014, 2016; Sargianou et al., 2013; Fouskis et al., 2018).

New technology has entered the study of infectious diseases in humans in recent years. This technology is that of drones, which can provide information at a much better resolution than that available from aerial and/or satellite remote sensing (about 23.000 articles and books have been written until today focused on this interdisciplinary process). There are now many applications for the use of drones ranging from simple mapping to assistance (Tsatsaris and Miliaresis, 2011, Fornace et al., 2014; Schootman et al., 2016; Li et al., 2016; Hardy et al., 2017; Carrasco-Escobar et al., 2019; Laksham, 2019; Schenkel et al., 2020; Aragão et al., 2020).

The models used are based on various population-related assumptions and are subject to assumptions regarding the dynamics of colonization (Hess et al., 2002; Hagenaars et al., 2004). This is an intermediate solution, in which the use of maps is not required. However, as far as models applied to infectious diseases are concerned, it is necessary to use maps showing various spatial phenomena such as example clusters, heterogeneities, and potential spatial patterns regarding transmission and spread (Elliott & Wartenberg, 2004; Pergantas et al., 2017, 2021; Riley, 2007). Space and time are important in these types of diseases, as their distribution and dynamics are critical for immediate decisions related to controlling and monitoring. For these reasons, the study of infectious diseases has found an ideal ally in the fight against them, none other than GIS and remote sensing (Hay et al., 2000; Beck et al., 2000; Huh & Malone, 2001; Rogers and Randolph, 2003; Graham et al., 2004; Herbreteau et al., 2007), and spatial models (Robinson, 2000; Graham et al., 2004; Riley, 2007; Pfeiffer et al., 2008).

1.5 Conclusions

The quality of results from the use of the technological tools of Geographic Information Science improves according to the quality of the data used. The possibility of combining GIS and remote sensing technology with spatial statistical analysis can lead to the adoption of appropriate policies for the management of public health-related crises.

This chapter has been set within a framework that refers to environmental components that have geographical characteristics and influence the mechanisms responsible for the emergence, re-emergence, and spread of pathogenic diseases. As mentioned, the dimension of this concern is transnational, as the continuous human interventions in the natural environment, have undoubtedly created local and global climate changes, (Tsatsaris et al., 2021), although they indicate the direction in which research into the causes of the re-development of these diseases should be directed, do not ensure comprehensive knowledge at present of the causes of such events. This is precisely where the social dimension of the issue comes in, as there is now an international awareness of the need to prevent adverse human health events from the effects of environmental changes on human health and to avoid major social and economic crises. Similar actions can be perceived if one observes other actions that are taking place at the European and international level, that operate in the context of prevention of socio-economic crises and concern various manifestations of natural or man-made activities. These are international networks and observatories of all kinds which collect, record, classify, and manage specialized geographical information, always to inform decision-makers to prevent and avoid events that are harmful to people.

The main issue that arises is therefore how to make practical use of the procedures to provide solutions to issues of early warning and response, when a disease is endemic, when it becomes an epidemic, or when a natural disaster exacerbates the problem by imposing tight deadlines for response.

References

Albert, D. P., Gisler, W. M., & Levergood, B. (2000). *Spatial analysis, GIS and remote sensing. Applications in the health sciences.* Ann Arbor Press Inc.

Anderson, R. M., & May, R. (1991). *Infectious diseases of humans.* Oxford University Press.

Antoniou, M., Christodoulou, V., Mazeris, A., Davies, C., Tsatsaris, A., Ready, P., Cox, J., & Tselentis, Y. (2010). *Validation of a novel approach to estimating prevalence of Canine Leishmaniasis in Cyprus. International Conference of Emerging Vector-borne Diseases in a Changing European Environment Proceedings, France, Montpellier* (p. 38).

Antoniou, M., Psaroulaki, A., Mazeris, A., Ioannou, I., Papaprodromou, M., Georgiou, I., Hristofi, N., Patsias, A., Moschandreas, J., Tsatsaris, A., & Tselentis, Y. (2010). Rats as indicators of the presence and dispersal of pathogens in Cyprus: Ectoparasites, parasitic helminths, enteric bacteria, and Encephalomyocarditis virus. *Vector Borne and Zoonotic Diseases (Larchmont, N.Y.).* Available from http://doi.org/10.1089/vbz.2009.0123.

Aragão, F., Zola, F., Marinho, L., de Genaro Chiroli, D., Junior, A., & Colmenero, J. (2020). Choice of unmanned aerial vehicles for identification of mosquito breeding sites. *Geospatial Health, 15*.

Beck, L. R., Lobitz, B. M., & Wood, B. L. (2000). Remote sensing and human health: New sensors and new opportunities. *Emerging Infectious Diseases, 6*, 217. Available from http://doi.org/10.3201/eid0603.000301.

Brooker, S., Hay, S. I., & Bundy, D. A. (2002). Tools from ecology: Useful for evaluating infection risk models? *Trends in Parasitology, 18*, 70–74. Available from http://doi.org/10.1016/S1471-4922(01)02223-1.

Brownlea, A. (1967). An urban ecology of infectious disease: City of Greater Wollongong-Shell Harbour. *Australian Geographer, 10*, 169–187.

Bucheton, B., Kheir, M., El-Safi, H. S., Hammad, A., Mergani, A., Mary, C., Abel, L., & Dessein, A. (2002). The interplay between environmental and host factors during an outbreak of visceral leishmaniasis in eastern Sudan. *Microbes and Infection, 4*, 1449–1457.

Carrasco-Escobar, G., Manrique, E., Ruiz-Cabrejos, J., Saavedra, M., Alava, F., Bickersmith, S., Prussing, C., Vinetz, J. M., Conn, J. E., Moreno, M., Gamboa, D., & Costantini, C. (2019). High-accuracy detection of malaria vector larval habitats using drone-based multispectral imagery. *PLOS Neglected Tropical Diseases, 13*(1), e0007105. Available from http://doi.org/10.1371/journal.pntd.0007105.

Cleaveland, S., Laurenson, M. K., & Taylor, L. H. (2001). Diseases of humans and their domestic mammals: Pathogen characteristics, host range and the risk of emergence. *Philosophical Transactions of the Royal Society B, 356*, 991–999. Available from http://doi.org/10.1098/rstb.2001.0889.

Cromley, E. K., & McLafferty, S. L. (2002). *GIS and public health*. New York, NY: The Guildford Press.

Dietz, K. (1988). The first epidemic model: A historical note on P.D. En'ko. *Australian Journal of Statistics, 30*(1). Available from http://doi.org/10.1111/j.1467-842X.1988.tb00464.x.

Dobell, C. (1960). *Antony Van Leeuwenhoek and his "Little Animals" Clifford Dobell 1960 Dover PB*. Dover PB.

Elliot, P. J., Cuzick, J., English, D., & Stern, R. (1992). *Geographical and environmental epidemiology: Methods for small area studies* (pp. 40–49). Oxford: Oxford University Press.

Elliot, P., Cuzick, J., English, D., & Stern, R. (1997). *Geographical & environmental epidemiology: Methods for small-area studies. London*: Oxford University Press.

Elliott, P., & Wartenberg, D. (2004). Spatial epidemiology: Current approaches and future challenges. *Environmental Health Perspectives, 112*, 998–1006. Available from http://doi.org/10.1289/ehp.6735.

Elnaiem, D. E., et al. (2003). Risk mapping of visceral leishmaniasis: The role of local variation in rainfall and altitude on the presence and incidence of kala-azar in eastern Sudan. *The American Journal of Tropical Medicine And Hygiene, 68*(1), 10–17.

Eyles, J. (1993). From disease ecology and spatial analysis to ... ? The challenges of medical geography in Canada. *Health and Canadian Society, 1*, 113–146.

Eyles, J., & Woods, K. J. (1983). *The social geography of medicine and health*. London: Croom Helm.

Ferguson, N. M., Cummings, D. A. T., Cauchemez, S., Fraser, C., Riley, S., Meeyai, A., Iamsirithaworn, S., & Burke, D. S. (2005). Strategies for containing an emerging influenza pandemic in Southeast Asia. *Nature, 437*, 209–214. Available from http://doi.org/10.1038/nature04017.

Feyerabend, P. K. (1981). *Realism, Rationalism and Scientific Method*. Cambridge University Press.

Fornace, K. M., Drakeley, C. J., William, T., Espino, F., & Cox, J. (2014). Mapping infectious disease landscapes: Unmanned aerial vehicles and epidemiology. *Trends in Parasitology, 30* (11), 514–519. Available from http://doi.org/10.1016/j.pt.2014.09.001.

Fouskis, J., Sandalakis, V., Cristidou, A., Tsatsaris, A., Tzanakis, N., Tselentis, Y., & Psaroulaki, A. (2018). The epidemiology of Brucellosis in Greece, 2007-2012: A "One Health" approach. *Transactions of the Royal Society of Tropical Medicine and Hygiene*. Available from http://doi.org/10.1093/trstmh/try031.

Gebgre, M. T., et al. (2004). Mapping the potential distribution of Phlebotomus martini and P. orientalis (Diptera: Psychodidae), vectors of kala-azar in East Africa by use of geographic information systems. *Acta Tropica, 90*, 73–86.

Gething, P. W., Noor, A. M., Gikandi, P. W., Ogara, E. A. A., Hay, S. I., Nixon, M. S., Snow, R. W., & Atkinson, P. M. (2006). Improving imperfect data from health management information systems in Africa using space-time geostatistics. *PLoS Med, 3*, e271. Available from http://doi.org/10.1371/journal.pmed.0030271.

Graham, A. J., Atkinson, P. M., & Danson, F. M. (2004). Spatial analysis for epidemiology. *Acta Tropica, 91*, 219–225. Available from http://doi.org/10.1016/j.actatropica.2004.05.001.

Hagenaars, T. J., Donnelly, C. A., & Ferguson, N. M. (2004). Spatial heterogeneity and the persistence of infectious diseases. *Journal of Theoretical Biology, 229*, 349–359. Available from http://doi.org/10.1016/j.jtbi.2004.04.002.

Hardy, A., Makame, M., Cross, D., Majambere, S., & Msellem, M. (2017). Using low-cost drones to map malaria vector habitats. *Parasites & Vectors, 10*(1), 29.

Hay, S. I. (2000). An overview of remote sensing and geodesy for epidemiology and public health application. *Advances in Parasitology, 47*, 1–35. Available from http://doi.org/10.1016/S0065-308X(00)47005-3.

Hay, S. I., Okiro, E. A., Gething, P. W., Patil, A. P., Tatem, A. J., Guerra, C. A., & Snow, R. W. (2010). Estimating the global clinical burden of *Plasmodium falciparum* malaria in 2007. *PLOS Medicine, 7*, e100029.

Hay, S. I., Randolph, S. E., & Rogers, D. J. (2000). *Remote sensing and geographical information systems in epidemiology*. Oxford, UK: Academic Press.

Herbreteau, V., Salem, G., Souris, M., Hugot, J. P., & Gonzalez, J. P. (2007). Thirty years of use and improvement of remote sensing, applied to epidemiology: From early promises to lasting frustration. *Health Place, 13*, 400–403. Available from http://doi.org/10.1016/j.healthplace.2006.03.003.

Hess, G., Randolph, S., Arneberg, P., Chemini, C., Furlanello, C., Harwood, J., Roberts, M., & Swinton, J. (2002). *Spatial aspects of disease dynamics. The ecology of wildlife diseases* (pp. 102–118). Oxford University Press.

Howe, G. M. (1970). Some Recent Developments in Disease Mapping. *The Royal Society of Health Journal, 90*, 16–20.

Howe, M. (1979). Death in London. *The Geographical Magazine, LI*(4), 284–289.

Huh, O. K., & Malone, J. B. (2001). New tools: Potential medical applications of data from new and old environmental satellites. *Acta Tropica, 79*, 35–47. Available from http://doi.org/10.1016/S0001-706X(01)00101-2.

Iliopoulou, P., Tsatsaris, A., & Katsios, I. (2009). *Disease risk maps: The case of leishmaniasis in the Greater Athens Region, Greece, EUGEO. Congress: Challenges for the European Geography in the 21st century*, Bratislava Slovakia August 13–16 2009.

Iliopoulou, P., Tsatsaris, A., Katsios, I., Panagiotopoulou, A., Romaliadis, S., Papadopoulos, B., & Tselentis, Y. (2018). Risk mapping of Visceral Leishmaniasis: A spatial regression model for Attica Region, Greece. *Tropical Medicine and Infectious Diseases, 3*(3), 1–12. Available from http://doi.org/10.3390/tropicalmed3030083.

Iliopoulou, P., Tsatsaris, A., Katsios, I., Panagiotopoulou, A., & Tselentis, Y. (2007). *Urban expansions in the Greater Athens Region and the spread of Zoonoses: The case of Leishmaniasis* http://www.openarchives.gr/view/2572353 *Federation International dés Geometrés (FIG) Com3-WPLA-CHLM International Joint Workshop, Athens, Greece*.

Joseph, A. E., & Bantock, P. R. (1982). Measuring potential physical accessibility to general practitioners in rural areas. *Social Science & Medicine, 16*, 85–90.

Joseph, A. E., & Phillips, D. (1984). *Accessibility and utilization*. London: Harper & Row.

Koopman, J. (1996). *Epidemiology seen more broadly*. In Epidemiology Monitor. Roswell.

Kubiak, R. J., Arinaminpathy, N., & McLean, A. R. (2010). Insights into the evolution and emergence of a novel infectious disease. *PLOS Computational Biology, 6*, e1000947. Available from http://doi.org/10.1371/journal.pcbi.1000947.

Laksham, K. B. (2019). Unmanned aerial vehicle (drones) in public health: A SWOT analysis. *Journal of Family Medicine and Primary Care, 8*(2), 342–346.

Lawson, A. (2001). *Statistical methods in spatial epidemiology*. New York: Wiley.

Learmonth, A. (1957). Some contrasts in the regional geography of malaria in India and Pakistan. *Transactions of the Institute of British Geographers, 23*, 37–59.

Learmonth, A. (1972). Atlases in medical geography 1950-1970: A review. In N. D. McGlashan (Ed.), *Medical geography: Techniques and field studies*. London: Methuen.

Leontidou, L. (1992). *Contemporary Anthropogeography: from Fragmentation to Interdisciplinarity. In Development and Planning: An Interdisciplinary Approach*. Athens: Papazisis Editions.

Leontidou, L., & Sklias, P. (2001). *General geography, anthropogeography and material culture of Europe*. Patra: Hellenic Open University.

Li, C. X., Zhang, Y. M., Dong, Y. D., Zhou, M. H., Zhang, H. D., Chen, H. N., et al. (2016). An unmanned aerial vehicle-mounted cold mist spray of permethrin and tetramethylfluthrin targeting Aedes albopictus in China. *Journal of the American Mosquito Control Association, 32*(1), 59–62.

Mansell, R. P. (1965). *Migrants and malaria*. New York: John Wiley & Sons.

Marbury, M. C., et al. (1996). The indoor air and children's health study: Methods and incidence rates. *Epidemiology, 7*(2), 166–174.

Mayer, J., & Meade, M. S. (1994). A reformed medical geography reconsidered. *Professional Geographer, 46*, 103–106.

Mazeris, A., Soteriadou, K., Dedet, J. P., Haralambous, Ch., Tsatsaris, A., Moschandreas, J., Messaritakis, I., Christodoulou, V., Papadopoulos, B., Ivović, V., Pratlong, F., Loucaides, F., & Antoniou, M. (2010). Leishmaniases and the Cyprus paradox. *The American Journal of Tropical Medicine & Hygiene (AJTMH), 82*(3), 441–448. Available from http://doi.org/10.4269/ajtmh.2010.09-0282.

McGlashan, N. D., & Harrington, J. S. (1976). Some techniques for mapping mortality. *The South African Geographical Journal, 58*, 18–24.

Mollison, D., & Levin, S. A. (1995). Spatial dynamics of parasitism. In B. T. Grenfell, & A. P. Dobson (Eds.), *Ecology of infectious diseases in natural populations* (pp. 384–398). Cambridge University Press.

Murray, C. J. L., Lopez, A. D., & Wibulpolprasert, S. (2004). Monitoring global health: Time for new solutions. *British Medical Journal, 329*, 1096–1100. Available from http://doi.org/10.1136/bmj.329.7474.1096.

O' Hear, A. (1996). *Karl Popper: Philosophy and problems*. Cambridge University Press.

Ostfeld, R. S., Glass, G. E., & Keesing, F. (2005). Spatial epidemiology: An emerging (or re-emerging) discipline. *Trends in Ecology & Evolution, 20*, 328–336. Available from http://doi.org/10.1016/j.tree.2005.03.009.

Papa, A., Chaligiannis, I., Kontana, N., Sourba, T., Tsioka, A., Tsatsaris, A., & Sotiraki, S. (2014). A novel AP92-like Crimean-Congo haemorrhagic fever virus strain, Greece. *Ticks and Tick-borne Diseases*, *5*(5). Available from http://doi.org/10.1016/j.ttbdis.2014.04.008.

Papa, A., Sidira, P., Kallia, S., Ntouska, M., Zotos, N., Doumbali, E., Maltezou, C. H., Demiris, N., & Tsatsaris, A. (2013). Factors associated with IgG positivity to Crimean-Congo hemorrhagic fever virus in the area with the highest seroprevalence in Greece. *Ticks and Tick-borne Diseases*. Available from http://doi.org/10.1016/j.ttbdis.2013.04.003.

Papa, A., Sidira, P., & Tsatsaris, A. (2016). Spatial cluster analysis of Crimean-Congo hemorrhagic fever virus seroprevalence in humans, Greece. *Parasite Epidemiology and Control*, *1*(3), 211–218. Available from http://doi.org/10.1016/j.parepi.2016.08.002.

Papadimitriou, K., Kastania, A., & Tsatsaris, A. (2006). *Medical geography and social medicine. The example of Megisti. 12th Regional Conference of WoncaEurope, August 2006, Florence*.

Papadopoulos, B., & Tselentis, Y. (1994). Sandflies in the greater Athens region, Greece. *Parasite*, *1*, 131–140.

Pergantas, P., Papanikolaou, N., Malesios, C., Tsatsaris, A., Kondakis, M., Perganta, I., Tselentis, Y., & Demiris, N. (2021). Towards a semi-automatic early warning system for vector-borne diseases. *International Journal of Environmental Research and Public Health*, *10*(2). Available from http://doi.org/10.3390/ijerph18041823.

Pergantas, P., Tsatsaris, A., Demiris, N., Moraitis, J., & Tselentis, Y. (2012). *Spatial predictive model for malaria resurgence: Integrates entomological. Environmental and Social Data in Central Greece, 18th E-sove Conference, Montpellier, France*.

Pergantas, P., Tsatsaris, A., Malesios, C., Kriparakou, G., Demiris, N., & Tselentis, Y. (2017). A spatial predictive model for malaria resurgence in Central Greece integrating entomological, environmental, and social data. *Plos One*. Available from http://doi.org/10.1371/journal.pone.0178836.

Pfeiffer, D. U., Robinson, T. P., Stevenson, M., Stevens, K. B., Rogers, D. J., & Clements, A. C. A. (2008). *Spatial analysis in epidemiology*. New York: Oxford University Press.

Platt, J. R. (1964). Strong inference. *Science*, *146*, 347–353.

Pyle, G. F. (1979). *Applied medical geography*. Holt Rinehart and Winston.

Riley, S. (2007). Large-scale spatial-transmission models of infectious disease. *Science*, *316*, 1298. Available from http://doi.org/10.1126/science.1134695.

Robinson, A. G. (1982). *Early thematic mapping in the history of cartography*. University of Chicago Press; London.

Robinson, T. P. (2000). Spatial statistics and geographical information systems in epidemiology and public health. *Advances in Parasitology*, *47*, 81–128.

Rogers, D. J., & Randolph, S. E. (2003). Studying the global distribution of infectious diseases using GIS and RS. *Nature Reviews Microbiology*, *1*, 231–236. Available from http://doi.org/10.1038/nrmicro776.

Rogers, D. J., Randolph, S. E., Snow, R. W., & Hay, S. I. (2002). Satellite imagery in the study and forecast of malaria. *Nature*, *415*, 710–715. Available from http://doi.org/10.1038/415710a.

Ross, N. A., Rosenberg, M. W., Pross, D. C., & Bass, B. (1994). Contradiction in women's health care provision: A case study of attendance for breast cancer screening. *Social Science & Medicine*, *39*. Available from http://doi.org/10.1016/0277-9536(94)90373-5.

Sandalakis, V., Pasparaki, E., Chochlakis, D., Svirinaki, E., Sifaki-Pistolla, D., Tsafantakis, E., Tsatsaris, A., Antoniou, M., Psaroulaki, A., & Tselentis, I. (2012). *Timeless surveillance of seven zoonosis in a livestock region of Crete. Proceedings of 2nd Panhellenic Forum of Public Health.*

Sanson, R. L., Morris, R. S., & Stern, M. W. (1999). EpiMAN-FMD: A decision support system for managing epidemics of vesicular disease. *Revue Scientifique et Technique Office International des Epizooties*, *18*, 593–605.

Sargianou, M., Panos, G., Tsatsaris, A., Gogos, Ch., & Papa, A. (2013). Crimean-Congo hemorrhagic fever: seroprevalence and risk factors among humans in Achaia, western Greece. *International Journal of Infectious Diseases (IJID)*. Available from http://doi.org/10.1016/j.ijid.2013.07.015.

Schenkel, J., Taele, P., Goldberg, D., Horney, J., & Hammond, T. (2020). identifying potential mosquito breeding grounds: Assessing the efficiency of UAV technology in public health. *Robotics*, *9*(4), 91. Available from http://doi.org/10.3390/robotics9040091.

Schootman, M., Nelson, E. J., Werner, K., Shacham, E., Elliott, M., Ratnapradipa, K., et al. (2016). Emerging technologies to measure neighborhood conditions in public health: Implications for interventions and next steps. *International Journal of Health Geographics*, *15*(1), 20.

Sergeant, E. S. G., Cameron, A. R., Baldock, F. C., & Vongthilath, S. (1997). An animal health information system for Laos. In S. More (Ed.), *Epidemiology Programme of the 10th Federation of Asian Veterinary Associations Congress*. Proceedings.

Shacklette, H. T., Sauer, H. I., & Miesch, A. T. (1972). Distribution of trace elements in the occurrence of heart disease in Georgia. *Geological Society of America Bulletin*, *83*, 1077–1082.

Taylor, L. H., Latham, S. M., & Woolhouse, M. E. (2001). Risk factors for human disease emergence. *Philosophical Transactions of the Royal Society B*, *356*, 983–989. Available from http://doi.org/10.1098/rstb.2001.0888.

Thomson, M. C., & Connor, S. J. (2000). Environmental information systems for the control of arthropod vector of disease. *Medical and Veterinary Entomology*, *14*, 227–244.

Tsatsaris, A., Kalogeropoulos, K., Stathopoulos, N., Louka, P., Tsanakas, K., Tsesmelis, D., Krassanakis, V., Petropoulos, G., Pappas, V., & Chalkias, Ch. (2021). Geoinformation technologies in support of environmental hazards monitoring under climate change: An extensive review. *International Journal of Geo-Information*, *10*(2). Available from http://doi.org/10.3390/ijgi10020094.

Tsatsaris, A., Chochlakis, D., Papadopoulos, B., Petsa, K., Georgalis, L., Angelakis, Em, Ioannou, I., Tselentis, Y., & Psaroulaki, A. (2016). Species composition, distribution, ecological preference and host association of ticks in Cyprus. *Experimental and Applied Acarology*. Springer International Publishing. https://link.springer.com/article/10.1007/s10493-016-0091-9.

Tsatsaris, A., Iliopoulou, P., Panagiotopoulou, A., & Tselentis, I. (2005). *A geographical database for the control of Leishmaniasis: The case of Greater Athens, Greece. 10th International Symposium on Health Information Management Research (ISHIMR2005), Thessalonica, Greece* (pp. 120–135).

Tsatsaris, A., & Miliaresis, G. (2011). Spatial correlation of tuberculosis (TB) incidents to the MODIS LST biophysical signature of African countries. *The International Journal of Environmental Protection*, *1*(1), 49–57.

Woolhouse, M. E., & Gowtage-Sequeria, S. (2005). Host range and emerging and reemerging pathogens. *Emerging Infectious Diseases*, *11*, 1842–1847. Available from http://doi.org/10.3201/eid1112.050997.

Chapter 2

Quality of life in Athens, Greece, using geonformatics

Antigoni Faka[1], Kleomenis Kalogeropoulos[2] and Christos Chalkias[1]
[1]Department of Geography, Harokopio University of Athens, Athens, Greece, [2]Department of Surveying and Geoinformatics Engineering, University of West Attica (UniWA), Athens, Greece

2.1 Introduction

Quality of life (QoL) has been studied by scientists and researchers in various fields, such as environment, sociology, geography, health-sciences, psychology, and economics. It is a multidimensional concept, including aspects of social, economic, and environmental conditions, as well as physical and mental health, and deals with the overall QoL of society (Farquhar, 1995; Schalock, 2000). For many years, QoL research was limited to health-related issues, investigating individuals' physical and mental health. In the middle of the 20th century, QoL was primarily recognized as the best possible living conditions and related to high levels of consumption and property possession. Afterwards, researchers started to study QoL taking under consideration the general living framework which contains concepts such as education and the natural environment. Nowadays, QoL incorporates many aspects including health, psychology, environment, economy, and social issues (Cabello Eras et al., 2014; Faka, 2020; Faka et al., 2021a, 2021b; Faka et al., 2022; Farquhar, 1995; Garau & Pavan, 2018; Maggino & Zumbo, 2012; Mizgajski et al., 2014; Sirgy et al., 2006; Ventegodt et al., 2003).

The improvement of QoL is often included in the policy agendas of governments and global organizations (Eurofound, 2017; European Commission & Directorate-General for Regional and Urban Policy 2020; OECD, 2020; UN-Habitat, 2016). European Union has planned and accomplished a series of studies about QoL. These studies focus on all the parameters that play a crucial role in shaping QoL such as health, education, economy, employment, services, security, environmental quality, culture, living and housing conditions, and family and social life (Eurofound, 2004, 2009, 2012, 2017).

The last decades urban environments are subject to major transformations due to the rapid population growth combined with the high rate of

Geoinformatics for Geosciences. DOI: https://doi.org/10.1016/B978-0-323-98983-1.00003-X

urbanization, affecting the urban QoL (Cramer et al., 2004; European Commission, 2016). Nowadays, urban areas encounter problems such as social dissimilarities, ecological deprivation, and crime (Eurofound, 2012), reducing the QoL (Eurofound, 2017; European Commission, 2016). Consequently, exploring spatial inequalities of urban QoL and its determinants are crucial for urban planning and can be used as a policy tool.

According to the international literature, QoL has been assessed using a wide variety of variables related to the quality of natural environment, the socioeconomic conditions, the quality of housing, the adequacy of public schools and health infrastructures, the leisure facilities, etc. (Din et al., 2013; Gill, 1994; Kazemzadeh-Zow et al., 2018; Lee & Guest, 1983; Murgaš & Klobučník, 2016, 2018; Najafpour et al., 2014; Peach & Petach, 2016; Vukmirovic et al., 2019; Weziak-Białowolska, 2016). In many studies, the above variables have been used to examine either the objective or the subjective approach of QoL. Some researchers support that the measurement of QoL presupposes the use of statistical and geospatial data to evaluate the objective environmental conditions independently of the individuals' perceptions (Apparicio et al., 2008; Murgaš & Klobučník; 2016). On the contrary, other researchers focus on the perceived QoL and the subjective way people evaluate the geographic features (Sirgy & Cornwell, 2002; Sirgy et al., 2010). However, the perception of the QoL across city neighborhoods is influenced by the environmental conditions prevailing in Greenberg and Crossney (2007).

Perceived QoL and neighborhood satisfaction is generally affected by various characteristics of the geographical environment. It has been widely argued that natural, built, and socioeconomic environment influence on how individuals assess their neighborhood and evaluate their overall satisfaction with the area's environmental conditions (Dumith et al., 2022; Gandelman et al., 2012; Mouratidis & Yiannakou, 2022; Permentier et al., 2011; Weckroth et al., 2022). The positive association between perceived QoL and the natural environment, as well as the socioeconomic status has been extensively documented (Gandelman et al., 2012; Mouratidis & Yiannakou, 2022; Permentier et al., 2011; Weckroth et al., 2022; Welsch, 2006). On the other hand, high neighborhood density, long distances to health, and other services seem to have a negative effect on perceived QoL (Florida et al., 2013; Mouratidis & Yiannakou, 2022; Weckroth et al., 2022). However, the effect of some environmental characteristics on QoL, such as density and built environment, has been revealed insignificant in other case studies (Arifwidodo, 2012; Hua et al., 2022).

The above results may vary due to the considerable spatial heterogeneity among different regions, which concerns the differentiation of natural and built environment, as well as socioeconomic and housing conditions. The effect of various geographical elements and spatial disparities of urban areas on residential QoL perceptions has different impact in different regions (Li & Liu, 2021).

The development of spatial statistics tools, such as Geographically Weighted Regression (GWR) (Fotheringham & Brunsdon, 1999; Fotheringham et al., 2002), allows the investigation of the local relationships between geographical data. In spite the fact that the geospatial research on the above associations is limited (Li & Liu, 2021; Ogneva-Himmelberger et al., 2013), GWR has been widely used to investigate various geographical phenomena and examine the spatial variations in relationships between predictors and outcome variables (among others Adamiak et al., 2021; Chen et al., 2018; Faka et al., 2019; Luo et al., 2022). This provides a powerful tool to analyze spatial patterns of perceived QoL and its determinants.

This research aims at investigating the relationship between the urban environment characteristics and perceived QoL, among neighborhoods in Municipality of Athens (MoA). The effect of MoA's environmental conditions on QoL was investigated using the perceived QoL of MoA's residents and the estimated quality of six environmental criteria; natural, built and socioeconomic environment, housing conditions and access to public services and infrastructures, and cultural and recreational facilities, as evaluated in a previous study on the assessment of the QoL in the study area (Faka et al., 2022), by implementing GWR.

2.2 Study area

Athens is the capital of Greece, located in Attica prefecture, in the central part of Greece which concentrates almost the half population of the country (4,000,000 of almost 11,000,000 inhabitants). Attica prefecture has 58 municipalities, where the biggest is the MoA with 664,046 inhabitants based on the last Population and Housing Census in 2011. MoA is divided into 494 urban analysis units (URANUs); neighborhoods of similar population size (1000 inhabitants each) which correspond to the spatial unit of analysis (Fig. 2.1).

2.3 Materials and methods

2.3.1 Criteria and variables

The effect of MoA's environmental conditions on QoL was investigated using the perceived QoL of MoA's residents and the estimated quality of six environmental criteria, as evaluated in a previous study on the assessment of the QoL in the study area (Faka et al., 2022). The criteria were based on basic domains of QoL, including natural, built and socioeconomic environment, housing conditions and access to public services and infrastructures, and cultural and recreational facilities. Each criterion was composed by a set of mappable subcriteria/variables that evaluate the domains

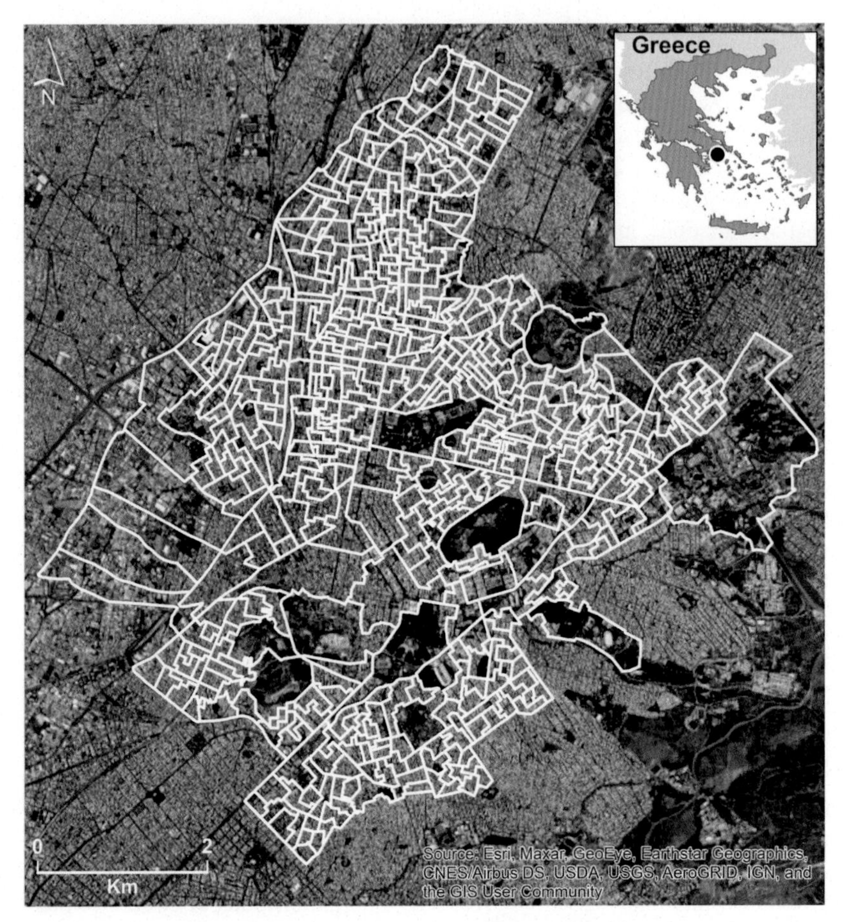

FIGURE 2.1 The study area and the spatial units of analysis (URANUs).

described previously (Table 2.1). The variables were estimated using both spatial and tabular data. Weighted cartographic overlay was implemented to assess each criterion, ranking QoL on a five-category ordinal scale, from very low (1) to very high value (5) (Fig. 2.2). Further details about the criteria and the variables that compose them could be found at Faka et al. (2022).

The residents' perception on overall QoL was obtained using questionnaires. The questionnaire survey was conducted during a five-month period (September 2020 to February 2021) on a total sample of 181 residents/participants through online questionnaires. The residents were asked to rate from 1 (very low) to 5 (very high) the geographical environment and QoL of their residence area. The questionnaire contained closed-ended questions regarding

TABLE 2.1 Quality of life criteria and variables.

Criteria	Subcriteria/variables
Built environment	Population density (habitants/km^2)
	Open spaces (%)
Natural environment	Mean distance to industries (m)
	Density of high-traffic roads and highways (km/km^2)
	Urban green (%)
Socioeconomic environment	Higher educated population (%)
	Mean income (euro)
Housing conditions	Detached houses (%)
	New buildings (%)
Public services and infrastructures	Accessibility to medical services/hospitals (min)
	Accessibility to schools (min)
	Accessibility to sport facilities (min)
Cultural and recreational facilities	Accessibility to cultural facilities (min)
	Accessibility to recreational facilities (min)

the score of QoL criteria, the factors valuing them, as well as overall QoL. The mean value of overall QoL scored in each spatial unit was used to estimate the relation between QoL and the six environmental criteria.

2.3.2 Geographically Weighted Regression

GWR was applied to explore the spatial variations in the relationships between perceived QoL and the six environmental criteria in MoA. GWR allows for local parameters to be estimated (Fotheringham & Brunsdon, 1999) and investigate the existence of spatial nonstationarity in the relationships between a phenomenon and its determinants (Fotheringham et al., 2002). The general GWR equation is defined as:

$$y_i u = \beta_{0i} u + \beta_{1i} u x_{1i} + \beta_{2i} u x_{2i} + \ldots + \beta_{mi} u x_{mi} \tag{2.1}$$

where the dependent variable y at a location (u) is regressed on a set (m) of independent variables (x) at the same location. β describes a relationship around the location (u), and it is specific to that location. GWR constructs a separate equation for every spatial unit (i) of the area that is being studied, incorporating the dependent and explanatory variables.

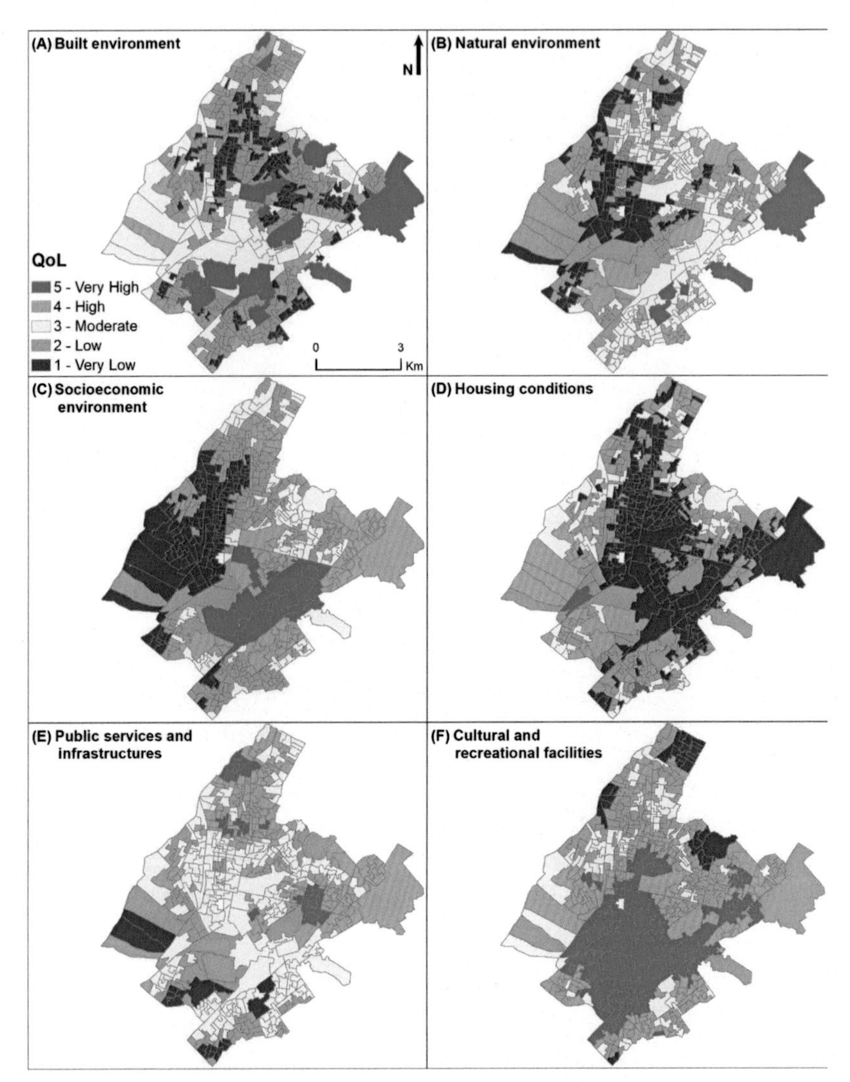

FIGURE 2.2 The criteria of QoL in MoA. (A) Built environment, (B) natural environment, (C) socioeconomic environment, (D) housing conditions, (E) public services and infrastructures, (F) cultural and recreational facilities. *MoA*, Municipality of Athens; *QoL*, quality of life.

Local models were implemented to examine the spatial heterogeneity of the relationship between QoL and the six environmental criteria, across the study area. Each of the criteria was defined as the explanatory variable and perceived QoL as the dependent variable. The fixed Gaussian kernel function was employed for the estimation of bandwidth in the GWR models, using a fixed distance to solve each local regression analysis. The optimal distance was determined using the corrected Akaike Information Criterion (AICc). The AICc is tested at various distances (maximum to minimum) incrementally between them, concluding to the best distance value based on the model with the minimum AICc value which provides better fit to the observed data (Fotheringham et al., 2002; Hurvich et al., 1998).

GWR analysis provide a set of statistics which denote local relationships and can be used to map the spatial distribution for the local parameter estimates. Mapping the regression coefficient distribution illustrated how spatially consistent the relationships between QoL and each criterion were across MoA. GWR analysis and mapping were performed using the ArcGIS 10.2 commercial package (ESRI Inc., Redlands, California, USA).

2.4 Results

Fig. 2.3 illustrates the spatial distribution of each QoL criterion coefficient values. The variation in the regression coefficients for the GWR models shows the nature and the strength of the nonstationarity between QoL criteria and perceived overall QoL. The analysis revealed either weak or strong but nonstationary association between perceived overall QoL and the six environmental criteria across the MoA.

The greatest positive effect of the built environment on perceived QoL was revealed in some central and south-eastern neighborhoods (Fig. 2.3A). Positive but weaker relationship was noticed in most neighborhoods of the study area, whereas negative association was observed in some north-central, south, and south-western neighborhoods. In the latter neighborhoods, negative association between the natural environment and the perceived QoL was also revealed (Fig. 2.3B). This negative relation was also noticed in north MoA. However, the positive effect of natural environment on perceived QoL seemed to dominate the rest of the neighborhoods. The strong positive association of the socioeconomic environment with perceived QoL was observed in most neighborhoods of MoA, except a limited number of neighborhoods, mainly in the west MoA, where a negative association was noticed (Fig. 2.3C). The housing conditions appeared to strongly positively influence the perceived QoL in a group of neighborhoods in the central study area (Fig. 2.3D). Positive but weaker relationship was revealed in most neighborhoods of the MoA, whereas weak negative association was noticed in the south-eastern and some western neighborhoods. The negative effect of public services and infrastructures on perceived QoL was mainly observed in north-eastern and west-central neighborhoods (Fig. 2.3E). On the contrary, the public services and infrastructures coefficient values across the rest neighborhoods indicated a positive association with the perceived QoL. Finally, the access to cultural and recreational facilities affects negatively the perceived QoL in the north and central neighborhoods, whereas, in the west, south, and east neighborhoods, this association is positive (Fig. 2.3F).

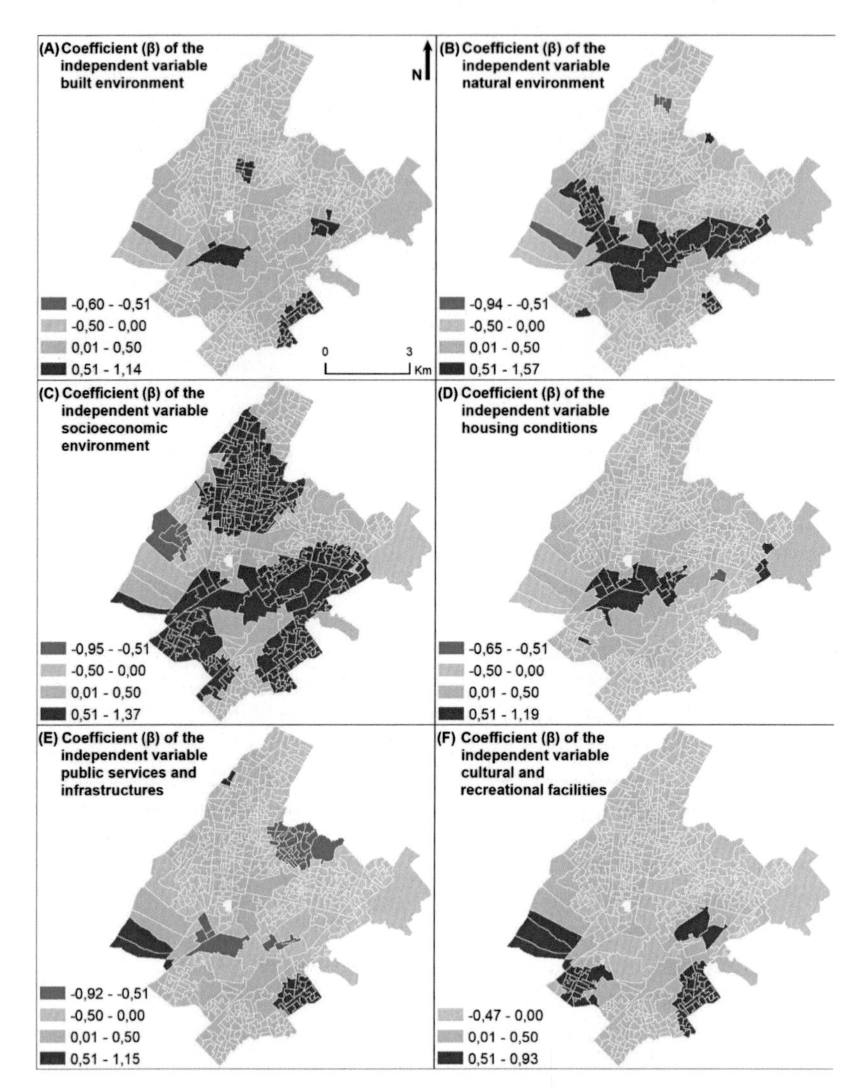

FIGURE 2.3 The spatial variation of the QoL criteria coefficient from GWR modeling. (A) Built environment, (B) natural environment, (C) socioeconomic environment, (D) housing conditions, (E) public services and infrastructures, (F) cultural and recreational facilities. *GWR*, Geographically Weighted Regression; *QoL*, quality of life.

2.5 Discussion

This study investigates the spatial heterogeneity in the relationship of perceived QoL and the urban environment characteristics, in MoA. The study area is characterized by extensive geographical disparities (Faka et al., 2022), and this study revealed that these inequalities are also related to spatial heterogeneity in the relationship of perceived QoL and its determinants.

The results of local regression analysis revealed nonstationary associations between perceived overall QoL and the six environmental criteria, across the neighborhoods of MoA. The positive effect of all criteria on perceived QoL prevailed in most MoA's neighborhoods, although inverse relationships was noticed as well.

The positive association of the good quality of the built environment (low population density and high percentages of open spaces) and the perceived QoL was confirmed in most neighborhoods. This relationship appears to be consistent with previous studies, stating that neighborhood density is negatively linked to perceived QoL and satisfaction (Weckroth et al., 2022; Streimikiene, 2014) and proximity to open spaces is positively (Sugiyama et al., 2009). However, the inverse association between the built environment and perceived QoL was noticed in some neighborhoods. This nonstationary relationship may exist due to interrelationships between neighborhood characteristics. For instance, high neighborhood density contributes to a larger number of amenities in the neighborhood (Mouratidis & Yiannakou, 2022), a factor that enhance QoL in urban areas.

The positive correlation between QoL and the natural environment has been confirmed in the international literature (Gandelman et al., 2012; Li & Liu, 2021; Ogneva-Himmelberger et al., 2013; Streimikiene, 2014; Zhang et al., 2017). In this study, although the positive relationship was noticed between perceived QoL and high levels of natural environment (high distance to industries, low density of highways and high percentages of urban green), this association was negative in some north and south neighborhoods. This may be due to road network variable, which seem to affect in two ways QoL. High road network density is associated with high levels of noise pollution that has negative impact on QoL (Gandelman et al., 2012; Streimikiene, 2014). On the other hand, high street connectivity, linked to road density, has been advocated for its benefits to neighborhoods environments, and poor street connectivity is detrimental to perceived QoL (Cao, 2016).

The influence of socioeconomic environment, such as education and economic level, has been shown in previous researches (Permentier et al., 2011; Weckroth et al., 2022; Li & Liu, 2021; Ogneva-Himmelberger et al., 2013). In this study, the high socioeconomic status (high percentages of higher educated population and high mean income) seems to be a strong predictor of perceived QoL, but the positive correlation was not revealed across all MoA. In a limited number of neighborhoods, mainly in the west MoA, this association was negative. In spite the fact that in these neighborhoods the lower socioeconomic strata are highly concentrated, this relationship was negative, indicating that various factors may interact and multicomponents need to be examined.

The privileged housing conditions (high percentages of detached and newly built houses) were associated positively with perceived QoL in an extensive number of the study area's neighborhoods, whereas weak negative correlation was noticed in the south-eastern and some western neighborhoods. Previous studies have found the quality of the housing stock in a neighborhood to have positive impact on neighborhood satisfaction (Basolo & Strong, 2002; Lee & Guest, 1983). These findings may be related to the detached houses variable, which may serves as an adequate proxy for housing space sufficiency, although in some areas detached houses are abandoned buildings in poor condition that attract crime activities.

Proximity to various services, facilities, and amenities have been found to be associated with greater residents' satisfaction and perception of QoL (Mouratidis & Yiannakou, 2022; Basolo & Strong, 2002; Hoogerbrugge et al., 2022; Park et al., 2021). On the other hand, Kyttä et al. (2016) found that closeness to services positively contributed to higher perceived environmental quality in urban neighborhoods and negatively in suburban neighborhoods. The nonstationarity of this relationship was also revealed in this study. The negative correlation between accessibility to public services and infrastructures (hospitals, schools, and sports facilities) and perceived QoL was extending to more neighborhoods than the corresponding relationships of the other local models, indicating the processes vary between different areas.

The spatial distribution of cultural and recreational facilities coefficient values from GWR analysis revealed that the relationship between the accessibility to them and perceived QoL was not spatially consistent across the study area, as well. One possibility of increasing understanding the reason that good accessibility to services and facilities does not everywhere contribute to the better perceived QoL would be to examine the quality of the services and facilities, alongside the closeness.

The implementation of local modeling in this study proved to be a valuable method for investigating local variations in the relationship between perceived QoL and urban environmental conditions. Mapping the explanatory variable's coefficient values from GWR analysis contributed to the better understanding and explanation of the results across the study area. An important limitation that has to be acknowledged is the synthesis of the proposed criteria. The selected variables were strongly based on the related international literature, but more determinants should be examined in future work. Another limitation is related to the boundary spatial units which neighbor other areas of the city's continuous urban fabric, not included in the study area. Limiting the analysis in the MoA, QoL was not explored in the neighboring areas and the contiguity with these neighborhoods was not taken into account.

The local regression analysis applied to the present study revealed that the relationship between the urban environmental conditions and perceived QoL was not spatially consistent across the study area. The aforementioned findings underline the need of investigating the processes and the interaction among various factors that determine QoL, which vary across different areas. The identification of the nonstationary relationships between the determinants of QoL and the residents' perceived QoL provided by the environment of their neighborhood is a powerful tool for decision-makers to design future targeted actions and strategies, focusing on the improvement of QoL. In future work, more variables will be tested to improve local models. Furthermore, the temporal dimension will be included to identify the examined relationships trends in the study area.

References

Adamiak, M., Jażdżewska, I., & Nalej, M. (2021). Analysis of built-up areas of small polish cities with the use of deep learning and geographically weighted regression. *Geosciences (Switzerland), 11*(5), 223.

Arifwidodo, S. D. (2012). Exploring the effect of compact development policy to urban quality of life in Bandung, Indonesia City. *Culture and Society, 3*(4), 303–311.

Apparicio, P., Séguin, A.-M., & Naud, D. (2008). The quality of the urban environment around public housing buildings in Montréal: An objective approach based on GIS and multivariate statistical analysis. *Social Indicators Research, 86*, 355–380.

Basolo, V., & Strong, D. (2002). Understanding the neighborhood: from residents' perceptions and needs to action. *Housing Policy Debate, 13*, 83–105.

Cabello Eras, J. J., Covas Varela, D., Hernández Pérez, G. D., Sagastume Gutiérrez, A., García Lorenzo, D., Vandecasteele, C., & Hens, L. (2014). Comparative study of the urban quality of life in Cuban first-Level cities from an objective dimension. *Environment, Development and Sustainability, 16*, 195–215.

Cao, X. J. (2016). How does neighborhood design affect life satisfaction? Evidence from twin cities. *Travel Behaviour and Society, 5*, 68–76.

Chen, J., Zhou, C., Wang, S., & Hu, J. (2018). Identifying the socioeconomic determinants of population exposure to particulate matter (PM2.5) in China using geographically weighted regression modeling. *Environmental Pollution, 241*, 494–503.

Cramer, V., Torgersen, S., & Kringlen, E. (2004). Quality of life in a city: The effect of population density. *Social Indicators Research, 69*, 103–116.

Din, H. S. E., Shalaby, A., Farouh, H. E., & Elariane, S. A. (2013). Principles of urban quality of life for a neighborhood. *HBRC Journal, 9*, 86–92.

Dumith, S. C., Leite, J. S., Fernandes, S. S., Sanchez, É. F., & Demenech, L. M. (2022). Social determinants of quality of life in a developing country: Evidence from a Brazilian sample. *Journal of Public Health (Germany), 30*(6), 1465–1472.

Eurofound. (2004). Quality of life in Europe: First European Quality of Life Survey 2002. Office for Official Publications of the European Communities.

Eurofound. (2009). Quality of life in Europe 2003–2007: Second European Quality of Life Survey. Office for Official Publications of the European Communities.

Eurofound. (2012). Third European Quality of Life Survey—Quality of Life in Europe: Impacts of the Crisis. Publications Office of the European Union.

Eurofound. (2017). European Quality of Life Survey 2016: Quality of Life, Quality of Public Services, and Quality of Society. Publications Office of the European Union.

European Commission. (2016). The State of European Cities 2016: Cities leading the way to a better future. European Commission.

European Commission & Directorate-General for Regional and Urban Policy. (2020). Report on the Quality of Life in European Cities 2020. Publications Office of the European Union

Faka, A. (2020). Assessing quality of life inequalities. A geographical approach. *ISPRS International Journal of Geo-Information*, *9*, 600.

Faka, A., Chalkias, C., Georgousopoulou, E. N., Tripitsidis, A., Pitsavos, C., & Panagiotakos, D. B. (2019). Identifying determinants of obesity in Athens, Greece through global and local statistical models. *Spatial and Spatio-temporal Epidemiology*, *29*, 31–41.

Faka, A., Kalogeropoulos, K., Maloutas, T., & Chalkias, C. (2021a). Urban quality of life: Spatial modeling and indexing in Athens Metropolitan Area, Greece. *ISPRS International Journal of Geo-Information*, *10*, 347.

Faka, A., Kalogeropoulos, K., Maloutas, T., & Chalkias, C. (2021b). Assessing spatial inequalities in urban quality of life—Empirical evidences from analysis in Athens, Greece. *Proceedings of the 28th APDR Congress: Green and Inclusive Transitions in Southern European Regions: What Can We Do Better? UTAD*, 16–17.

Faka, A., Kalogeropoulos, K., Maloutas, T., & Chalkias, C. (2022). Spatial variability and clustering of quality of life at local level: A geographical analysis in Athens, Greece. *ISPRS International Journal of Geo-Information*, *11*(5), 276.

Farquhar, M. (1995). Definitions of quality of life: A taxonomy. *Journal of Advanced Nursing*, *22*(3), 502–508.

Florida, R., Mellander, C., & Rentfrow, P. J. (2013). The happiness of cities. *Regional Studies*, *47*(4), 613–627.

Fotheringham, A. S., & Brunsdon, C. (1999). Local forms of spatial analysis. *Geographical Analysis*, *31*(4), 340–358.

Fotheringham, A. S., Brunsdon, C., & Charlton, M. (2002). *Geographically weighted regression: The analysis of spatially varying relationships*. Wiley and Sons.

Gandelman, N., Piani, G., & Ferre, Z. (2012). Neighborhood determinants of quality of life. *Journal of Happiness Studies*, *13*(3), 547–563.

Garau, C., & Pavan, V. M. (2018). Evaluating urban quality: Indicators and assessment tools for smart sustainable cities. *Sustainability*, *10*, 575.

Gill, T. M. (1994). A critical appraisal of the quality of quality-of-life measurements. *JAMA The Journal of the American Medical Association*, *272*, 619.

Greenberg, M., & Crossney, K. (2007). Perceived neighborhood quality in the United States: Measuring outdoor, housing and jurisdictional influences. *Socio-Economic Planning Sciences*, *41*, 181–194.

Hoogerbrugge, M. M., Burger, M. J., & Van Oort, F. G. (2022). Spatial structure and subjective well-being in north-west Europe. *Regional Studies*, *56*(1), 75–86.

Hua, J, Mendoza-Vasconez, A. S., Chrisinger, B. W., Conway, T. L., Todd, M., Adams, M. A., Sallis, J. F., Cain, K. L., Saelens, B. E., Frank, L. D., & King, A. C. (2022). Associations of social cohesion and quality of life with objective and perceived built environments: A latent profile analysis among seniors. *Journal of Public Health*, *44*(1), 138–147.

Hurvich, C. M., Simonoff, J. S., & Tsai, C. L. (1998). Smoothing parameter selection in nonparametric regression using an improved Akaike information criterion. *Journal of the Royal Statistical Society Series B (Statistical Methodology)*, *60*(2), 271–293.

Kazemzadeh-Zow, A., Darvishi Boloorani, A., Samany, N. N., Toomanian, A., & Pourahmad, A. (2018). Spatiotemporal modelling of urban quality of life (UQoL) using satellite images and GIS. *International Journal of Remote Sensing*, *39*, 6095–6116.

Kyttä, M., Broberg, A., Haybatollahi, M., & Schmidt-Thomé, K. (2016). Urban happiness: Context-sensitive study of the social sustainability of urban settings. *Environment and Planning B: Planning and Design*, *43*(1), 34–57.

Lee, B. A., & Guest, A. M. (1983). Determinants of neighborhood satisfaction: A Metropolitan-level analysis. *The Sociological Quarterly*, *24*, 287–303.

Li, X., & Liu, H. (2021). The influence of subjective and objective characteristics of urban human settlements on residents' life satisfaction in China. *Land*, *10*(12), 1400.

Luo, Y., Yan, J., McClure, S. C., & Li, F. (2022). Socioeconomic and environmental factors of poverty in China using geographically weighted random forest regression model. *Environmental Science and Pollution Research*, *29*(22), 33205–33217.

Maggino, F., & Zumbo, B. (2012). *Measuring the quality of life and the construction of social indicators. Handbook of social indicators and quality of life research* (pp. 201–238). Springer.

Mizgajski, A., Walaszek, M., & Kaczmarek, T. (2014). Determinants of the quality of life in the communes of the Poznań agglomeration: A quantitative approach. *Quaestiones Geographicae*, *33*, 67–80.

Mouratidis, K., & Yiannakou, A. (2022). What makes cities livable? Determinants of neighborhood satisfaction and neighborhood happiness in different contexts. *Land Use Policy*, *112*, 105855.

Murgaš, F., & Klobučník, M. (2016). Municipalities and regions as good places to live: Index of quality of life in the Czech Republic. *Applied Research in Quality of Life*, *11*, 553–570.

Murgaš, F., & Klobučník, M. (2018). Quality of life in the city, quality of urban life or well-being in the city: Conceptualization and case study. *Ekologia Bratislava*, *37*, 183–200.

Najafpour, H., Bigdeli Rad, V., Lamit, H., & Bin Rosley, M. S. (2014). The systematic review on quality of life in urban neighborhoods. *Life Science Journal*, *11*, 355–364.

OECD. (2020). *How's life? 2020: Measuring well-being*. OECD Publishing.

Ogneva-Himmelberger, Y., Rakshit, R., & Pearsall, H. (2013). Examining the impact of environmental factors on quality of life across Massachusetts. *Professional Geographer*, *65*(2), 187–204.

Park, Y., Kim, M., & Seong, K. (2021). Happy neighborhoods: Investigating neighborhood conditions and sentiments of a shrinking city with Twitter data. *Growth Change*, *52*(1), 539–566.

Peach, N. D., & Petach, L. A. (2016). Development and quality of life in cities. *Economic Development Quarterly*, *30*, 32–45.

Permentier, M., Bolt, G., & van Ham, M. (2011). Determinants of neighbourhood satisfaction and perception of neighbourhood reputation. *Urban Studies*, *48*(5), 977–996.

Schalock, R. L. (2000). Three decades of quality of life. *Focus on Autism and Other Developmental Disabilities*, *15*, 116–127.

Sirgy, M. J., & Cornwell, T. (2002). How neighborhood features affect quality of life. *Social Indicators Research*, *59*, 79–114.

Sirgy, M. J., Michalos, A. C., Ferriss, A. L., Easterlin, R. A., Patrick, D., & Pavot, W. (2006). The quality-of-life (QOL) research movement: Past, present, and future. *Social Indicators Research*, *76*, 343–466.

Sirgy, M. J., Widgery, R. N., & Lee, D. J. (2010). Developing a measure of community well-being based on perceptions of impact in various life domains. *Social Indicators Research*, *96*, 295–311.

Streimikiene, D. (2014). Natural and built environments and quality of life in EU member states. *Journal of International Studies*, *7*(3), 9–19.

Sugiyama, T., Thompson, C. W., & Alves, S. (2009). Associations between neighborhood open space attributes and quality of life for older people in Britain. *Environment and Behavior*, *41*, 3–21.

UN Habitat. (2016). *Measurement of city prosperity: Methodology and metadata*. United Nations Human Settlements Programme.

Ventegodt, S., Hilden, J., & Merrick, J. (2003). Measurement of quality of life I. A methodological framework. *Scientific World Journal*, *3*, 950–961.

Vukmirovic, M., Gavrilovic, S., & Stojanovic, D. (2019). The improvement of the comfort of public spaces as a local initiative in coping with climate change. *Sustainability*, *11*, 6546.

Weckroth, M., Ala-Mantila, S., Ballas, D., Ziogas, T., & Ikonen, J. (2022). Urbanity, neighbourhood characteristics and perceived quality of life (QoL): Analysis of individual and contextual determinants for perceived QoL in 3300 Postal Code Areas in Finland. *Social Indicators Research*.

Welsch, H. (2006). Environment and happiness: Valuation of air pollution using life satisfaction data. *Ecological Economics*, *58*, 801–813.

Weziak-Białowolska, D. (2016). Quality of life in cities—Empirical evidence in comparative European perspective. *Cities*, *58*, 87–96.

Zhang, Y., Van den Berg, A. E., Van Dijk, T., & Weitkamp, G. (2017). Quality over quantity: Contribution of urban green space to neighborhood satisfaction. *International Journal of Environmental Research and Public Health*, *14*(5), 535.

Chapter 3

A new kind of GeoInformatics built on living structure and on the organic view of space

Bin Jiang
Urban Governance and Design Thrust, Society Hub, The Hong Kong University of Science and Technology (Guangzhou), Guangzhou, P.R. China

3.1 Introduction

The great architect Christopher Alexander had some remarkable insights about architecture, which applies equally to geography. Over the past century, geography (or architecture) has always been a minor science, seeking application of the physical sciences such as physics and anthropology. In the next two centuries, geography (or architecture) might become a major science, a sort of complexity science, when the deep question of space has been properly understood (Grabow, 1983). The deep question touches the very nature of space or the organic or third view of space, as formulated by Alexander (2002–2005, 1999), that space is neither lifeless nor neutral but a living structure capable of being more living or less living. Living structure is such a structure that consists of far more small things (or substructures) than large ones across all scales ranging from the smallest to the largest (scaling law, Jiang, 2015), yet with more or less similar-sized things (or substructures) on each of the scales (Tobler's law, Tobler, 1970). It is initially the deep question or the very notion of a living structure or the organic view of space that triggered us to develop this chapter.

The third view of space differs fundamentally from the first two views of space: Newtonian absolute space and Leibnizian relational space, which are framed under Cartesian mechanical worldview (Descartes, 1637/1954). The mechanical worldview is so dominated in science and in our thinking as if it were the only mental model, or even worse it may be considered to the world itself. It is a powerful model about our world, for what we human beings have achieved in science over the past 100 years is largely attributed to the mental model. However, the mechanical mental model is limited when

Geoinformatics for Geosciences. DOI: https://doi.org/10.1016/B978-0-323-98983-1.00004-1

comes to design or creation, as the goodness of designed or created things is sidelined as an opinion or personal preference rather than a matter of fact (Alexander, 2002–2005). Under Newtonian absolute and Leibnizian relational views of space—a geographic space is represented as a collection of geometric primitives such as points, lines, polygons, and pixels (c.f., Fig. 3.1 for illustration), which tend to be "*cold and dry*" (Mandelbrot, 1982), so it is not seen as a living structure.

The mechanical world picture has two devastating results according to Alexander (2002–2005). The first was that the "I" went out of the world picture and the inner experience of being a person is not part of this picture. The second was that the mechanical world picture no longer has any definite feeling of value in it, or value has become sidelined as a matter of opinion

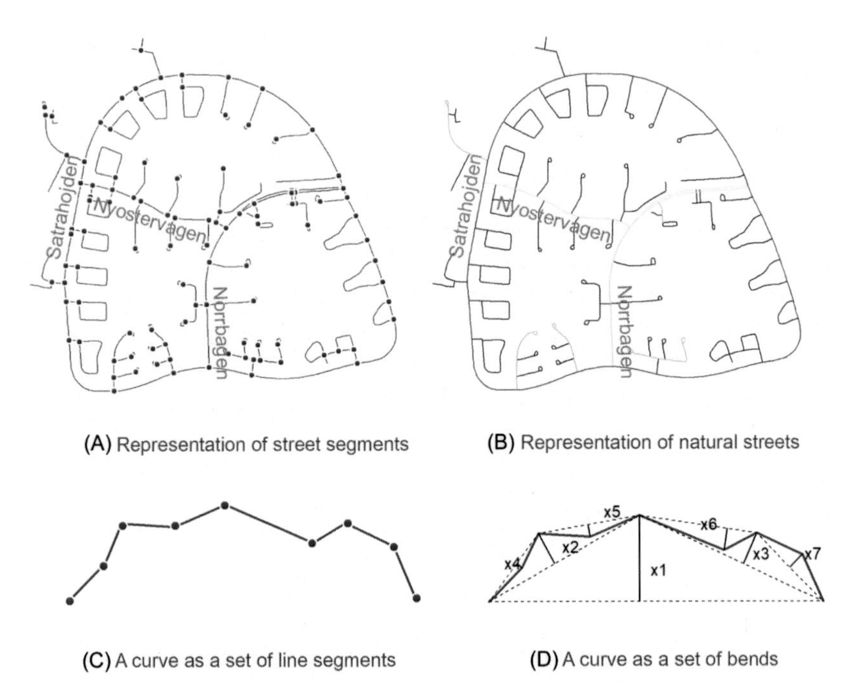

FIGURE 3.1 Nonliving versus living structure views of geographic features. *Note*: Conventionally, a street network is represented as a set of geometric primitives, which are not the right things or substructures (A), whereas it is more correctly perceived as a collection of named streets, which are the right things or substructures for seeing the street network as a living structure (B). Each street is colored as one of the four levels of scale: blue for the least connected streets, red for the most connected street (only one), and yellow and turquoise for those between the most and the least connected. A coastline is conventionally represented as a set of line segments, which are not the right things or substructures (C), but it is more correctly perceived as a collection of far more small bends than large ones, which are the right things or substructures for seeing the coastline as a living structure (D). It is because the notion of far more small bends than large ones occurs twice: (1) $x_1 + x_2 + x_3 > x_4 + x_5 + x_6 + x_7$, and (2) $x_1 > x_2 + x_3$.

rather than as something intrinsic to the nature of the world. The organic world picture first conceived by Whitehead (1929) extends the mechanical world picture to include human beings as part of the organic world picture. The same worldview has been advocated by quantum physicist Bohm (1980) among many others. Under the organic worldview, we human beings are part of the world rather than separated from the world. In other words, the physical world or the universe is organ-like rather than machine-like (Whitehead, 1929). It is under the organic world picture that Alexander (2002–2005, 1999) formulated the third view of space. Under the third view of space, value lies on the underlying configuration of space and the goodness of space is no longer conceived as a matter of opinion, but a matter of fact. The shift from the opinion view to the fact view or from the mechanical worldview to the organic worldview represents something fundamental (Kuhn, 1970) in our thinking about geography, for design or how to make living or more living space is at the forefront of geographic inquiry.

We in this chapter attempt to setup a new kind of GeoInformatics on the notion of living structure and on the third or organic view of space. Living structure is said to be governed by two fundamental laws: the scaling law and Tobler's law. Among the two laws, the scaling law is the first, or dominant law, as it is universal, global, and across scales, whereas Tobler's law is available locally or on each of the scales. Conventionally, GeoInformatics has been viewed as a minor science or an applied science that seeks to use or apply major sciences for understanding geographic forms and processes. In this chapter, we argue that the new kind of GeoInformatics is a major science, a science of living structure, not only for better understanding geographic forms and processes but also for better making and remaking geographic space or the Earth's surface toward a living or more living structure.

The remainder of this chapter is organized as follows. Section 3.2 introduces the two fundamental laws of living structure that favor statistics over exactitude. Section 3.3 illustrates how living structure differs from nonliving one under two different worldviews. Section 3.4 presents two design principles—differentiation and adaptation—to make or transform a space to be living or more living. Section 3.5 further discusses the new kind of GeoInformatics and its deep implications. Finally in Section 3.6, the chapter concludes with a summary pointing to a prosperous future of the new GeoInformatics.

3.2 Two statistical laws together for characterizing the living structure

The notion of living structure applies to all organic and inorganic phenomena in the scales ranging from the smallest Planck's length to the largest scale of the universe (Alexander, 2002–2005, 2003), so do the scaling law and

Tobler's law. The applicability implies that there are far more small particles than large ones, far more rats than elephants, far more small stars than large ones, far more small galaxies than large ones, and so on. This chapter deals with a range of scales of the Earth's surface between 10^{-2} and 10^{6} m. Table 3.1 shows how these two laws complement rather than contradict to each other from various perspectives (Jiang & Slocum, 2020). It is wise to keep the scaling law as the dominant one, as it is global or across scales, whereas Tobler's law is local or on each scale. In conventional GeoInformatics, Tobler's law is usually overstated as the first law of geography, and it implies that the Earth's surface is in a simple and well-balanced equilibrium state. However, we know that the Earth's surface is unbalanced and very heterogeneous and every place is unique (Goodchild, 2004). Dominated by the scaling law or the nonequilibrium character, the new kind of GeoInformatics aims not only to better understand the complexity of the Earth's surface but also to make the Earth's surface a living or more living structure. For creating living structures, two design principles—differentiation and adaptation—will be introduced later on.

Unlike many other laws in science, these two laws are statistical rather than exact. The statistical nature is more powerful than the exactitude one. Below, we cite three sets of evidence in science and art to make it clear why exactitude is less important. First, Zipf's law (Zipf, 1949) is also statistical rather than exact. It states that in terms of city sizes, the largest city is about twice as big as the second largest, approximately three times as big as the third largest, and so on. Here twice, three times, and so on are not exact but statistical or roughly. Among the two sets for example: [1, 1/2, 1/3, ..., 1/10] and [$1 + e_1$, $1/2 + e_2$, $1/3 + e_3$, ..., $1/10 + e_{10}$] (where e_1, e_2, e_3, ... e_{10} are very small values), the first dataset does not follow Zipf's law, whereas the second does. Zipf's law is a major source of inspirations of fractal geometry (Mandelbrot, 1982). In his autobiography, Mandelbrot (2012) made the

TABLE 3.1 Two complementary laws of geography or living structure.

Scaling law	Tobler's law
There are far more small things than large ones across all scales, and the ratio of smalls to larges is disproportional (80/20).	There are more or less similar things available at each scale, and the ratio of smalls to larges is closer to proportional (50/50).
Globally, there is no characteristic scale, so exhibiting Pareto distribution, or a heavy-tailed distribution, due to spatial heterogeneity or hierarchy, indicating complex and nonequilibrium character.	Locally, there is a characteristic scale, so exhibiting a Gauss-like distribution, due to spatial homogeneity or dependence, indicating simple and equilibrium character.

following remark while describing the first time he was introduced to a book review on Zipf's law: "*I became hooked: first deeply mystified, next totally incredulous, and then hopelessly smitten ... to this day. I saw right away that, as stated, Zipf's formula could not conceivably be exact*." A dataset following Zipf's law meets the scaling law, but not vice versa, which means that the scaling law is even more statistical than Zipf's law. Zipf's law requires a power law, whereas the scaling law does not.

The second evidence is not only statistical but also geometrical. The leaf vein shown in Fig. 3.2 (Jiang & Huang, 2021) apparently has far more small substructures than large ones from the largest square to the smallest white spots. Carefully examining the structure of the leaf vein, it is not difficult to find that there are four different levels of scale according to the thickness of their outlines. In contrast, the Sierpinski carpet also has far more smalls than larges; that is, far more small squares than large ones, exactly rather than statistically (Sierpínski, 1915). Let us carefully examine the exactitude of the carpet. The largest square is in the middle of the carpet of size 1/3, which is surrounded by eight squares of size 1/9, each of which is surrounded by eight squares of size 1/27, each of which is surrounded by eight squares of size 1/81. Thus there are two exponential data series, each of which is controlled by some exact number. The size of squares is exponentially decreased by the exact number 1/3 (1/3, 1/9, 1/27, 1/81), whereas the number of squares is exponentially increased by the exact number 8 (1, 8, 64, 512). Clearly there are far more small squares than large ones exactly rather than statistically.

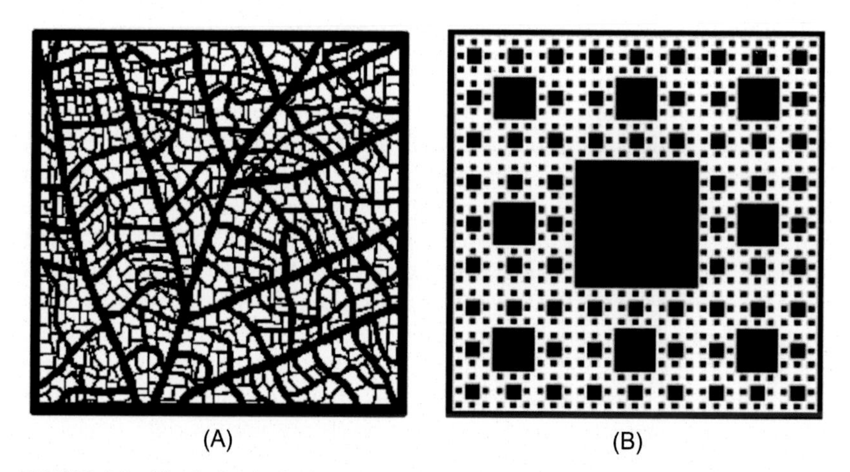

FIGURE 3.2 The leaf vein looks more living or more structurally beautiful than the stiff Sierpinski carpet. *Note*: The leaf vein (A) and the Sierpinski carpet (B) both meet the scaling law and Tobler's law, but the leaf vein is more living than the Sierpinski carpet. This is because the squares of the Sierpinski carpet on each scale are precisely the same rather than more or less similar, thus violating Tobler's law to some extent.

Because of the exactitude, the Sierpinski carpet is less living structurally than the leaf vein.

French painter Matisse (1947) made a famous statement about the essence of art: "*Exactitude is not truth.*" In terms of exactitude, a photograph is far better than a painting. However, the value of a painting lies not in its exactitude, but in something else, which is not only inexact but also distorted or exaggerated. The distorted or exaggerated nature is often used in drawing a cartoon. A human face is a living structure governed by the scaling law and Tobler's law, with the recurring notion of far more smalls than larges. The eyes, nose, mouth, and ears are the largest features and are therefore the most salient; each of them—if examined carefully—is a living structure again, with the recurring notion of far more smalls than larges. All human faces are universally beautiful in terms of the underlying living structure, despite some tiny cultural effects on their beauty.

The scaling law and Tobler's law are really two fundamental laws about livingness or beauty. They can be used to examine many patterns or structures (e.g., Wade, 2006; Wichmann & Wade, 2017) for understanding not only why they are beautiful but also how beautiful they are. For example, the leaf vein is living or beautiful because of the recurring notion of far more small structures than large ones. This way, through these two laws, the livingness or beauty of a structure or pattern can be objectively judged. Importantly, the livingness judged through these two laws can be well reflected in the human mind and heart, thus evoking a sense of beauty. This point will be further discussed in the following.

3.3 Living versus nonliving structure: the "things" the two laws refer to

The two laws introduced above have a common keyword—"things": (1) more or less similar things on each scale and (2) far more small things than large ones across all scales. What are the "things" the two laws refer to? In general terms, the things that collectively constitute a living structure are the right things, whereas the things that collectively do not constitute a living structure are not the right things. For example, if the leaf vein was saved as a gray-scale image with 1024 by 1024 pixels, each of which has a gray scale between 0 and 255, careful examination of these pixel values would show that they do not have far more light (or dark) pixels than dark (or light) ones. This way, we would end up with an absurd conclusion that the leaf vein is not a living structure. In fact, the pixels are not the right things, or the pixel perspective is not the right perspective for seeing the living structure.

In addition to the perspective discussed above, the scope also matters in seeing a living structure. A tree has surely far more small branches than large ones across scales from the largest to the smallest, whereas branches on each scale are more or less similar. Thus the tree is no doubt a living structure, not biologically but in terms of the underlying structure. However, its leaves can be both living and nonliving structures depending on the scope we see them. It is a living structure, if we go down to the scope or scale of intra-leaves, each of them has multiple scales (as shown in Fig. 3.2). It is a nonliving structure, if we on the other hand concentrate on inter-leaves, they are all more or less similar sized, being the smallest scale of the tree. In addition, the leaf vein shown in Fig. 3.2 is not a complete leaf, but part of it, with the large enough scope for us to see the living structure. All geographic features are living structures, if they are seen correctly with the right perspective and scope.

Let us further clarify the term "things" or substructures through two working examples: a street network and a coastline (Fig. 3.1, Jiang & Slocum, 2020). Conventionally, in geography or GeoInformatics, the things often refer to geometric primitives such as pixels, points, lines, and polygons. There is little wonder that Tobler's law is seen pervasively, as there are more or less similar-sized things seen from the perspective of geometric primitives. For example, a street network has more or less similar street segments, or all the street junctions have more or less similar numbers of connections (1–4) (Fig. 3.1A). A coastline consists of a set of more or less similar line segments (Fig. 3.1C). Unfortunately, all these geometric primitives are not the right things for seeing the street network or coastline as a living structure. There is little wonder, constrained by the geometric primitives, that living structure was not a formal concept in geography or GeoInformatics.

A street network is more correctly conceived of as a set of far more short streets than long ones or a set of far more less connected streets than well connected ones (Fig. 3.1B). The street network has four levels of scale, indicated by the four colors, far more short streets than long ones across the scales, and more or less similar streets on each of the four scales. A coastline is more correctly represented as a set of far more small bends than large ones (Fig. 3.1D). The coastline has three levels of scale, indicated by three sets of bends: [x_1], [x_2, x_3], and [x_4, x_5, x_6, x_7]. The notion—or recurring notion—of far more smalls than larges should be the major criteria for whether things are the right things that enable us to see a living structure or whether we have the right perspective and scope for seeing a living structure.

The "things" that collectively constitute a living structure are also called centers (Alexander, 2002–2005), a term that was initially inspired by the notion of organisms conceived by Whitehead (1929). Centers or organisms are the building blocks of a living structure, and their definitions are somewhat obscure. Instead, in this chapter, we use substructures to refer to the right things for seeing a living structure. This way, a living structure can be stated—in a recursive manner—as the structure of the structure of the structure, and so on. The things or substructures constitute an iterative system. To make the point clear, it is necessary to introduce the head/tail breaks (Jiang, 2013), a classification scheme for data with a heavy-tailed distribution.

For the sake of simplicity, we use the 10 numbers [1, 1/2, 1/3, . . ., 1/10] to show how they are classified through the head/tail breaks (Fig. 3.3, Jiang & Slocum, 2020). The dataset is a whole, and its average is about 0.29, which partitions the whole into two subwholes: those greater than the average are called the head [1, 1/2, 1/3] and those less than the average are called the tail [1/4, . . ., 1/10]. The average of the head subwhole is about 0.61, and it partitions the head subwhole into two subwholes again: those greater than the average are called the head [1] and those less than the average are called the tail [1/2, 1/3]. Instead of expressing the dataset as a set of numbers, we state the 10 numbers as an iterative system consisting of three subwholes recursively defined: [1], [1, 1/2, 1/3], and [1, 1/2, 1/3, . . ., 1/10]. Instead of perceiving these numbers as a set of 10 numbers, we consider them as a coherent whole, consisting of three subwholes including the whole itself. Or alternatively, these numbers as a coherent structure consists of three substructures including the structure itself. The dataset [1, 1/2, 1/3, . . ., 1/10], because of its inherent hierarchy of 3, is more living than the other dataset [1, 2, 3, . . ., 10] that is without any inherent hierarchy, or violates the notion of far more smalls than larges.

Now let us apply the recursive way of stating a whole or structure into the street network illustrated in Fig. 3.1. Seen from above, the sample street network consists of 50 streets at four hierarchical levels indicated by the four colors: red (r), yellow (y), turquoise (t), and blue (b). Instead of stating the street network as a set or as four classes, we state it as an iterative system consisting of four subwholes or substructures that are recursively defined:

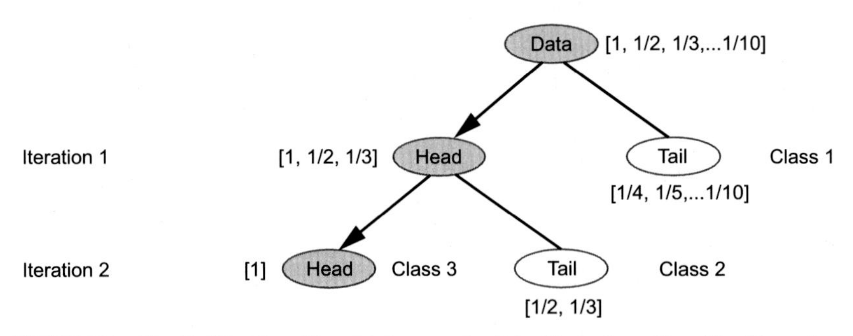

FIGURE 3.3 Head/tail breaks with a simple example of the 10 numbers. *Note*: The 10 numbers [1, 1/2, 1/3, . . ., 1/10] are classified into three classes: [1/4, 1/5, . . ., 1/10], [1/2, 1/3], and [1], which can be said to have three inherent hierarchical levels. The dataset, due to its inherent hierarchy, is therefore more living or more structurally beautiful than another dataset [1, 2, 3, . . ., 10], which lacks any inherent hierarchy or violates the scaling law.

[r], [r, y_1, y_2], [r, y_1, y_2, t_1, t_2, t_3, t_4, t_5], and [r, y_1, y_2, t_1, t_2, t_3, t_4, t_5, b_1, b_2, b_3, ..., b_{42}]. In the same way, it is not difficult to figure out the three recursively defined subwholes for the coastline: [x_1], [x_1, x_2, x_3], and [x_1, x_2, x_3, ..., x_7]. This living structure representation is recursive and holistic, so it differs fundamentally from existing representations that tend to focus on segmented individuals or mechanical pieces. An advantage of the living structure representation is that the inherent hierarchy of space is obvious. To this point, we have seen clearly how the right things constitute an iterative system, being a living structure consisting of far more smalls than larges.

3.4 Two design principles: differentiation and adaptation

In line with the two laws of living structure, there are two design principles—differentiation and adaptation—for transforming a space or structure to be living or more living. The purpose of the differentiation principle is to create far more small substructures than large ones, whereas the adaptation principle ensures that the created substructures are well adapted to each other, for example, nearby substructures are more or less similar. These two design principles ensure that any geographic space would become living or more living from the current status. Importantly, goodness of a geographic space is considered as a fact rather than an opinion, as mentioned above. These two design principles are what underlie the 15 structural properties (Fig. 3.4) distilled by Alexander (2002–2005) from traditional buildings,

FIGURE 3.4 Fifteen properties in natural and human-made things. *Note*: The fifteen properties exist pervasively in physical space, not only nature but also in what we human beings make and build. The two fundamental laws and the two design principles are distilled from these 15 properties.

cities, and artifacts. The 15 structural properties can be used to transform a space or structure into living or more living structure. Interested readers should refer to Alexander (2002–2005), specifically Volumes 2 and 3, for numerous examples. In this section, we use two working examples—two paintings and two city plans—to clarify these two design principles.

The two paintings shown in Fig. 3.5 are not very living, as they meet only the minimum condition of being a living structure with three or four inherent hierarchical levels. Painting (A) by Dutch painter Piet Mondrian (1872–1944) is entitled *Composition II*, with the three colors of red, yellow, and blue, whereas painting (B) is modified slightly from painting (A) by the author (Jiang & Huang, 2021). Fig. 3.5 demonstrates that how these two paintings are evolved—in a step-by-step fashion—from an empty square. Structurally speaking, painting (B) is more living than painting (A). It can equally be said that structure (G) is more living than structure (F), which is more living than (E), which is more living than structure (D), which is more living than structure (C). Thus, among all these structures or substructures, the empty square is the deadest, while structure (G) is the most living. On the one hand, there is the recurring notion of far more newborn substructures than old ones; on the other hand, within each iteration, there are far more

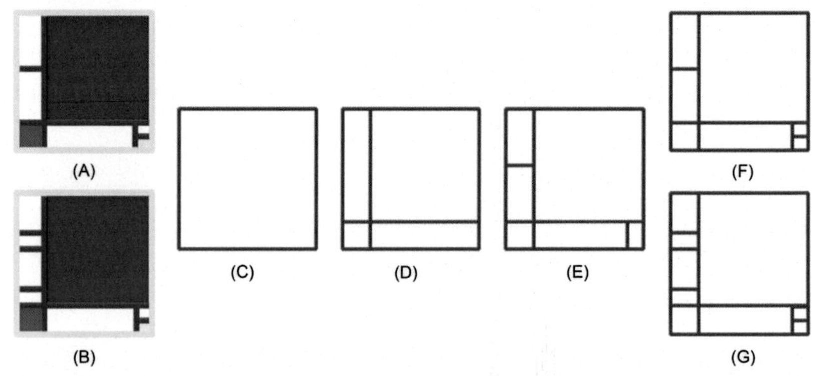

FIGURE 3.5 Living and less-living structures and their differentiation processes. *Note*: The two paintings—*Composition* (A) by the Dutch painter Piet Mondrian (1872–1944) and *Configuration* (B) modified from *Composition* by the author of this chapter—meet the minimum condition of being a living structure. Both paintings can be viewed to be differentiated like cell division from the empty square (C), so they are featured by the recurring notion of far more newborn (newly generated) substructures than old ones. More specifically, there are far more newborns than old one from (C) to (D), and again from (D) to (E), except from (E) to (F) in which there is a violation of far more newborns than old ones. However, there is again far more newborns than old one from (E) to (G). On the other hand, in each iteration, there are far more small substructures than large ones. Thus the painting *Configuration* is more living or more beautiful—structurally—than the painting *Composition*. If the reader prefers *Composition* over *Configuration*, do not be panic and your preference is likely to be dominated by nonstructural factors such as cultures, faiths, and ethnicities. However, the kind of beauty determined by the underlying living structure accounts for the feelings shared by most people or peoples.

small substructures than large ones. Seen from the comparison, it is not hard to understand that one structure is—objectively—more living than another.

The goodness or livingness of a space—or a city in particular—is a matter of fact rather than an opinion or personal preference, based on the underlying living structure. More specifically, the goodness of a space depends on substructures within the space, as we have already seen in the above discussion. The goodness also depends on larger space that contains the particular space. This way of judging goodness or order of things is universal across all cultures, faiths, and ethnicities, not only for natural things but also for what we make or build. This is probably the single most important message in the masterful work *The Nature of Order* (Alexander 2002–2003, 2005). This is a radical departure from the current view of space in terms of its goodness, judged by various technical parameters such as density, accessibility, and greenness. The living structure constitutes the foundation of the new kind of GeoInformatics this chapter seeks to advocate and promote.

The living structure perspective implies that a geographic space is in a constant evolution from less living to living or more living. Importantly, a geographic space or its design and planning process is an embryo-like evolution rather than LEGO-like assembly of prefabricated elements (Alexander, 2002–2005; Jiang & Huang, 2021). Note that the evolution view differs fundamentally from the assembly view, with the former being organic or natural, while the latter being mechanical or less natural. The living structure perspective implies also that a structure or substructures must be seen recursively. For example, conventionally painting (A) is seen as composed of seven pieces, but it is more correct to say it consists of 18 (1 + 4 + 6 + 7) recursively defined structures or substructures (Fig. 3.5). Instead of being nine pieces for painting (B), it is more correct to say that it consists of 20 (1 + 4 + 6 + 9) recursively defined structures.

Using the recursive perspective, it is not hard to understand why traditional city plans are usually more living than modernist counterparts. For example, with the city of London plan, the notion of far more small substructures than large ones recurs five times, so there are six hierarchical levels, whereas for the Manhattan one, the notion of far more small substructures than large ones recurs twice, so there are only three hierarchical levels (Fig. 3.6). Thus the city of London city plan is more living—structurally or objectively—than the Manhattan one. There have been many human perception tests supporting the conclusion that traditional city plans are more living than modernist counterparts (e.g., Alexander, 2002–2005; Wu, 2015), indicating over 75% agreement between the human perception and the reasoning based on the two laws. There may be some people (fewer than 25%) who prefer modernist buildings because they look new and luminous or for whatever personal reasons. A recent biometric investigation (Salingaros & Sussman, 2020) has provided further neuroscientific evidence that traditional façades are more "engaging" with people than contemporary façades.

FIGURE 3.6 Why the city of London plan is more living than the Manhattan one. *Note*: The city of London plan (the left) is obviously a living structure, for it meets scaling law, or the recurring notion of far more small substructures than large ones across the six hierarchical levels, shown in colors in those reduced panels to the left (Jiang & Huang, 2021). The part of Manhattan plan (the right) is less living, with only three inherent hierarchical levels to the right. Additionally, the number of substructures for the city of London is almost twice that of Manhattan, which is another reason why the left plan is more living than the right one.

Space has a healing effect, and this insight into space has been well established in the literature (e.g., Ulrich, 1984). Human beings have an innate nature of loving lifelike things and processes such as forests and weathering. This affinity to nature is termed by the eminent biologist Wilson (1984) as biophilia. The biophilia effect has been used to help create living environments by integrating lifelike things such as light, water, and trees (Kellert et al., 2008). It should be noted that a true biophilia goes beyond the simple integration of natural things but to create things that look like nature structurally (Salingaros, 2015). Jackson Pollock (1912–56) once said that he was not interested in mimicking nature, yet his poured paintings capture the order of nature. In this connection, living structure or the recurring notion of far more smalls than larges, as Alexander (2002–2005) has argued, appears to be the order that exists not only in nature, but also in what we make or build. The order—or living structure—constitutes the core of the new kind of GeoInformatics.

3.5 The new kind of GeoInformatics, its implications, and future works

The new kind of GeoInformatics laid down in the chapter is established on the third view of space or on the solid foundation of living structure. The new kind of GeoInformatics is inclusive of a wide range of conventional disciplines, including for example architecture, urban design and planning, urban science, and regional science, all to do with how to transform our cities and communities to be more livable, more living or more beautiful. Thus the new kind of GeoInformatics is a science of living structure, not only for better understanding geographic forms and processes but also—more importantly—for better making and remaking geographic space to be living or

TABLE 3.2 Differences between the conventional and new GeoInformatics.

Conventional GeoInformatics	New kind of GeoInformatics
Mechanical worldview of Descartes	Organic worldview of Whitehead
First and second views of space of Newton and Leibniz	Third view of space of Alexander
Understanding geographic forms and processes	Understanding + making living structures
Tobler's law dominated	Scaling law dominated
A minor science or application of other major sciences	A major science or a science of living structure

more living (i.e., sustainable spatial planning or design). Table 3.2 lists the differences between the conventional GeoInformatics and the new kind. The new kind of GeoInformatics goes beyond the two cultures under which science is separated from art (Snow, 1959), toward the third culture (Brockman, 1995) under which science and art is one. In the rest of this section, we further discuss on implications of the new kind of GeoInformatics and future works to be done.

It is important to note that the concept of living structure is part of physics, part of mathematics, and part of psychology. As a physical phenomenon, living structure pervasively exists in physical space or in any part of space or matter, and the physical phenomenon constitutes part of physics, or part of quantum physics to be more precise rather than that of classic physics. In this connection, living structure has another name called wholeness that is essentially the same as implicate order (Bohm, 1980). Living structure can be defined mathematically, but the mathematics is a nonlinear mathematics rather than a linear mathematics. The physical or mathematical structure can be psychologically or cognitively reflected in the human mind and heart, triggering a sense of livingness or beauty. Living structure is to livingness or beauty what temperature is to warmth. Given this, human-related research such as spatial cognition, mental map, human way-finding, and even perception of beauty must consider the underlying living structure.

The new kind of GeoInformatics has huge implications on design and art, because goodness of art or design is no longer considered to be an arbitrary opinion or personal preference, but a matter of fact. It is essentially the underlying living structure that evokes a sense of goodness or beauty in the human mind and heart. Thus there is a shared notion of quality or goodness of art among people or different peoples regardless of our culture, gender, and races. Goodness can be measured and quantified mathematically, and the outcome has over 70% agreement with people perception (e.g., Salingaros & Sussman, 2020; Wu, 2015). In this regard, the mirror-of-the-self experiment (Alexander, 2002–2005) provides an effective measure for testing people on their judgement on goodness of things. In this experiment, two things or pictures (e.g., those pairs in Figs. 3.2 and 3.6) are put side by side and human subjects are asked to provide their personal judgment to which one they have a higher degree of belonging or wholeness. The experiment is not kind of psychological or cognitive tests that seek intersubjective agreement, but rather on degree of livingness, something objective or structural. This kind of experiment, as well as eye-tracking and other biometrics data (Sussman & Hollander, 2015), will provide neuroscientific evidence for living structure, thus being an important future work in the new kind of GeoInformatics.

The new kind of GeoInformatics is a science of living structure, substantially based on living structure that resembles yet exceeds fractal geometry (Mandelbrot, 1982). Like conventional GeoInformatics, fractal geometry belongs to the camp of mechanical thought. For example, the commonly used box-counting method for calculating fractal dimension is too mechanical, as the boxes defined at different levels of scale are not the right things (or the right perspective) for seeing living structure (c.f., Section 3.3). As we have illustrated in Figs. 3.1 and 3.2, we adopt an organic rather than mechanical way of seeing living structures. Fractals emerge from an iterative process, but the iterative process is often too strict or too exact. The real world is indeed evolved iteratively, but it is not as simple as fractals, neither classic fractals nor statistical fractals. Nature—naturally occurring things—has its own geometry, which is neither Euclidean nor fractal, but a living geometry that "*follows the rules, constraints, and contingent conditions that are inevitably encountered in the real world*" (Alexander, 2002–2005). The major difference between fractal and living geometries lies probably on the two different worldviews. More importantly, goodness of a shape is not what fractal geometry concerned about, but it is the primary issue of living geometry.

Geographic information gathered through geographic information technologies has provided rich data sources for studying living structures on the Earth's surface from the perspectives of space, time, and human activities. This is particularly true for big data emerging from social media or the Internet. The big data are better than government owned or defined data for revealing the underlying living structure for two main reasons. First, big data have high resolution (like GPS locations of a couple of meters) and finer time scales (down to minutes and seconds for social media location data). Thus they are better than government data for revealing living structure at different levels of scale. Second, government-defined spatial units, such as census tracts, are too rough or too arbitrary for seeing living structure. Instead, we should use naturally defined spatial units such as natural cities and auto-generated substructures (Jiang & Huang, 2021; Jiang, 2018), which are all defined from the bottom up, rather than imposed from the top down, thus making it easy to see living structures. While working with big data, we should try to avoid using grid-like approaches such as the digital elevation model. Although the digital elevation model has far more low elevations than high ones, the grid approach is not the right perspective for seeing living structures. Instead, we should use watersheds or water streams which are naturally or structurally defined. All these topics will be studied in the future for the new kind of GeoInformatics.

3.6 Conclusion

This chapter is intended to help set GeoInformatics on the firm foundation of living structure, based on the belief that how to make and remake livable spaces—or living structures in general—should remain at the core of the new kind GeoInformatics. Considering a room, for example, we should first diagnose whether it is a living structure. If not, try to make it a living structure; if it is already, try to make it more living. This pursuit of living or more living structure extends from our rooms, gardens, buildings to streets, cities, and even the entire Earth's surface. The new kind of GeoInformatics should not just be a minor science—as currently conceived under the Cartesian mechanical worldview—that seeks to apply other major sciences or technology for understanding geographic forms and processes (or city structure and dynamics in particular). This is because these major sciences have not yet solved the problem of how to do an effective making or creation. Instead, the problem of making or creating is commonly left to art, design, or engineering, where there is a lack of criteria for judging the quality or goodness of the created things. In this chapter, the new kind of GeoInformatics is built on the criteria of living structure, not only for understanding geographic forms and processes but also for transforming geographic space to be living or more living.

The new kind of GeoInformatics is founded on the third or organic view of space, under which space is conceived as neither lifeless nor neutral, but a living structure capable of being more living or less living. The third view of space reveals that the nature of geographic space is a living structure or coherent whole, and its livingness or the degree of coherence can be quantified by the inherent hierarchy or the recurring notion of far more smalls than larges. Throughout this chapter, we have attempted to argue that the scaling law should play a dominant role for it is universal, global, and across scales, whereas Tobler's law is available on each of these scales. These two laws are the two fundamental laws of living structure. To make a space living or more living, we must follow the two design principles or, more specifically, a series of biophilia design principles or the 15 structural properties. There are three fundamental issues about a geographic space (or a city in particular): (1) how it looks, (2) how it works, and (3) what it ought to be. A short response to these three issues is that a geographic space should look and work like a living structure and ought to become living or more living. Facing various challenges of our cities and environments, the new kind of GeoInformatics provides new concepts, questions, and solutions to tackle problems and to make and remake cities and communities to be more livable and more beautiful toward a sustainable society. It is time to transform conventional GeoInformatics into the new kind of GeoInformatics, a science of living structure for the Earth's surface.

Acknowledgment

This chapter was condensed from the open-access one (Jiang, 2021). I would like to thank the anonymous referees and the editor A-Xing Zhu for their constructive comments. In addition, Yichun Xie, Jia Lu, and Ge Lin read an earlier version of this chapter, and Chris de Rijke helped with part of the figures. Thanks to you all. This project is partially supported by the Swedish Research Council FORMAS through the ALEXANDER project with grant number 2017-00824.

References

Alexander, C. (1999). The origins of pattern theory: The future of the theory, and the generation of a living world. *IEEE Software*, *16*(5), 71–82.

Alexander, C. (2002–2005). *The nature of order: An essay on the art of building and the nature of the universe*. Berkeley, CA: Center for Environmental Structure.

Alexander C. (2003). *New concepts in complexity theory: Arising from studies in the field of architecture*, http://natureoforder.com/library/scientific-introduction.pdf.

Bohm, D. (1980). *Wholeness and the implicate order*. London and New York: Routledge.

Brockman, J. (1995). *The third culture: Beyond the scientific revolution*. New York: Touchstone.

Descartes, R. (1637/1954). In D. E. Smith, & M. L. Latham (Eds.), *The geometry of Rene Descartes*. New York: Dover Publications.

Goodchild, M. (2004). The validity and usefulness of laws in geographic information science and geography. *Annals of the Association of American Geographers*, *94*(2), 300–303.

Grabow, S. (1983). *Christopher Alexander: The search for a new paradigm in architecture*. Stocksfield: Oriel Press.

Jiang, B. (2013). Head/tail breaks: A new classification scheme for data with a heavy-tailed distribution. *The Professional Geographer*, *65*(3), 482–494.

Jiang, B. (2015). Geospatial analysis requires a different way of thinking: The problem of spatial heterogeneity. *GeoJournal*, *80*(1), 1–13.

Jiang, B. (2018). A topological representation for taking cities as a coherent whole. *Geographical Analysis*, *50*(3), 298–313.

Jiang, B. (2021). Geography as a science of the Earth's surface founded on the third view of space. *Annals of GIS*, *x*(x), xx–xx.

Jiang, B., & Huang, J. (2021). A new approach to detecting and designing living structure of urban environments. *Computers, Environment and Urban Systems*, *88*, 1–10.

Jiang, B., & Slocum, T. (2020). A map is a living structure with the recurring notion of far more smalls than larges. *ISPRS International Journal of Geo-Information*, *9*(6), 388.

Kellert, S. R., Heerwagen, J., & Mador, M. (2008). *Biophilic design: The theory, science and practice of bringing buildings to life*. Hoboken, New Jersey: John Wiley & Sons, Inc.

Kuhn, T. S. (1970). *The structure of scientific revolutions* (second edition). Chicago: The University of Chicago Press.

Mandelbrot, B. B. (1982). *The fractal geometry of nature*. New York: W. H. Freeman and Co.

Mandelbrot, B. B. (2012). *The fractalist: Memoir of a scientific maverick*. New York: Pantheon Books.

Matisse, H. (1947). Exactitude is not truth. In J. D. Flam (Ed.), *Matisse on art* (pp. 117–119). New York: E. P. Dutton, 1978.

Salingaros, N. A. (2015). *Biophilia and healing environments: Healthy principles for designing the built world*. New York: Terrain Bright Green, LLC.

Salingaros, N. A., & Sussman, A. (2020). Biometric pilot-studies reveal the arrangement and shape of windows on a traditional façade to be implicitly "engaging", whereas contemporary façades are not. *Urban Science*, *4*(2), 26. https://www.mdpi.com/2413-8851/4/2/26.

Sierpínski, W. (1915). Sur une courbe dont tout point est un point de ramification. *Comptes rendus hebdomadaires des séances de l'Académie des Sciences*, *160*, 302–305.

Snow, C. P. (1959). *The two cultures and the scientific revolution*. New York: Cambridge University Press.

Sussman, A., & Hollander, J. B. (2015). *Cognitive architecture: Designing for how we respond to the built environment*. London: Routledge.

Tobler, W. (1970). A computer movie simulating urban growth in the Detroit region. *Economic Geography*, *46*(2), 234–240.

Ulrich, R. S. (1984). View through a window may influence recovery from surgery. *Science (New York, N.Y.)*, *224*, 420–422.

Wade, D. (2006). *Symmetry: The ordering principle*. Bloomsbury, USA: Wooden Books.

Whitehead, A. N. (1929). *Process and reality: An essay in cosmology*. New York: The Free Press.

Wichmann, B., & Wade, D. (2017). *Islamic design: A mathematical approach*. Cham, Switzerland: Birkhauser.

Wilson, E. O. (1984). *Biophilia*. Cambridge, MA: Harvard University Press.

Wu J. (2015). *Examining the New Kind of Beauty Using Human Beings as a Measuring Instrument*, Master Thesis at the University of Gävle.

Zipf, G. K. (1949). *Human behaviour and the principles of least effort*. Cambridge, MA: Addison Wesley.

Further reading

Bak, P. (1996). *How nature works: The science of self-organized criticality*. New York: Springer-Verlag.

Christaller, W. (1933/1966). *Central places in Southern Germany*. Englewood Cliffs, N. J.: Prentice Hall.

Jiang, B. (2019). Living structure down to earth and up to heaven: Christopher Alexander. *Urban Science*, *3*(3), 96, Reprinted as the cover story in the magazine *Coordinates*, March and April issues, 29–38, 12–17, 2020.

Salingaros, N. A. (1995). The laws of architecture from a physicist's perspective. *Physics Essays*, *8*, 638–643.

Simon, H. A. (1996). *The sciences of the artificial* (*Third edition*). Cambridge, Massachusetts: The MIT Press.

Chapter 4

Geospatial modeling of invasive *Aedes* vectors in Europe and the diseases they transmit: a review of best practices

Mina Petrić, Cedric Marsboom, Gaëlle Nicolas, Emilie Hendrickx, Roger Venail and Guy Hendrickx
Avia-GIS NV, Zoersel, Belgium

4.1 Introduction

Awareness regarding the introduction and spread of *Aedes* vectors in Europe has increased following the rapid spread of *Aedes albopictus*, the Asian tiger mosquito, and the pending invasion of *Aedes aegypti*, the two most dangerous invasive mosquito species globally. It is predicted that vector-borne diseases (VBDs) will continue to move from endemic to nonendemic areas with increased globalization and climate change. The timing and location of these changes can be predicted using different modeling approaches.

Aedes (Stegomyia) *albopictus* (Skuse, 1894) (Diptera: Culicidae), the Asian tiger mosquito, is acknowledged as a foreign invading arthropod species of significant concern. *Ae. albopictus* is thought to be a potential vector for a number of pathogens that cause VBDs that are important for human and/or animal health, including the West Nile and Zika viruses, Rift Valley fever, chikungunya, dengue, and Japanese encephalitis (McKenzie et al., 2019; Schaffner et al., 2013). The species is known to be the primary carrier of the dengue and chikungunya viruses in continental Europe. Since 2007, sporadic dengue, chikungunya, and Zika virus transmissions have been reported in southern and western Europe (European Centre for Disease Prevention & Control, 2017, 2019, 2020).

During the transatlantic slave trade in the 16th century, *Ae. aegypti* was first introduced to America from Africa (Shocket et al., 2019). It currently lives in the majority of tropical and subtropical climates due to its worldwide expansion. It is not established in more northern latitudes, mainly due to the

Geoinformatics for Geosciences. DOI: https://doi.org/10.1016/B978-0-323-98983-1.00005-3

fact that it did not evolve a diapause mechanism, unlike *Ae. albopictus* and is thus not adapted to overwinter in the colder, higher-latitude regions.

Rainfall and temperature are the main climatic factors that have been associated with the mosquito vector's capacity to survive and be active in a particular location (Brady et al., 2013; Caminade et al., 2012; ECDC, 2009; Kobayashi et al., 2002; Medlock et al., 2006; Nawrocki et al., 1987; Proestos et al., 2015; Roiz et al., 2010; Scott et al., 2000; Thomas et al., 2012). In addition, it is noted that host availability (Burkett-Cadena et al., 2013; Liu et al., 2018), proximity to water bodies (Ferraguti et al., 2016; Zhou et al., 2012), relative humidity (Yamana & Eltahir, 2013), NDVI, and land cover (Ferraguti et al., 2016; Liu et al., 2018; Steiger et al., 2016; Vanwambeke et al., 2007) all affect vector activity and spread. *Ae. albopictus* has developed an overwintering mechanism that enables species survival in more northern latitudes, which is the main distinction between this species and *Ae. aegypti*. *Ae. albopictus* has also been discovered to be generally more tolerant of lower temperatures (Vinogradova, 2000). Although the commencement of each year's activity has been connected to springtime temperatures and photoperiods, the duration and magnitude of the active period has largely been connected to air temperature in the months of June, July, and August (T_{jja}) and the amount of annual precipitation (Medlock et al., 2006). The temperature of the coldest month and yearly precipitation are factors limiting the northern distribution of both *Ae. aegypti* and *Ae. albopictus* vectors. While *Ae. aegypti* cannot survive winters below $T_{\text{jan}} = 10°\text{C}$ (Eisen & Moore, 2013; Eisen et al., 2014; Marsboom et al., 2019; Valdez et al., 2018), *Ae. albopictus* has been seen to overwinter in regions with mean January temperatures (T_{jan}) as low as $-4°\text{C}$ (Petrić et al., 2021).

For *Ae. albopictus*, an annual mean temperature of 11°C is thought to be the temperature threshold for establishment (Kobayashi et al., 2002). Temperatures between 15°C and 35°C are in the ideal range for population growth; outside this range, mobility and lifespan are seen to be lower (Brady et al., 2013). Egg mortality increases dramatically if temperatures are below 0°C, and the expected progeny is significantly reduced when annual precipitation is below 500 mm, which have been proven to be the limiting factors for the survival of diapausing eggs in the northern hemisphere (Kobayashi et al., 2002; Nawrocki et al., 1987). Temperature and photoperiod in temperate regions have a major role in determining when diapausing eggs are produced in the fall and when they hatch in the spring (Focks et al., 1993). When the photoperiod is longer than 11.25 hours and the spring temperature is above 10.5°C, spring hatching is thought to begin (Medlock et al., 2006). When the temperature drops below 9.5°C and the photoperiod falls below 13.5 hours, female activity is found to cease (star of diapausing). On the other hand, for *Ae. aegypti*, it has been discovered that 10°C is the minimum temperature below which adults can no longer fly and enter a torpid state (Christophers, 1960; Couret et al., 2014; Reinhold et al., 2018; Rowley & Graham, 1968).

Meteorology impacts mosquito populations on several scales; climatic conditions influence the niche in which the vector may live and the anticipated yearly window of activity, while daily and weekly weather determine interannual population dynamics and affect the development and death rates of different mosquito life-cycle phases. The vector develops through four major stages: (1) the egg, (2) the larva, (3) the pupa, and (4) the adult. Most *Aedes* species, notably the anthropophilic *Ae. aegypti*, breed in artificial containers in urban environments. On the other hand, the common house mosquito (*Cx. pipiens*) may reproduce in a variety of natural and artificial locations in both urban and rural settings (Vinogradova, 2000; Werblow et al., 2014).

When modeling VBDs, climate change is an important subject to consider since it is a source of vector range expansion and disease transmission (Khormi & Kumar, 2014). By combining climate models with climate scenarios, it is possible to analyze future spread of vectors and illnesses under different scenarios of climate change. This can help with proactive planning and decision-making by identifying areas where circumstances are most likely to favor the cohabitation of vectors, host, and pathogens (Rocklöv et al., 2016).

The scaling and coupling of the components of an infectious disease system in a practical way can be performed by identifying the expected vector population *spatial* distribution and *temporal* dynamics and feedback loops through which this influences the disease transmission within the SIR vector cycle (susceptible, infections, and recovered) and the SEIR (susceptible, exposed, infectious, and recovered) host cycle.

Vector modeling techniques can firstly be differentiated into two broad categories: (1) spatial distribution models and (2) vector population dynamics (VPDs) models.

Spatial distribution models can be further considered into two subcategories: climatic suitability models and species distribution models (SDMs). Climatic suitability modeling is employed to examine the amicability of specific region for the establishment and annual activity of a vector species. This is critical for assessing potential risk and identifying possible hotspots for vector growth and disease transmission. This form of research, in particular, can help to estimate the sensitivity of a specific region to the arrival of a new invasive vector species.

On the other hand, in SDMs, the goal is to simulate the distribution and possible niches of vectors, hosts, and infections under present and future eco-climatic circumstances based on observed vector data. Various modeling techniques are applied to modeling the geographical distribution of these three key components (GLM, GAM, MAXENT, ENM, MARS, GBM, BRT, RF, SRE, CTA, ANN, etc.). These pipelines employ both machine learning and statistical methods. SDMs employ abiotic and biotic parameters and covariate layers to explain the spatio-temporal distribution of a vector species in a specific area. As a result, they can give valuable information on key habitat, connectivity within and between ecosystems, and the effects of anthropogenic influences on species (Robinson et al., 2017; Rosenberg, 2004). SDMs are also being used more often to address the challenge of anticipating how climate change and other variables may alter range extents, create distributional shifts, or affect the timing of host migrations (Brown et al., 2016; Edwards & Richardson, 2004; Hazen et al., 2013; Thorson et al., 2017). The versatility of SDMs has resulted in a plethora of modeling frameworks and parameterization choices. SDMs are broadly classified into correlative and mechanistic techniques (Connolly et al., 2017; Dormann et al., 2012; Robertson et al., 2003). Parameter responses in correlative models are not predefined and are instead modeled implicitly, resulting in responses that are not necessarily ecologically appropriate. Mechanistic models, on the other hand, employ explicit functions to characterize interactions between different components of the ecosystem (Connolly et al., 2017; Dormann et al., 2012). Correlative models, in contrast to mechanistic models, are frequently conceptually simple, capable of performing as well as or better than simple mechanistic models in estimation or short-term forecasting applications (Muhling et al., 2017; Robertson et al., 2003), and their independence from explicit assumptions can avoid confirmation biases (Connolly et al., 2017). Correlative models, on the other hand, are hampered in their ability to estimate conditions when system states change or display nonlinearity, such as under climate change (Lurgi et al., 2012; Plagányi et al., 2011, 2014). There has been a recent increase in the number of correlative tools available to model species distributions (Elith et al., 2006; Lawler et al., 2006), and there is a need to thoroughly analyze the performance, accuracy, and possible biases of these applications.

It is possible to anticipate seasonal risk and potential VBD outbreaks using mathematical models to simulate the temporal dynamics of a vector population. A single equation or set of equations that simulates, describes, and predicts the behavior and dynamics of a system is known as a mathematical model (Garira & Chirove, 2020). To model the temporal dynamics of disease vectors, numerical VPDs models are employed. These models can be used to simulate daily vector population dynamics of all stages in the mosquito life-cycle using a set of coupled dynamic equations, as well as incorporate the effect of long-range and seasonal forecasts on vector abundance. This forecasting necessitates the integration of ground-measured and modeled data at many scales with numerical weather prediction models. Because of their reliance on accurate, local micro-meteorological data, VPDs are most often run as dynamic time-series models for a specific location and are not readily scalable. These models have a wide range of applications in disease and pest management, such as assessing the impact of (integrated) control methods and forecasting vector activity.

Specific research problems regarding data manipulation and processing include: working with data at multiple scales, and combining NUTS level, vectorized and gridded data. Moreover, collecting presence/absence as well as abundance data is extremely costly and time-consuming, often resulting in sparse, low-frequency, and irregular data. A persistent problem is also the lack of absence data which is needed as input to different types of SDM models.

The challenges of abundance monitoring and its impact on human and/or animal health were emphasized by ECDC (European Centre for Disease Prevention & Control and European Food Safety Authority, 2018). Season and seasonality can change from year to year and can be affected by a wide range of environmental and anthropological factors, including climate, vegetation, and host availability. Temperature, photoperiodicity, and precipitation appear to be the most important abiotic variables. Threshold values for *Aedes* mosquitoes have been calculated for these variables in a number of surveys conducted in Asia (Kobayashi et al., 2002; Nawrocki et al., 1987), the Americas (Nawrocki et al., 1987), and Europe (ECDC, 2009).

There is a growing need to summarize trends, identify emerging questions, debunk controversies, and explain contradictory results in light of the ever-expanding body of scientific literature (Grames et al., 2019; Haddaway et al., 2018; Sutherland & Wordley, 2018). Meta-analyses, which statistically evaluate data identified via systematic and scoping reviews, which examine the existing literature for evidence with which to address a research issue, are two key methods for synthesizing evidence (Grames et al., 2019).

The goal of this chapter is to perform a scoping assessment to identify and analyze prediction and importation models for mosquito-borne diseases with a specific focus on *Aedes* mosquitoes in Europe, with the aim of outlining best practices and potential bottlenecks in the data-to-model chain.

4.2 Materials and methods

The goal of this study was to answer the question: "What is the current state of the art in geospatial modeling of invasive *Aedes* vectors in Europe and the diseases that they transmit?"

To improve the body of data around health risk concerns and provide answers to ecological and epidemiological challenges, systematic review, meta-analysis, and other types of evidence synthesis are essential. However, due to time and resource costs, literature synthesis often lags behind the rate of scientific publication. Additionally, the approaches used today might be biased in favor of studies that the researchers are already acquainted with and research teams may mistakenly reject studies from the review in areas without standardized terminology.

The literature review presented in this chapter was carried out by applying the generated search string (given below) to the PubMed database. We utilize the method developed by Grames et al (2019), as a comprehensive literature review approach implemented in R. The screening was centered on peer-reviewed studies and reviews that focus on modeling the spatio-temporal distribution of *Aedes* mosquitoes in Europe. The search string was run on the 21st of July 2022. Only publications written in English were considered, and nonprimary research was omitted. Because there is a lot of variability in the methodology and parameters used in the quickly evolving field of vector and VBD modeling, a scoping review was used to gather and characterize the pertinent modeling literature.

We performed several review iterations by employing (1) network DFM analysis, (2) pruning, and (3) grouping to run and refine the search string.

A network analysis was performed on the search phrase to identify outliers and refine the initial search string. The reasoning behind this is that words are connected to one another since they are found in the same publications. By identifying the terms that frequently appear together in a single article, we may identify groups of terms that are likely all referring to the same subject and exclude terms that do not frequently appear with any of the major groups of terms (Grames et al., 2019).

The final search string is given below:

((\"GIS model\" OR \"GIS simulation\" OR \"GIS analysis\" OR \"Geographical model\" OR \"Geographical simulation\" OR \"Geographical analysis\" OR \"Spatial model\" OR \"Spatial simulation\" OR \"Spatial analysis\" OR Simulation OR Analysis OR \"Species distribution model\" OR \"Spatial distribution model\" OR \"Distribution model\" OR \"climatic suitability\" OR spatiotemporal OR model OR \"risk maps\") AND (Aedes OR albopictus OR aegypti OR japonicus OR astropalpus OR koreicus OR Dengue OR Chikungunya OR Zika) AND (Aedes) AND (Europe) AND (Europe OR Albania OR Andorra OR Austria OR Belarus OR Belgium OR \"Bosnia and Herzegovina\" OR Bulgaria OR Croatia OR Cyprus OR Czechia OR Denmark OR Estonia OR Finland OR France OR Georgia OR Germany OR Greece OR Hungary OR Iceland OR Ireland OR Italy OR Kosovo OR Latvia OR Liechtenstein OR Lithuania OR Luxembourg OR Malta OR Monaco OR Montenegro OR Netherlands OR \"North Macedonia\" OR Norway OR Poland OR Portugal OR Romania OR \"San Marino\" OR Serbia OR Slovakia OR Slovenia OR Spain OR Sweden OR Switzerland OR Ukraine OR \"United Kingdom\"))

Finally, the collected literature was validated by confirming that the selected publications contain a subset of specific relevant results which were treated as golden-standard.

The following parameters were examined in the analysis: (1) year of publication; (2) domain of application; (3) author affiliation; (4) modeled species; (5) modeled disease; (6) climate change scenarios; (7) source and list of covariates used; and (8) source and list of entomological data used.

To efficiently find prospective keywords without depending on a possibly biased collection of preselected articles, we employed text mining and keyword cooccurrence networks (Grames et al., 2019).

4.3 Results

In total, 148 full text papers were screened, of which 83 were retained. In Fig. 4.1, a graphic overview of the reference history (from 2006 to 2021) as well as the target species and modeling approach is outlined. The references are grouped by geographic region over which the models were employed. Europe is categorized into four domains: (1) southern Europe; (2) eastern Europe; (3) western Europe; and (4) northern Europe following the EuroVoc standard (Publications Office of the European Union, 2022). Box shapes represent the model type: mathematical and mechanistic models (parallelogram), statistical models (curved rectangle), SDMs (trapezoid), vector population dynamics models (ellipse), and SEIR-SIR disease models (rectangle with wavy bottom edge). The box colors represent the species included in the analysis: *Ae. albopictus* (red), *Ae. aegypti* (black), *Ae. japonicus* (green), *Ae. koreicus* (purple), and *Ae. sticticus* (blue). Studies focusing on both *Ae. albopictus* and *Ae. egypti* (dashed black line), and studies focusing on multiple species including *Ae. caspius*, *Ae. Cinereus*, and *Ae. geminus* (finer black dashed line). Boxes that are shaded gray indicate that the study considers the effect of climate change and expected future projected distribution of *Aedes* mosquitoes in Europe.

4.3.1 Historic profile

We see a positive trend in the number of studies exploring the spatio-temporal distribution of *Aedes* species in Europe, with the steepest increase observed from 2014 to 2016 (Fig. 4.2).

4.3.2 Geographical representation

The domain of application of the different modeling techniques is predominantly focused on the continental (Europe, $n = 18$) and global scale ($n = 15$). The studies focusing on individual European countries were predominantly located in Italy ($n = 17$), followed by France ($n = 8$), Germany ($n = 8$), Netherlands ($n = 3$), United Kingdom ($n = 2$), Sweden ($n = 2$), Switzerland ($n = 1$), Spain ($n = 1$), Portugal ($n = 1$), and Hungary ($n = 1$). Two studies focused on the broader Balkan ($n = 1$) and eastern Mediterranean region ($n = 1$). The authors affiliated institutions behind these publications are mostly located in Italy ($n = 22$), Germany ($n = 20$), France ($n = 10$), Sweden ($n = 10$), and Belgium ($n = 7$).

4.3.3 Model type

The majority of the models employed are SDMs ($n = 27$) (Fig. 4.3), followed by mathematical and mechanistic models ($n = 19$), and statistical models ($n = 18$). Nine studies modeled the temporal dynamics of *Aedes* mosquitoes (population dynamics models, $n = 9$), and eight studies considered the

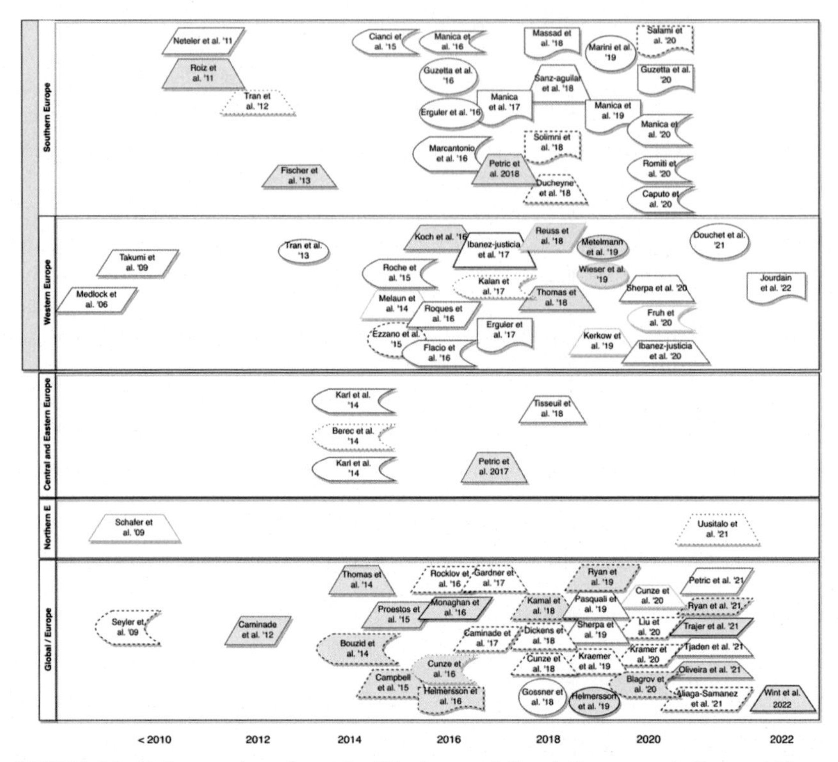

FIGURE 4.1 Reference chronology of publications modeling Aedes vectors in Europe (Aliaga-Samanez et al., 2021; Berec et al., 2014; Blagrove et al., 2020; Bouzid et al., 2014; Caminade et al., 2012; Caminade et al., 2017; Campbell et al., 2015; Cianci et al., 2015; Cunze et al., 2016; Cunze et al., 2018; Cunze et al., 2020; Dickens et al., 2018; Douchet et al., 2021; Ducheyne et al., 2018; Erguler et al., 2016; Erguler et al., 2017; Fischer et al., 2014; Flacio et al., 2016; Früh et al., 2020; Gardner et al., 2017; Gossner et al., 2018; Guzzetta et al., 2016; Guzzetta et al., 2020; Ibañez-Justicia et al., 2017; Ibáñez-Justicia et al., 2020; Jourdain et al., 2022; Kalan et al., 2017; Kamal et al., 2018; Karl et al., 2014; Kerkow et al., 2019; Koch et al., 2016; Kraemer et al., 2019; Kramer et al., 2020; Liu et al., 2020; Liu-Helmersson et al., 2016; Liu-Helmersson et al., 2019; Manica et al., 2016; Manica et al., 2017; Manica et al., 2019; Manica et al., 2020; Marcantonio et al., 2016; Marini et al., 2019; Massad et al., 2018; Medlock et al., 2006; Melaun et al., 2015; Metelmann et al., 2019; Monaghan et al., 2016; Neteler et al., 2011; Oliveira et al., 2021; Pasquali et al., 2020; Petrić et al., 2017; Petrić et al., 2018; Petrić et al., 2021; PrReuss et al., 2018; Rocklöv et al., 2016; Roiz et al., 2011; Romiti et al., 2021; Roques & Bonnefon, 2016; Ryan et al., 2019; Ryan et al., 2021; oeSalami et al., 2020; Sanz-Aguilar et al., 2018; Schäfer & Lundström, 2009; Seyler et al., 2009; Sherpa et al., 2019; Sherpa et al., 2020; stSolimini et al., 2018; Takumi et al., 2009; Thomas et al., 2018; Tisseuil et al., 2018; Tjaden et al., 2021; Trájer, 2021; Tran et al., 2013; Uusitalo et al., 2021; Wieser et al., 2019; Wint et al., 2022os et al., 2015). The box colors represent the species included in the analysis: *Ae. albopictus* (red), *Ae. aegypti* (black), *Ae. japonicus* (green), *Ae. koreicus* (purple), and *Ae. sticticus* (blue). Studies focusing on both *Ae. albopictus* and *Ae. egypti* (dashed black line), and studies focusing on multiple species including *Ae. caspius*, *Ae. cinereus*, and *Ae. geminus* (finer black dashed line). Boxes that are shaded gray indicate that the study considers the effect of climate change and expected future projected distribution of *Aedes* mosquitoes in Europe.

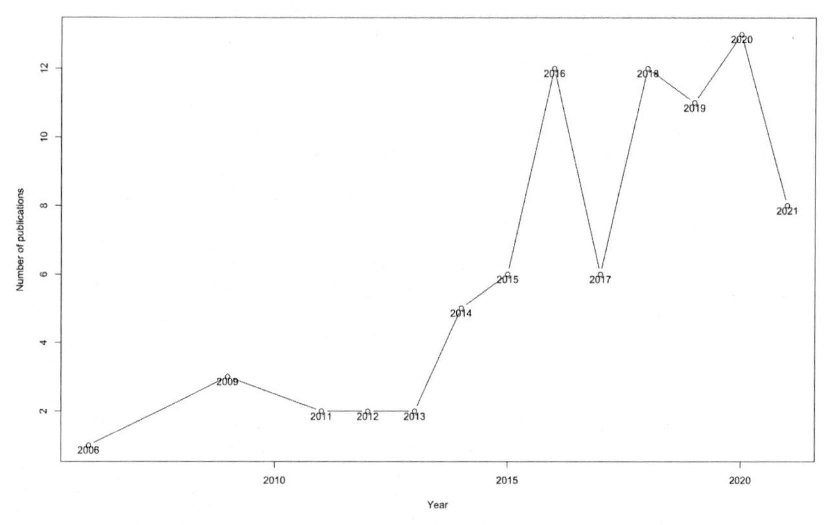

FIGURE 4.2 Number of publications by year (2006–2021).

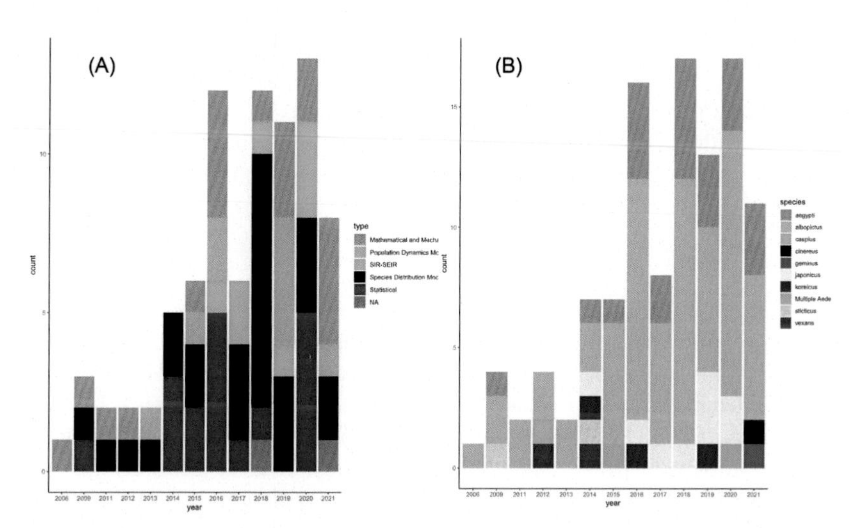

FIGURE 4.3 Frequency of the 83 relevant publications by year, separated by model type (A) and modeled species (B).

coupled vector-VBD system employing a SIR-SEIR framework ($n = 8$). Most of the SDM models used were ecological niche models, of which most used the Maxent software (Phillips et al., 2017). There were five studies employing an ensemble approach combining ecological niche models, GLM, GAM, BRT, RF, and GBM models ($n = 5$).

4.3.4 Presence/absence and abundance data

The highest utilized vector database was the global compendium of *Ae. aegypti* and *Ae. albopictus* occurrence (Kramer et al., 2020), followed by ECDC data collected though Vectornet and Vbornet (Blagrove et al., 2020). Other sources include data from Culbase (Leibniz Centre for Agricultural Landscape Research ZALF e.V., 2022), EID (EID, 2022), GBIF (GBIF, 2022), INaturalist (National Geographic Society & California Academy of Science, 2022), Mosquito Alert (Južnič-Zonta et al., 2022), Survnet (Department of Bioinformatics & Computational Biology, 2022), Vectorbase (Amos et al., 2022), and Vectormap (National Museum of Natural History, 2000).

Chikungunya and dengue were the most frequently studied illnesses. Globally, dengue fever is among the most commonly studied *Aedes* VBDs. This might be due to the enormous global burden, as well as their worldwide prominence, which has sponsored research for decades. As a result, there is a substantial but still not homogenized, quantity of data, which is essential for appropriate model parameterization.

4.3.5 Species decomposition

The highest number of papers focused on modeling *Ae. albopictus* ($n = 63$), followed by *Ae. aegypti* ($n = 23$), *Ae. japonicus* ($n = 9$), *Ae. koreicus* ($n = 3$), *Ae. sticticus* ($n = 2$), *Ae. vexans* ($n = 2$), *Ae. cinereus* ($n = 1$), and *Ae. caspius* ($n = 1$), with several papers focusing on multiple species. Chikungunya ($n = 11$) is most represented in terms of diseases, followed by dengue ($n = 11$), zika ($n = 7$), and yellow fever ($n = 2$) (Fig. 4.3).

4.3.6 Meteorological data

The meteorological variables used were temperature ($n = 56$), bioclimatic variables ($n = 14$), elevation ($n = 3$), cargo transport data ($n = 1$), flight data ($n = 2$), length of growing season ($n = 1$), human mobility ($n = 1$), land cover ($n = 10$), NDVI ($n = 3$), NDWI ($n = 1$), photoperiod ($n = 5$), human population density ($n = 9$), precipitation ($n = 11$), relative humidity ($n = 4$), road network ($n = 1$), snow depth ($n = 1$), solar radiation ($n = 1$), and wind speed ($n = 1$). The most used sources for the meteorological covariates were worldclim ($n = 34$) (Fick & Hijmans, 2017), ground observations (WSN, synoptic, logger, $n = 20$), MODIS ($n = 11$) (Wan, 2015), EOBS ($n = 10$) (van der Schrier et al., 2013), Corine Land Cover ($n = 7$) (Büttner et al., 2004), and Chelsa ($n = 6$) (Karger et al., 2016). Satellite data was utilized in 13% of the publications.

Despite the fact that many of the models included climatic parameters, the majority of the models included in this review did not explicitly investigate climate change ($n = 59$), as they did not use a climate model to explore outcomes under projected climate scenarios, despite the potential relevance to disease transmission, vector range, and distribution.

The majority of articles listed the parameters employed in their model, while 11% of publications did not reveal their parameters, either in the paper or in extra material.

In Fig. 4.4, we provide an overlay analysis of the current and expected tendencies for *Ae. albopictus* based on available literature. A quantitative analysis was not possible because of the differences in model-type, covariates and climate model used, emission scenarios, projected coordinate system as well as spatial and temporal resolution. Therefore the comparison and combined suitability analysis shown in Fig. 4.4. must be seen as a qualitative and schematic representation of the current end expected trends.

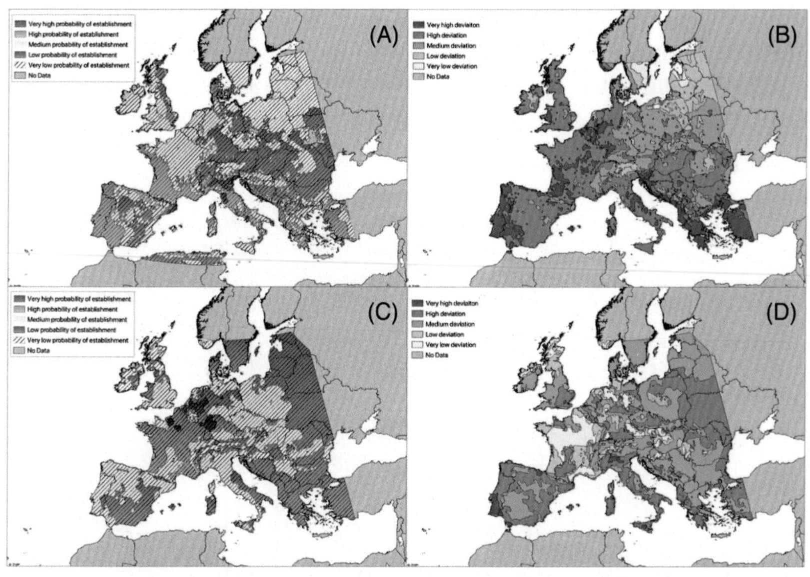

FIGURE 4.4 Comparison of current (A) and expected (C) climatic suitability for the establishment of *Ae. albopictus* toward the end of the 21st century. Outputs from multiple models (Aliaga-Samanez et al., 2021; Caminade et al., 2012; Kamal et al., 2018; Oliveira et al., 2021; Pasquali et al., 2020; Proestos et al., 2015; Ryan et al., 2019; Sherpa et al., 2019; Thomas et al., 2014) were used in an ensemble overlay analysis in which the different metrics were categorically grouped in five classes of climatic suitability (from stripped-white, very low probability, to stripped-red, corresponding to very high probability of establishment). Panels (C) and (D) indicate the degree of deviation between the models simulating current (B) and future (D) conditions for vector establishment (with white corresponding to the biggest agreement, that is, the lowest deviation between models, to dark green indicating very high differences in the modeled outputs). Due to differences in modeling approach, climatic projections, time periods and spatial resolution, this analysis represents a schematic abstraction of the mean simulated suitability (A, C) and divergence between the different models (B, D).

Nonetheless, certain common trends in the simulated climatic suitability for *Ae. albopictus* in Europe can be observed. The highest agreement between models can be seen in the low-suitability areas across the northern parts of central and eastern Europe (shown in white and blue on Fig. 4.4A). The highest deviation between models simulating the current climatic suitability of *Ae. albopictus* is on the other hand observed over most of the Mediterranean. Future projections (Fig. 4.4C) indicate that the vector is likely to become active in many regions which are currently unfavorable for its establishment. Certain areas in western Europe (Belgium, Netherlands and parts of Germany) are expected to become highly suitable for the establishment of *Ae. albopictus* toward the end of the century with fairly high agreement between models. As a general trend visible across all models, it can be assumed that most of western Europe will become increasingly suitable for *Ae. albopictus* in the coming decades. Interestingly, the whole of the Mediterranean is expected to become less favorable for the mosquito by the end of the century, and this is most likely a result of the expected drop in total annual precipitation which limits the activity and number of available breeding sites needed for the mosquito to reproduce.

Fig. 4.5 illustrates the speed of invasion of *Ae. albopictus* in Europe based on VectorNet data (Braks et al., 2022). The areas given in dark-red

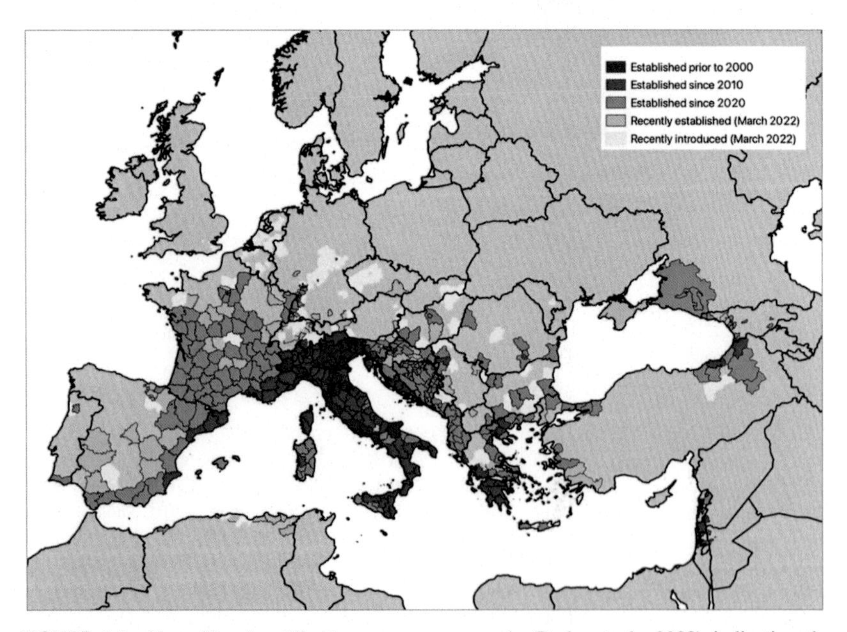

FIGURE 4.5 VectorNet *Ae. Albopictus* presence records (Braks et al., 2022) indicating the speed of its invasion of Europe, with the areas given in dark-red representing the first administrative units (prior to 2000) in which the vector was able to overwinter for multiple years, while the most recent areas in which *Ae. albopictus* was found but is not yet proven to be established are given in yellow.

representing the first administrative units (prior to 2000) which the vector had colonized, while the most recent areas in which *Ae. albopictus* was found, but is not yet proven to be established, are given in yellow.

4.4 Discussion

Given the wide range of scales of vector and VBD modeling, it is possible that the most significant lesson learned from 16 years of modeling *Aedes* mosquitoes in Europe is that a diverse suite of modeling approaches is required. Because of the broad geographical and temporal scales or uncertainty arising from ecological processes, some distribution models rely on less precision and complexity to be scalable. However, models with comparable goals have converged over time to incorporate the majority of the same crucial elements. For instance, more modern machine learning models include up to 50 different covariate layers, however, maintain the focus on primary factors (and their derivatives) that govern development and mortality rates such as temperature and precipitation.

By suggesting putative causative pathways, environmental variables in SDMs provide insight into the fundamental processes driving species distributions. Temperature, for example, can impact a vector species' distribution either directly by influencing its ability to become established on a certain area or by influencing mating, mortality and development, thus limiting the number of generations, and expected seasonal abundance. Models can be poorly fit if crucial environmental processes are not accounted for in the modeling framework. However, because environmental variables are frequently geographically and temporally autocorrelated, space and time covariates can operate as proxies for these missing variables (Legendre, 1993). This being said, the calibration of SDMs using environmental and spatiotemporal covariates should always be done with the model aim in mind (García-Díaz et al., 2019; Guillera-Arroita et al., 2015). Comparative assessments of models based on both simulated and observed data can aid in identifying model performance biases and further guide parameters selection.

While the spatial distribution and seasonality of *Aedes* mosquitoes can be simulated on a European scale, estimating and modeling vector abundance in a quantitative way on the same scale is difficult. The volume and quality of breeding and resting sites, as well as local temperature trends, which have an impact on the life cycles of populations, are eco-climatic factors with a highly local character and significant small-scale spatial variability. Moreover, the accessibility of these breeding and resting locations will consequently be influenced by a wide range of variables, including rainfall and proximity to water bodies, vegetation and land cover, artificial watering, and the presence of human-made containers. The fact that micro-environmental conditions in potential habitats are currently not assessed in a standardized fashion and that there is no exact link between results derived in laboratory settings and the same parameters under variable environmental conditions is one of the main problems in effective abundance modeling. The relationship between the numbers of observed egg/larval/adult stages and the total size of the vector population throughout the season must also be further studied. Finally, only locally calibrated parameters can provide precise information on crucial life-history qualities, like survival and development, but cannot be comprehensively applied to continental-scale analysis. Currently, research networks such as AIM-COST are making an attempt to develop a standardized database for *Aedes* mosquitoes in Europe that should support abundance modeling efforts. There are currently no models that accurately predict abundance on a wider scale, and the research is primarily focused on predicting activity and relative abundance within a presence area, producing locally calibrated VPDs that cannot be readily extrapolated to a wider area.

Despite existing differences across models identified in this review, some broad trends about the changing environmental suitability for *Ae. albopictus* throughout Europe may be deduced. Most projections show that many areas where the species is not yet established will see an improvement in climatic suitability. Areas that are now characterized by a relatively low or moderate suitability in northern Europe may become suitable by the end of the century, while areas in southern and central Europe would become suitable for the establishment of *Ae. aegypti* within the same timeframe. On the other hand, parts of the Mediterranean coast are predicted to become less suitable for *Ae. albopictus*. This is most likely a result of expected drier weather conditions over the summer which are projected for most of the area. Precipitation projections and derived variables often reflect higher uncertainty than temperature projections, which is especially noticeable in southern Europe. It is unclear if the climate will remain favorable or deteriorate in the future in the core regions of the Iberian Peninsula as well as parts of Italy and Greece, and the projections from different studies vary in the spatial representation of the current and expected shift in suitability for *Aedes* mosquitoes.

There is a strong correlation between the simulated climatic suitability of *Aedes albopictus* for current climate and the occurrence of related VBDs in Europe, particularly in France (Giron et al., 2019; Succo et al., 2018), and Italy (Angelini et al., 2008; Rezza et al., 2007; Venturi et al., 2017).

The assumption of a northward spread throughout Europe and higher-altitude areas up to the middle of the century is based on the overall trend of improving climatic suitability in places that are now already conducive for the establishment of *Ae. albopictus*, however not yet suitable for the establishment of *Ae. aegypti*.

Interestingly, *Ae. aegypti* was widespread in the Mediterranean in the first half of the 20th century (Wint et al., 2022) but has been absent since. Until recently, with its latest detection in Cyprus 2022, it has failed to re-establish in Europe. This could be due to a number of potential factors such as interspecies competition, unsuitable points of entry, reduced availability of breeding sites, and unfavorable overwintering conditions (Wint et al., 2022). However, the exact reasons are still not certain.

The kinetics of vector invasion processes are expected to change with the climate. As a result, monitoring invasive mosquito species in Europe that have a strong propensity to colonize new areas, such as *Aedes* mosquitoes, presents a significant concern. A thorough approach to integrating climate change and variable vulnerability analysis in an interdisciplinary context is highly necessary for future research on invasive species of societal value. One of the primary questions that should be addressed in future research is the effect of human activity on breeding site availability since this has proven to override the climatic constraints in the south of Spain. Meaning that human activity and water storage in applicable containers could provide enough potential breeding sites despite the drier conditions that will be exasperated by climate change. Moreover, when estimating the potential spread of disease vectors, anthropogenic infrastructure and transit cannot be neglected. Specifically, points of import such as ports and distribution networks must be examined. Invasive arthropod disease vectors typically have low habitat needs, for the larval and pupal development of *Aedes* mosquitoes, any type of still-water body even at modest volumes are sufficient. This suggests that human traffic and passive long-distance trading are an important channel for VBD. Taking transport routes into consideration will facilitate efficient monitoring and surveillance operations. Thus risk analysis of the impact of climate change on emerging vector diseases must take into account changes in societal behavior as a whole.

As an initial step in early-warning and risk assessment, climatic suitability studies should always be taken into consideration to identify potential target-areas and support vector surveillance. Climate change is also anticipated to encourage the formation of nondiapausing populations of *Ae. albopictus* in the Mediterranean, presuming an increase in temperature in the short- to medium-term (Collantes et al., 2015). Southern Spain and Italy have previously recorded continuous egg-laying (Bonacci et al., 2015; Collantes et al., 2015). According to Lacour (Lacour et al., 2015), this adaptive selection is probably motivated by the advantage of extending its time of activity, but it may also be connected to the suppression of the diapause cost (Scheiner, 1993). All these elements make it difficult to accurately forecast when the vector activity season will begin and end across a broader region since localized small-scale environmental and climatic conditions may have a significant impact on the species seasonality and abundance peaks.

4.5 Conclusions

The outlined recent studies present the state-of-the art in modeling the current and future distribution and seasonal dynamics of *Aedes* mosquitoes in Europe, however with existing uncertainties regarding climate change, expected extremes and biotic interactions, the very process of species invasion could change in the future and require novel considerations and modeling approaches to capture and anticipate this risk (Fischer et al., 2014). A thorough approach to integrating climate risk calculations in a larger scientific framework, under the One Health umbrella, is necessary for future research on invasive vector species and their effect on human and animal health.

There are noted uncertainties regarding mosquito vectors and climate change (Fischer et al., 2014). Applying an ensemble modeling approach to both the vector and climate models should be explored in future research to make more accurate and robust assessments of possible variations in the climate-driven expansion of *Aedes* mosquitoes in Europe.

This scoping review may be used by researchers developing their own models to rapidly find published material and techniques or references of interest and assess which parameters have been utilized in those models. In areas where outbreaks of endemic *Aedes*-borne illnesses or emerging diseases are more likely to occur, these models can be valuable in anticipating when and where changes in disease and vector distribution may take place, allowing for early warning and planning. VBD and vector population dynamics models operating in a quickly changing environment must be supplied with the most recent suitable methodologies in the face of advancing globalization and climate change.

Acknowledgments

We thank Alisa Aliaga-Samanez, Cyril Caminade, Mahmoud Kamal, Sandra Oliveira, Pasquali Sara, Yiannis Proestos, Sadie Ryan, Stéphanie Sherpa, and Stephanie Thomas for their contribution in providing data on the simulated distribution and climatic suitability for *Aedes albopictus* in Europe.

References

Aliaga-Samanez, A., et al. (2021). Worldwide dynamic biogeography of zoonotic and anthroponotic dengue. *PLoS Neglected Tropical Diseases, 15*(6), e0009496. Available from https://doi.org/10.1371/journal.pntd.0009496.

Amos, B., et al. (2022). VEuPathDB: The eukaryotic pathogen, vector and host bioinformatics resource center. *Nucleic Acids Research, 50*(D1), D898–D911.

Angelini, P., et al. (2008). Chikungunya epidemic outbreak in Emilia-Romagna (Italy) during summer 2007. *Parassitologia, 50*(1–2), 97–98.

Berec, L., Gelbic, I., & Sebesta, O. (2014). Worthy of their name: how floods drive outbreaks of two major floodwater mosquitoes (Diptera: Culicidae. *Journal of Medical Entomology, 51* (1), 76–88. Available from https://doi.org/10.1603/me12255.

Blagrove, M. S. C., et al. (2020). Potential for Zika virus transmission by mosquitoes in temperate climates. *Proceedings. Biological Sciences/The Royal Society, 287*(1930), 20200119. Available from https://doi.org/10.1098/rspb.2020.0119.

Bonacci, T., Mazzei, A., Hristova, V. K., & Ayaz Ahmad, M. (2015). Monitoring of *Aedes albopictus* (Diptera, Cilicidae) in Calabria, Southern Italy. *International Journal of Scientific & Engineering Research*, *6*(5), 1186–1189.

Bouzid, M., Colón-González, F. J., Lung, T., Lake, I. R., & Hunter, P. R. (2014). Climate change and the emergence of vector-borne diseases in Europe: Case study of dengue fever. *BMC Public Health*, *14*, 781. Available from https://doi.org/10.1186/1471-2458-14-781.

Brady, O. J., et al. (2013). Modelling adult *Aedes aegypti* and *Aedes albopictus* survival at different temperatures in laboratory and field settings. *Parasites & Vectors*, *6*, 351. Available from https://doi.org/10.1186/1756-3305-6-351.

Braks, M., et al. (2022). VectorNet: Putting vectors on the map. *Frontiers in Public Health*, 549.

Brown, C. J., et al. (2016). Ecological and methodological drivers of species' distribution and phenology responses to climate change. *Global Change Biology*, *22*(4), 1548–1560.

Burkett-Cadena, N. D., McClure, C. J., Estep, L. K., & Eubanks, M. D. (2013). Hosts or habitats: What drives the spatial distribution of mosquitoes? *Ecosphere*, *4*(2), 1–16.

Büttner, G., Feranec, J., Jaffrain, G., Mari, L., Maucha, G., & Soukup, T. (2004). The CORINE land cover 2000 project. *EARSeL eProceedings*, *3*(3), 331–346.

Caminade, C., et al. (2012). Suitability of European climate for the Asian tiger mosquito *Aedes albopictus*: Recent trends and future scenarios. *Journal of the Royal Society Interface*, *9*(75), 2708–2717.

Caminade, C., et al. (2017). Global risk model for vector-borne transmission of Zika virus reveals the role of El Niño 2015. *Proceedings of the National Academy of Sciences of the United States of America*, *114*(1), 119–124. Available from https://doi.org/10.1073/pnas.1614303114.

Campbell, L. P., Luther, C., Moo-Llanes, D., Ramsey, J. M., Danis-Lozano, R., & Peterson, A. T. (2015). Climate change influences on global distributions of dengue and chikungunya virus vectors. *Philosophical Transactions of the Royal Society of London. Series B, Biological Sciences*, *370*(1665). Available from https://doi.org/10.1098/rstb.2014.0135.

Christophers, S. R. (1960). *Aedes aegypti: The yellow fever mosquito*. CUP Archive.

Cianci, D., Hartemink, N., & Ibáñez-Justicia, A. (2015). Modelling the potential spatial distribution of mosquito species using three different techniques. *International Journal of Health Geographics*, *14*. Available from https://doi.org/10.1186/s12942-015-0001-0.

Collantes, F., et al. (2015). Review of ten-years presence of Aedes albopictu s in Spain 2004–2014: Known distribution and public health concerns. *Parasites & Vectors*, *8*(1), 1–11.

Connolly, S. R., Keith, S. A., Colwell, R. K., & Rahbek, C. (2017). Process, mechanism, and modeling in macroecology. *Trends in Ecology & Evolution*, *32*(11), 835–844.

Couret, J., Dotson, E., & Benedict, M. Q. (2014). Temperature, larval diet, and density effects on development rate and survival of *Aedes aegypti* (Diptera: Culicidae). *PLoS One*, *9*(2). Available from https://doi.org/10.1371/journal.pone.0087468.

Cunze, S., Kochmann, J., & Klimpel, S. (2020). Global occurrence data improve potential distribution models for *Aedes japonicus japonicus* in non-native regions. *Pest Management Science*, *76*(5), 1814–1822. Available from https://doi.org/10.1002/ps.5710.

Cunze, S., Kochmann, J., Koch, L. K., & Klimpel, S. (2016). *Aedes albopictus* and its environmental limits in Europe. *PLoS One*, *11*(9), e0162116. Available from https://doi.org/10.1371/journal.pone.0162116.

Cunze, S., Kochmann, J., Koch, L. K., & Klimpel, S. (2018). Niche conservatism of *Aedes albopictus* and *Aedes aegypti*—Two mosquito species with different invasion histories. *Scientific Reports*, *8*(1), 7733. Available from https://doi.org/10.1038/s41598-018-26092-2.

Department of Bioinformatics and Computational Biology, University of Texas, "SurvNet version 3.0.2." https://bioinformatics.mdanderson.org/public-software/survnet/. Accessed August 10, 2022.

Dickens, B. L., Sun, H., Jit, M., Cook, A. R., & Carrasco, L. R. (2018). Determining environmental and anthropogenic factors which explain the global distribution of *Aedes aegypti* and *Ae. albopictus*. *BMJ Global Health*, *3*(4), e000801. Available from https://doi.org/10.1136/bmjgh-2018-000801.

Dormann, C. F., et al. (2012). Correlation and process in species distribution models: Bridging a dichotomy. *Journal of Biogeography*, *39*(12), 2119–2131.

Douchet, L., et al. (2021). Comparing sterile male releases and other methods for integrated control of the tiger mosquito in temperate and tropical climates. *Scientific Reports*, *11*(1), 7354. Available from https://doi.org/10.1038/s41598-021-86798-8.

Ducheyne, E., et al. (2018). Current and future distribution of *Aedes aegypti* and *Aedes albopictus* (Diptera: Culicidae) in WHO Eastern Mediterranean Region. *International Journal of Health Geographics*, *17*(1), 4. Available from https://doi.org/10.1186/s12942-018-0125-0, 14.

ECDC. (2009). Development of *Aedes albopictus* risk maps. European Centre for Disease Prevention and Control.

Edwards, M., & Richardson, A. J. (2004). Impact of climate change on marine pelagic phenology and trophic mismatch. *Nature*, *430*(7002), 881–884.

EID. EID emoustication Rhone-Alpes. https://www.eid-rhonealpes.com/moustiques/les-especes-de-moustique-importees-en-france-que-l-on-retrouve-en-region-auvergne-rhone-alpes. Accessed August 10, 2022.

Eisen, L., Monaghan, A. J., Lozano-Fuentes, S., Steinhoff, D. F., Hayden, M. H., & Bieringer, P. E. (2014). The impact of temperature on the bionomics of *Aedes* (Stegomyia) *aegypti*, with special reference to the cool geographic range margins. *Journal of Medical Entomology*, *51*(3), 496–516.

Eisen, L., & Moore, C. G. (2013). *Aedes* (Stegomyia) *aegypti* in the continental United States: A vector at the cool margin of its geographic range. *Journal of Medical Entomology*, *50*(3), 467–478.

Elith, J., et al. (2006). Novel methods improve prediction of species' distributions from occurrence data. *Ecography*, *29*(2), 129–151.

Erguler, K., et al. (2016). Large-scale modelling of the environmentally-driven population dynamics of temperate *Aedes albopictus* (Skuse). *PLoS One*, *11*(2), e0149282. Available from https://doi.org/10.1371/journal.pone.0149282.

Erguler, K., Chandra, N. L., Proestos, Y., Lelieveld, J., Christophides, G. K., & Parham, P. E. (2017). A large-scale stochastic spatiotemporal model for *Aedes albopictus*-borne chikungunya epidemiology. *PLoS One*, *12*(3), e0174293. Available from https://doi.org/10.1371/journal.pone.0174293.

European Centre for Disease Prevention and Control. (2017). Clusters of autochthonous chikungunya cases in France. ECDC [Online]. Available 14-08-2017-RRA-Chikungunya-France (europa.eu). Accessed April 20, 2021.

European Centre for Disease Prevention and Control. (2019). Zika virus disease in Var department, France. ECDC [Online]. Available: https://www.ecdc.europa.eu/en/publications-data/rapid-risk-assessment-zika-virus-disease-var-department-france. Accessed April 20, 2021.

European Centre for Disease Prevention and Control. (2020). Autochthonous transmission of dengue virus in EU/EEA, 2010–2020. ECDC [Online]. Available: https://www.ecdc.europa.eu/sites/default/files/documents/RRA-dengue-in-Spain-France_1Oct2019.pdf. Accessed April 20, 2021.

European Centre for Disease Prevention and Control and European Food Safety Authority. (2018). The importance of vector abundance and seasonality—Results from an expert consultation. ECDC and EFSA [Online]. Available: https://www.ecdc.europa.eu/sites/default/files/documents/vector-abundance-and-seasonality.pdf. Accessed April 20, 2021.

Ferraguti, M., Martínez-de La Puente, J., Roiz, D., Ruiz, S., Soriguer, R., & Figuerola, J. (2016). Effects of landscape anthropization on mosquito community composition and abundance. *Scientific Reports*, *6*(1), 1–9.

Fick, S. E., & Hijmans, R. J. (2017). WorldClim 2: New 1-km spatial resolution climate surfaces for global land areas. *International Journal of Climatology*, *37*(12), 4302–4315.

Fischer, D., Thomas, S. M., Neteler, M., Tjaden, N. B., & Beierkuhnlein, C. (2014). Climatic suitability of *Aedes albopictus* in Europe referring to climate change projections: Comparison of mechanistic and correlative niche modelling approaches. *Eurosurveillance*, *19*(6), 20696. Available from https://doi.org/10.2807/1560-7917.ES2014.19.6.20696.

Flacio, E., Engeler, L., Tonolla, M., & Müller, P. (2016). Spread and establishment of *Aedes albopictus* in southern Switzerland between 2003 and 2014: an analysis of oviposition data and weather conditions. *Parasites & Vectors*, *9*(1), 304. Available from https://doi.org/10.1186/s13071-016-1577-3.

Focks, D. A., Haile, D. G., Daniels, E., & Mount, G. A. (1993). Dynamic life table model for *Aedes aegypti* (Diptera: Culicidae): Analysis of the literature and model development. *Journal of Medical Entomology*, *30*(6), 1003–1017.

Früh, L., Kampen, H., Koban, M. B., Pernat, N., Schaub, G. A., & Werner, D. (2020). Oviposition of *Aedes japonicus japonicus* (Diptera: Culicidae) and associated native species in relation to season, temperature and land use in western Germany. *Parasites & Vectors*, *13* (1), 623. Available from https://doi.org/10.1186/s13071-020-04461-z.

García-Díaz, P., Prowse, T. A., Anderson, D. P., Lurgi, M., Binny, R. N., & Cassey, P. (2019). A concise guide to developing and using quantitative models in conservation management. *Conservation Science and Practice*, *1*(2), e11.

Gardner, L., Chen, N., & Sarkar, S. (2017). Vector status of *Aedes* species determines geographical risk of autochthonous Zika virus establishment. *PLoS Neglected Tropical Diseases*, *11* (3), e0005487. Available from https://doi.org/10.1371/journal.pntd.0005487.

Garira, W., & Chirove, F. (2020). A general method for multiscale modelling of vector-borne disease systems. *Interface Focus*, *10*(1), 20190047.

GBIF. GBIF—the Global Biodiversity Information Facility—is an international network and data infrastructure funded by the world's governments and aimed at providing anyone, anywhere, open access to data about all types of life on Earth. Available from: https://www.gbif.org/. Accessed Aug. 10, 2022.

Giron, S., et al. (2019). Vector-borne transmission of Zika virus in Europe, southern France, August 2019. *Eurosurveillance*, *24*(45), 1900655.

Gossner, C. M., Ducheyne, E., & Schaffner, F. (2018). Increased risk for autochthonous vector-borne infections transmitted by *Aedes albopictus* in continental Europe. *Euro Surveillance: Bulletin Europeen sur les Maladies Transmissibles (European Communicable Disease Bulletin)*, *23*(24). Available from https://doi.org/10.2807/1560-7917.ES.2018.23.24.1800268.

Grames, E. M., Stillman, A. N., Tingley, M. W., & Elphick, C. S. (2019). An automated approach to identifying search terms for systematic reviews using keyword co-occurrence networks. *Methods in Ecology and Evolution*, *10*(10), 1645–1654.

Guillera-Arroita, G., et al. (2015). Is my species distribution model fit for purpose? Matching data and models to applications. *Global Ecology and Biogeography*, *24*(3), 276–292.

Guzzetta, G., et al. (2016). Potential risk of dengue and chikungunya outbreaks in northern Italy based on a population model of *Aedes albopictus* (Diptera: Culicidae). *PLoS Neglected Tropical Diseases, 10*(6), e0004762.

Guzzetta, G., et al. (2020). Spatial modes for transmission of chikungunya virus during a large chikungunya outbreak in Italy: A modeling analysis. *BMC Medicine, 18*(1), 226. Available from https://doi.org/10.1186/s12916-020-01674-y.

Haddaway, N. R., Macura, B., Whaley, P., & Pullin, A. S. (2018). ROSES RepOrting standards for systematic evidence syntheses: Pro forma, flow-diagram and descriptive summary of the plan and conduct of environmental systematic reviews and systematic maps. *Environmental Evidence, 7*(1), 1–8.

Hazen, E. L., et al. (2013). Predicted habitat shifts of Pacific top predators in a changing climate. *Nature Climate Change, 3*(3), 234–238.

Ibáñez-Justicia, A., et al. (2020). Habitat suitability modelling to assess the introductions of *Aedes albopictus* (Diptera: Culicidae) in the Netherlands. *Parasites & Vectors, 13*(1), 217. Available from https://doi.org/10.1186/s13071-020-04077-3.

Ibañez-Justicia, A., Gloria-Soria, A., den Hartog, W., Dik, M., Jacobs, F., & Stroo, A. (2017). The first detected airline introductions of yellow fever mosquitoes (*Aedes aegypti*) to Europe, at Schiphol International airport, the Netherlands. *Parasites & Vectors, 10*(1), 603. Available from https://doi.org/10.1186/s13071-017-2555-0.

Jourdain, F., et al. (2022). Estimating chikungunya virus transmission parameters and vector control effectiveness highlights key factors to mitigate arboviral disease outbreaks. *PLoS Neglected Tropical Diseases, 16*(3), e0010244. Available from https://doi.org/10.1371/journal.pntd.0010244.

Južnič-Zonta, Ž., et al. (2022). Mosquito alert: leveraging citizen science to create a GBIF mosquito occurrence dataset. *Gigabyte, 2022*, 1–11.

Kalan, K., Ivovic, V., Glasnovic, P., & Buzan, E. (2017). Presence and potential distribution of *Aedes albopictus* and *Aedes japonicus japonicus* (Diptera: Culicidae) in Slovenia. *Journal of Medical Entomology, 54*(6), 1510–1518. Available from https://doi.org/10.1093/jme/tjx150.

Kamal, M., Kenawy, M. A., Rady, M. H., Khaled, A. S., & Samy, A. M. (2018). Mapping the global potential distributions of two arboviral vectors *Aedes aegypti* and *Ae. albopictus* under changing climate. *PLoS One, 13*(12), e0210122. Available from https://doi.org/10.1371/journal.pone.0210122.

Karger D.N. *et al.* (2016). CHELSA climatologies at high resolution for the earth's land surface areas (Version 1.0).

Karl, S., Halder, N., Kelso, J. K., Ritchie, S. A., & Milne, G. J. (2014). A spatial simulation model for dengue virus infection in urban areas. *BMC Infectious Diseases, 14*(1), 1–17.

Kerkow, A., et al. (2019). What makes the Asian bush mosquito *Aedes japonicus japonicus* feel comfortable in Germany? A fuzzy modelling approach. *Parasites & Vectors, 12*(1), 106. Available from https://doi.org/10.1186/s13071-019-3368-0.

Khormi, H. M., & Kumar, L. (2014). Climate change and the potential global distribution of *Aedes aegypti*: spatial modelling using geographical information system and CLIMEX. *Geospatial Health, 8*(2), 405–415.

Kobayashi, M., Nihei, N., & Kurihara, T. (2002). Analysis of northern distribution of *Aedes albopictus* (Diptera: Culicidae) in Japan by geographical information system. *Journal of Medical Entomology, 39*(1), 4–11.

Koch, L. K., et al. (2016). Modeling the habitat suitability for the arbovirus vector *Aedes albopictus* (Diptera: Culicidae) in Germany. *Parasitology Research, 115*(3), 957–964.

Kraemer, M. U. G., et al. (2019). Past and future spread of the arbovirus vectors *Aedes aegypti* and *Aedes albopictus*. *Nature Microbiology*, 1. Available from https://doi.org/10.1038/s41564-019-0376-y.

Kramer, I. M., et al. (2020). Does winter cold really limit the dengue vector *Aedes aegypti* in Europe? *Parasites & Vectors*, *13*(1), 178. Available from https://doi.org/10.1186/s13071-020-04054-w.

Lacour, G., Chanaud, L., L'Ambert, G., & Hance, T. (2015). Seasonal synchronization of diapause phases in Aedes albopictus (Diptera: Culicidae). *PLoS One*, *10*(12), e0145311.

Lawler, J. J., White, D., Neilson, R. P., & Blaustein, A. R. (2006). Predicting climate-induced range shifts: Model differences and model reliability. *Global Change Biology*, *12*(8), 1568–1584.

Legendre, P. (1993). Spatial autocorrelation: Trouble or new paradigm? *Ecology*, *74*(6), 1659–1673.

Leibniz Centre for Agricultural Landscape Research (ZALF) e.V. 'Mückenatlas': A citizen science project for mosquito surveillance in Germany. Available from https://climate-adapt.eea.europa.eu/metadata/case-studies/2018muckenatlas2019-a-citizen-science-project-for-mosquito-surveillance-in-germany. Accessed August 10, 2022.

Liu, B., Gao, X., Ma, J., Jiao, Z., Xiao, J., & Wang, H. (2018). Influence of host and environmental factors on the distribution of the Japanese Encephalitis vector Culex tritaeniorhynchus in China. *International Journal of Environmental Research and Public Health*, *15* (9), 1848.

Liu, Y., Lillepold, K., Semenza, J. C., Tozan, Y., Quam, M. B. M., & Rocklöv, J. (2020). "Reviewing estimates of the basic reproduction number for dengue, Zika and chikungunya across global climate zones. *Environmental Research*, *182*, 109114. Available from https://doi.org/10.1016/j.envres.2020.109114.

Liu-Helmersson, J., et al. (2016). Climate change and *Aedes* vectors: 21st century projections for dengue transmission in Europe. *EBioMedicine*, *7*, 267–277. Available from https://doi.org/10.1016/j.ebiom.2016.03.046.

Liu-Helmersson, J., Rocklöv, J., Sewe, M., & Brännström, Å. (2019). Climate change may enable *Aedes aegypti* infestation in major European cities by 2100. *Environmental Research*, *172*, 693–699. Available from https://doi.org/10.1016/j.envres.2019.02.026.

Lurgi, M., López, B. C., & Montoya, J. M. (2012). Novel communities from climate change. *Philosophical Transactions of the Royal Society B: Biological Sciences*, *367*(1605), 2913–2922.

Manica, M., et al. (2016). Spatial and temporal hot spots of *Aedes albopictus* abundance inside and outside a South European Metropolitan area. *PLoS Neglected Tropical Diseases*, *10*(6), e0004758. Available from https://doi.org/10.1371/journal.pntd.0004758.

Manica, M., Rosà, R., Della Torre, A., & Caputo, B. (2017). From eggs to bites: do ovitrap data provide reliable estimates of *Aedes albopictus* biting females? *PeerJ*, *5*, e2998.

Manica, M., et al. (2019). Assessing the risk of autochthonous yellow fever transmission in Lazio, central Italy. *PLoS Neglected Tropical Diseases*, *13*(1), e0006970. Available from https://doi.org/10.1371/journal.pntd.0006970.

Manica, M., Riello, S., Scagnolari, C., & Caputo, B. (2020). Spatio-temporal distribution of *Aedes albopictus* and *Culex pipiens* along an urban-natural gradient in the Ventotene Island, Italy. *International Journal of Environmental Research and Public Health*, *17*(22). Available from https://doi.org/10.3390/ijerph17228300.

Marcantonio, M., et al. (2016). First assessment of potential distribution and dispersal capacity of the emerging invasive mosquito *Aedes koreicus* in Northeast Italy. *Parasites & Vectors*, *9*, 63. Available from https://doi.org/10.1186/s13071-016-1340-9.

Marini, G., et al. (2019). First report of the influence of temperature on the bionomics and population dynamics of *Aedes koreicus*, a new invasive alien species in Europe. *Parasites & Vectors*, *12*(1), 524. Available from https://doi.org/10.1186/s13071-019-3772-5.

Marsboom C. *et al.* (2019). A systematic literature review of mathematical models for *Aedes*-borne disease spread and control.

Massad, E., et al. (2018). Estimating the probability of dengue virus introduction and secondary autochthonous cases in Europe. *Scientific Reports*, *8*(1), 4629. Available from https://doi.org/10.1038/s41598-018-22590-5.

McKenzie, B. A., Wilson, A. E., & Zohdy, S. (2019). *Aedes albopictus* is a competent vector of Zika virus: A meta-analysis. *PLoS One*, *14*(5), e0216794.

Medlock, J. M., Avenell, D., Barrass, I., & Leach, S. (2006). Analysis of the potential for survival and seasonal activity of *Aedes albopictus* (Diptera: Culicidae) in the United Kingdom. *Journal of Vector Ecology*, *31*(2), 292–305.

Melaun, C., et al. (2015). Modeling of the putative distribution of the arbovirus vector *Ochlerotatus japonicus japonicus* (Diptera: Culicidae) in Germany. *Parasitology Research*, *114*(3), 1051–1061. Available from https://doi.org/10.1007/s00436-014-4274-1.

Metelmann, S., Caminade, C., Jones, A. E., Medlock, J. M., Baylis, M., & Morse, A. P. (2019). The UK's suitability for *Aedes albopictus* in current and future climates. *Journal of the Royal Society, Interface/the Royal Society*, *16*(152), 20180761. Available from https://doi.org/10.1098/rsif.2018.0761.

Monaghan, A. J., et al. (2016). The potential impacts of 21st century climatic and population changes on human exposure to the virus vector mosquito *Aedes aegypti*. *Climatic Change*, *146*(3–4), 487–500. Available from https://doi.org/10.1007/s10584-016-1679-0.

Muhling, B. A., et al. (2017). Projections of future habitat use by Atlantic bluefin tuna: mechanistic vs. correlative distribution models. *ICES Journal of Marine Science*, *74*(3), 698–716.

National Geographic Society and California Academy of Science. iNaturalist is a joint initiative of the California Academy of Sciences and the National Geographic Society. Available from: https://www.inaturalist.org. Accessed August 10, 2022.

National Museum of Natural History. Smithsonian Institution, Washington DC, University of Queensland Insect Collection, Brisbane, and Museum of Vertebrate Zoology, University of California, VectorMap, August 15, 2000. Available from: http://www.vectormap.si.edu. Accessed August 10, 2022.

Nawrocki, S. J., Hawley, W. A., et al. (1987). Estimation of the northern limits of distribution of *Aedes albopictus* in North America. *Journal of the American Mosquito Control Association*, *3*(2), 314–317.

Neteler, M., Roiz, D., Rocchini, D., Castellani, C., & Rizzoli, A. (2011). Terra and Aqua satellites track tiger mosquito invasion: Modelling the potential distribution of Aedes albopictus in north-eastern Italy. *International Journal of Health Geographics*, *10*, 49. Available from https://doi.org/10.1186/1476-072X-10-49.

Oliveira, S., Rocha, J., Sousa, C. A., & Capinha, C. (2021). Wide and increasing suitability for *Aedes albopictus* in Europe is congruent across distribution models. *Scientific Reports*, *11* (1), 9916. Available from https://doi.org/10.1038/s41598-021-89096-5.

Pasquali, S., et al. (2020). Development and calibration of a model for the potential establishment and impact of *Aedes albopictus* in Europe. *Acta Tropica*, *202*, 105228. Available from https://doi.org/10.1016/j.actatropica.2019.105228.

Petrić, M., Lalić, B., Ducheyne, E., Djurdjević, V., & Petrić, D. (2017). Modelling the regional impact of climate change on the suitability of the establishment of the Asian tiger mosquito (*Aedes albopictus*) in Serbia. *Climatic Change*, *142*(3–4), 361–374.

Petrić, M., Lalić, B., Pajović, I., Micev, S., Djurdjević, V., & Petrić, D. (2018). Expected changes of Montenegrin climate, impact on the establishment and spread of the Asian tiger mosquito (*Aedes albopictus*), and validation of the model and model-based field sampling. *Atmosphere*, *9*(11), 453.

Petrić, M., et al. (2021). Seasonality and timing of peak abundance of *Aedes albopictus* in Europe: Implications to public and animal health. *Geospatial Health*, *16*(1). Available from https://doi.org/10.4081/gh.2021.996, Art. no. 1.

Phillips, S. J., Anderson, R. P., Dudík, M., Schapire, R. E., & Blair, M. E. (2017). Opening the black box: An open-source release of Maxent. *Ecography*, *40*(7), 887–893.

Plagányi, É. E., et al. (2011). Modelling climate-change effects on Australian and Pacific aquatic ecosystems: a review of analytical tools and management implications. *Marine and Freshwater Research*, *62*(9), 1132–1147.

Plagányi, É. E., et al. (2014). Ecosystem modelling provides clues to understanding ecological tipping points. *Marine Ecology Progress Series*, *512*, 99–113.

Proestos, Y., Christophides, G. K., Ergüler, K., Tanarhte, M., Waldock, J., & Lelieveld, J. (2015). Present and future projections of habitat suitability of the Asian tiger mosquito, a vector of viral pathogens, from global climate simulation. *Philosophical Transactions of the Royal Society B: Biological Sciences*, *370*(1665), 20130554.

Publications Office of the European Union. (2022). EuroVoc: multilingual, multidisciplinary thesaurus covering the activities of the EU. Thesaurus 20220708-0, [Online]. Available from: http://publications.europa.eu/resource/dataset/eurovoc.

Reinhold, J. M., Lazzari, C. R., & Lahondère, C. (2018). Effects of the environmental temperature on *Aedes aegypti* and *Aedes albopictus* mosquitoes: a review. *Insects*, *9*(4), 158.

Reuss, F., et al. (2018). Thermal experiments with the Asian bush mosquito (*Aedes japonicus japonicus*) (Diptera: Culicidae) and implications for its distribution in Germany. *Parasites & Vectors*, *11*(1), 81. Available from https://doi.org/10.1186/s13071-018-2659-1.

Rezza, G., et al. (2007). Infection with chikungunya virus in Italy: An outbreak in a temperate region. *Lancet*, *370*(9602), 1840–1846. Available from https://doi.org/10.1016/S0140-6736(07)61779-6.

Robertson, M. P., Peter, C. I., Villet, M. H., & Ripley, B. S. (2003). Comparing models for predicting species' potential distributions: A case study using correlative and mechanistic predictive modelling techniques. *Ecological Modelling*, *164*(2–3), 153–167.

Robinson, N. M., Nelson, W. A., Costello, M. J., Sutherland, J. E., & Lundquist, C. J. (2017). A systematic review of marine-based species distribution models (SDMs) with recommendations for best practice. *Frontiers in Marine Science*, *4*, 421.

Rocklöv, J., et al. (2016). Assessing seasonal risks for the introduction and mosquito-borne spread of Zika virus in Europe. *EBioMedicine*, *9*, 250–256. Available from https://doi.org/10.1016/j.ebiom.2016.06.009.

Roiz, D., Neteler, M., Castellani, C., Arnoldi, D., & Rizzoli, A. (2011). Climatic factors driving invasion of the tiger mosquito (*Aedes albopictus*) into new areas of Trentino, northern Italy. *PLoS One*, *6*(4).

Roiz, D., Rosa, R., Arnoldi, D., & Rizzoli, A. (2010). Effects of temperature and rainfall on the activity and dynamics of host-seeking *Aedes albopictus* females in northern Italy. *Vector-Borne and Zoonotic Diseases*, *10*(8), 811–816.

Romiti, F., et al. (2021). *Aedes albopictus* (Diptera: Culicidae) monitoring in the Lazio region (Central Italy). *Journal of Medical Entomology*, *58*(2), 847–856. Available from https://doi.org/10.1093/jme/tjaa222.

Roques, L., & Bonnefon, O. (2016). Modelling population dynamics in realistic landscapes with linear elements: A mechanistic-statistical reaction-diffusion approach. *PLoS One*, *11*(3), e0151217. Available from https://doi.org/10.1371/journal.pone.0151217.

Rosenberg, M. S. (2004). Wavelet analysis for detecting anisotropy in point patterns. *Journal of Vegetation Science*, *15*(2), 277–284.

Rowley, W. A., & Graham, C. L. (1968). The effect of temperature and relative humidity on the flight performance of female *Aedes aegypti*. *Journal of Insect Physiology*, *14*(9), 1251–1257. Available from https://doi.org/10.1016/0022-1910(68)90018-8.

Ryan, S. J., Carlson, C. J., Mordecai, E. A., & Johnson, L. R. (2019). Global expansion and redistribution of *Aedes*-borne virus transmission risk with climate change. *PLoS Neglected Tropical Diseases*, *13*(3), e0007213. Available from https://doi.org/10.1371/journal.pntd.0007213.

Ryan, S. J., et al. (2021). Warming temperatures could expose more than 1.3 billion new people to Zika virus risk by 2050. *Global Change Biology*, *27*(1), 84–93. Available from https://doi.org/10.1111/gcb.15384.

Salami, D., Capinha, C., Sousa, C. A., do, M., Martins, R. O., & Lord, C. (2020). Simulation models of dengue transmission in Funchal, Madeira Island: Influence of seasonality. *PLoS Neglected Tropical Diseases*, *14*(10), e0008679. Available from https://doi.org/10.1371/journal.pntd.0008679.

Sanz-Aguilar, A., et al. (2018). Water associated with residential areas and tourist resorts is the key predictor of Asian tiger mosquito presence on a Mediterranean island. *Medical and Veterinary Entomology*, *32*(4), 443–450. Available from https://doi.org/10.1111/mve.12317.

Schäfer, M. L., & Lundström, J. O. (2009). The present distribution and predicted geographic expansion of the floodwater mosquito *Aedes sticticus* in Sweden. *Journal of Vector Ecology: Journal of the Society for Vector Ecology*, *34*(1), 141–147. Available from https://doi.org/10.1111/j.1948-7134.2009.00017.x.

Schaffner, F., Medlock, J. M., & Van Bortel, W. (2013). Public health significance of invasive mosquitoes in Europe. *Clinical Microbiology and Infection*, *19*(8), 685–692.

Scheiner, S. M. (1993). Genetics and evolution of phenotypic plasticity. *Annual Review of Ecology and Systematics*, *24*(1), 35–68.

Scott, T. W., et al. (2000). Longitudinal studies of *Aedes aegypti* (Diptera: Culicidae) in Thailand and Puerto Rico: Blood feeding frequency. *Journal of Medical Entomology*, *37*(1), 89–101.

Seyler, T., Grandesso, F., Le Strat, Y., Tarantola, A., & Depoortere, E. (2009). Assessing the risk of importing dengue and chikungunya viruses to the European Union. *Epidemics*, *1*(3), 175–184. Available from https://doi.org/10.1016/j.epidem.2009.06.003.

Sherpa, S., et al. (2019). Predicting the success of an invader: Niche shift versus niche conservatism. *Ecology and Evolution*, *9*(22), 12658–12675.

Sherpa, S., et al. (2020). Landscape does matter: Disentangling founder effects from natural and human-aided post-introduction dispersal during an ongoing biological invasion. *The Journal of Animal Ecology*, *89*(9), 2027–2042. Available from https://doi.org/10.1111/1365-2656.13284.

Shocket M.S. *et al.* (2019). Transmission of West Nile virus and other temperate mosquito-borne viruses occurs at lower environmental temperatures than tropical diseases, *bioRxiv*, 597898.

Solimini, A. G., Manica, M., Rosà, R., Della Torre, A., & Caputo, B. (2018). Estimating the risk of Dengue, Chikungunya and Zika outbreaks in a large European city. *Scientific Reports*, *8*(1), 16435. Available from https://doi.org/10.1038/s41598-018-34664-5.

Steiger, D. B. M., Ritchie, S. A., & Laurance, S. G. (2016). Mosquito communities and disease risk influenced by land use change and seasonality in the Australian tropics. *Parasites & Vectors*, *9*(1), 387.

Succo, T., et al. (2018). Dengue serosurvey after a 2-month long outbreak in Nîmes, France, 2015: Was there more than met the eye. *Eurosurveillance*, *23*(23), 1700482. Available from https://doi.org/10.2807/1560-7917.ES.2018.23.23.1700482.

Sutherland, W. J., & Wordley, C. F. (2018). *A fresh approach to evidence synthesis*. Nature Publishing Group.

Takumi, K., Scholte, E.-J., Braks, M., Reusken, C., Avenell, D., & Medlock, J. M. (2009). Introduction, scenarios for establishment and seasonal activity of *Aedes albopictus* in The Netherlands. *Vector Borne and Zoonotic Diseases (Larchmont, N.Y.)*, *9*(2), 191–196. Available from https://doi.org/10.1089/vbz.2008.0038.

Thomas, S. M., Tjaden, N. B., van den Bos, S., & Beierkuhnlein, C. (2014). Implementing cargo movement into climate based risk assessment of vector-borne diseases. *International Journal of Environmental Research and Public Health*, *11*(3), 3360–3374. Available from https://doi.org/10.3390/ijerph110303360.

Thomas, S. M., Obermayr, U., Fischer, D., Kreyling, J., & Beierkuhnlein, C. (2012). Low-temperature threshold for egg survival of a post-diapause and non-diapause European aedine strain, *Aedes albopictus* (Diptera: Culicidae). *Parasites & Vectors*, *5*(1), 100.

Thomas, S. M., et al. (2018). Areas with high hazard potential for autochthonous transmission *of Aedes albopictus*-associated arboviruses in Germany. *International Journal of Environmental Research and Public Health*, *15*(6). Available from https://doi.org/10.3390/ijerph15061270.

Thorson, J. T., Ianelli, J. N., & Kotwicki, S. (2017). The relative influence of temperature and size-structure on fish distribution shifts: A case-study on Walleye pollock in the Bering Sea. *Fish and Fisheries*, *18*(6), 1073–1084.

Tisseuil, C., et al. (2018). Forecasting the spatial and seasonal dynamic of *Aedes albopictus* oviposition activity in Albania and Balkan countries. *PLoS Neglected Tropical Diseases*, *12*(2), e0006236. Available from https://doi.org/10.1371/journal.pntd.0006236.

Tjaden, N. B., Cheng, Y., Beierkuhnlein, C., & Thomas, S. M. (2021). Chikungunya beyond the tropics: Where and when do we expect disease transmission in Europe? *Viruses*, *13*(6). Available from https://doi.org/10.3390/v13061024.

Trájer, A. J. (2021). *Aedes aegypti* in the Mediterranean container ports at the time of climate change: A time bomb on the mosquito vector map of Europe. *Heliyon*, *7*(9), e07981. Available from https://doi.org/10.1016/j.heliyon.2021.e07981.

Tran, A., et al. (2013). A rainfall- and temperature-driven abundance model for *Aedes albopictus* populations. *International Journal of Environmental Research and Public Health*, *10*(5), 1698–1719. Available from https://doi.org/10.3390/ijerph10051698.

Uusitalo, R., et al. (2021). Predicting spatial patterns of Sindbis virus (SINV) infection risk in Finland using vector, host and environmental data. *International Journal of Environmental Research and Public Health*, *18*(13). Available from https://doi.org/10.3390/ijerph18137064.

Valdez, L. D., Sibona, G. J., & Condat, C. A. (2018). Impact of rainfall on *Aedes aegypti* populations. *Ecological Modelling*, *385*, 96–105.

van der Schrier, G., van den Besselaar, E. M., Klein Tank, A. M. G., & Verver, G. (2013). Monitoring European average temperature based on the E-OBS gridded data set. *Journal of Geophysical Research: Atmospheres*, *118*(11), 5120–5135.

Vanwambeke, S. O., et al. (2007). Landscape and land cover factors influence the presence of *Aedes* and *Anopheles* larvae. *Journal of Medical Entomology*, *44*(1), 133–144.

Venturi, G., et al. (2017). Detection of a chikungunya outbreak in Central Italy, August to September 2017. *Euro Surveillance: Bulletin Europeen sur les Maladies Transmissibles*

(European Communicable Disease Bulletin), *22*(39). Available from https://doi.org/10.2807/1560-7917.ES.2017.22.39.17-00646.

Vinogradova, E. B. (2000). *Culex pipiens pipiens mosquitoes: Taxonomy, distribution, ecology, physiology, genetics, applied importance and control, no. 2*. Pensoft Publishers.

Wan, S.H. Z. (2015). MOD11A1 MODIS/Terra Land Surface Temperature/Emissivity Daily L3 Global 1km SIN Grid V006. Available from https://doi.org/10.5067/MODIS/MOD11A1.006.

Werblow, A., Klimpel, S., Bolius, S., Dorresteijn, A. W., Sauer, J., & Melaun, C. (2014). Population structure and distribution patterns of the sibling mosquito species *Culex pipiens* and *Culex torrentium* (Diptera: Culicidae) reveal different evolutionary paths. *PLoS One*, *9*(7), e102158.

Wieser, A., Reuss, F., Niamir, A., Müller, R., O'Hara, R. B., & Pfenninger, M. (2019). Modelling seasonal dynamics, population stability, and pest control in *Aedes japonicus japonicus* (Diptera: Culicidae). *Parasites & Vectors*, *12*(1), 142. Available from https://doi.org/10.1186/s13071-019-3366-2.

Wint, W., Jones, P., Kraemer, M., Alexander, N., & Schaffner, F. (2022). Past, present and future distribution of the yellow fever mosquito Aedes aegypti: The European paradox. *Science of The Total Environment*, *847*, 157566.

Yamana, T. K., & Eltahir, E. A. B. (2013). Incorporating the effects of humidity in a mechanistic model of *Anopheles gambiae* mosquito population dynamics in the Sahel region of Africa. *Parasites & Vectors*, *6*, 235. Available from https://doi.org/10.1186/1756-3305-6-235.

Zhou, S., et al. (2012). Spatial correlation between malaria cases and water-bodies in *Anopheles sinensis* dominated areas of Huang-Huai plain, China. *Parasites & Vectors*, *5*(1), 1–7.

Section 2

Geospatial platforms and crowdsourced geospatial data

Chapter 5

Deposition and erosion dynamics in Axios and Aliakmonas river deltas (Greece) with the use of Google Earth Engine and geospatial analysis tools

Isidora Isis Demertzi[1,2], Spyridon E. Detsikas[2], Ioanna Tselka[1,2], George P. Petropoulos[2] and Efthimios Karymbalis[2]
[1]*School of Rural and Surveying Engineering, National Technical University of Athens, Zografou Campus, Athens, Greece,* [2]*Department of Geography, Harokopio University of Athens, Athens, Greece*

5.1 Introduction

River deltas are considered among the most diverse landscapes with a significant environmental and ecosystemic importance (Salimi et al., 2021). While being very important habitats for many endangered species and hotspots of biodiversity, river deltas are also very important human cultural and economic centers with more than 500 hundred million inhabitants living in deltas over the globe (Kuenzer et al., 2011). River deltas are amongst the most complex ecosystems, where the boundaries between land and water are continuously altering, have significant environmental and agricultural importance. These are vulnerable environments threatened by a range of factors related to climate, geology, and human activity (Dilalos et al., 2022; Vörösmarty et al., 2003). Activities such as agriculture, industrial development, and urbanization of the coastlines constitute the largest human-related threat to deltaic settings (Kapsimalis et al., 2005). Indeed, coastline movement due to erosion and deposition is a major concern for coastal zone management. Very dynamic coastlines pose considerable hazards to human use and development. Mapping the constant changes of the deltaic formations is of great significance to better

Geoinformatics for Geosciences. DOI: https://doi.org/10.1016/B978-0-323-98983-1.00006-5

understand the coastline alterations happening due to erosion and deposition in such formations. Given their importance, there is an urgent need for rapid, replicable techniques to map river deltas and their changes over time (White and El-Asmar, 1999; Petropoulos et al., 2015).

Traditional methods of estuaries monitoring, involving in situ visits and ground-based observations, are ineffective, cost, time, and labor intensive, as they demand frequent field visits to fully capture the highly dynamic nature of a river delta (Cai et al., 2022; Fragou et al., 2020a,2020b). On the other hand, Earth Observation (EO) has been widely used in monitoring river deltas and coastal diversity over the last decades (White and El-Asmar, 1999, Chen et al., 2005). In comparison to traditional mapping approaches such as field surveys, EO technology provides many advantages, such as rapid and cost-efficient mapping (Petropoulos et al., 2020; Tsatsaris et al., 2021). EO also provides access to unapproachable regions at regular time intervals which can be a frequent phenomenon in river deltas monitoring. Finally, geographic information systems (GIS) combined with EO data enable further processing through a wide variety of tools for spatial information extraction and analysis (Jiang et al., 2016; Sandric et al., 2022).

A pioneer satellite EO program, having a large contribution of the development of EO technology in general, is the Landsat program that provides the scientific community with a frequent and cost-free coverage of the planet since 1972. Landsat missions' data have been widely exploited in wetlands monitoring with high accuracies (e.g., Petropoulos et al., 2015). Landsat sensors provide a high potential in monitoring river deltas. In addition to the plethora of the available EO datasets, new software tools have been created to facilitate the analysis of large amount of EO data. In this context, cloud-based platforms such as Google Earth Engine (GEE) and Microsoft Planetary Computer have emerged as a very useful tool to handle the abundance of big EO datasets. Those cloud-based platforms allow a cost and computational efficient way to process big EO data on the cloud while also supporting the deployment of sophisticate algorithms such as machine learning (ML) techniques useful in pattern recognition and classification (Yang et al., 2022). Yet, to our knowledge, the use of cloud-based platforms such as GEE in the context of river deltas dynamics monitoring using EO data from sensors like Landsat series remains underexplored.

In purview of the above, the present study aims at developing a semiautomated GEE-based method for mapping and quantifying morphological changes that occur over time in river deltas. The use of the developed approach is demonstrated for mapping the morphological changes of the two largest in size river deltas located in Thermaikos Gulf, Northern Greece over a period of 36 years. In this context, the changes in soil erosion and deposition rates are also estimated in a GIS environment using geospatial analysis tools. To validate the GEE results, direct photo-interpretation to the same Landsat images is used.

5.2 Experimental setup

5.2.1 Study area description

For the purposes of this study, the deltas of two of the most important rivers of Greece and the southern Balkans, Axios and Aliakmonas, are used (Fig. 5.1). Those two river deltas are located closely to each other on the northwestern Aegean Sea and more specifically in Thermaikos Gulf near Thessaloniki. Axios, also known as Vardar, river springs are located in the mountain range of Sharr, in North Macedonia, near the borders of Albania and Kosovo. Axios has a total length of 380 km, with only 76 km of the total length lying within the Greek borders (Smardon, 2009). Axios delta is a fluvially dominated delta extending relatively around 40°31′ North and 22°43′ East and a southern aspect exposure (Poulos et al., 2009). On the other hand, Aliakmonas or Haliakmon river has a total length of 297 km extending relatively around 40°28′2 North and 22°39′ East. Haliakmon flows only within Greek borders, making it the longest river of the country. Like Axios deltas, Haliakmon has a fluvially dominated delta located with a South East aspect exposure (Poulos et al., 2009).

The two rivers create a fertile valley with significant environmental, economic, and agricultural importance for the Northern Greece. Particularly, Axios and Aliakmonas deltas region are accountable for more than half of

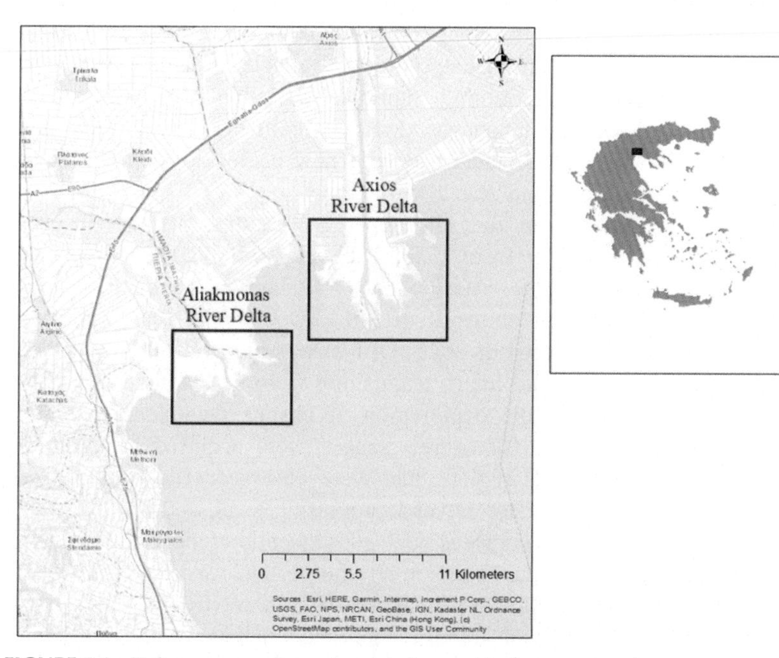

FIGURE 5.1 Reference map of experimental sites used in the present study.

the Greek rice and shellfish production (Petropoulos et al., 2015). However, the human activity and the intensification of the agricultural production have exacerbated the environmental conditions in the region (Kazantzidis & Goutner, 1996).

5.2.2 Datasets

To monitor the coastline deposition/erosion rates over the selected region and time, remotely sensed imagery from Landsat satellite program archive was used. Landsat is the longest existing EO program with an almost continuous archive from 1972. Landsat missions provide crucial data with a moderate pixel size of 30 m on the visible and shortwave electromagnetic spectrum and 100 m on the thermal electromagnetic spectrum coordinated by NASA and the US Geological Survey. Landsat data was found out ideal for this study, as the extensive and consistent archive has been widely used in environmental research with an emphasis on wetlands monitoring (Bishop-Taylor et al., 2021; Cao et al., 2020).

All Landsat images used in this study were obtained from the US Geological Survey (USGS, http://glovis.usgs.gov) archive using GEE JavaScript code editor. GEE is a publicly available, at no cost for research and education purposes, cloud-based platform that enables labor and time-efficient geospatial analysis at planetary scale. GEE exploits Google Cloud computational resources coupled with a constantly growing 50 Petabyte database integrating data from different EO missions (Gorelick et al., 2017). It provides an array of geospatial analysis tools available through a simple, yet powerful Application Programming Interface accessible both in JavaScript and Python programming languages. Landsat archive is being fully integrated and available to GEE database with new acquisitions of the active missions of Landsat being integrated within days.

In this study, data from four different Landsat missions was utilized. More specifically, imagery from Landsat 5 on April of 1986 (Thematic Mapper—TM), Landsat 7 on April of 2000 (Enhanced Thematic Mapper Plus—ETM +), Landsat 8 on April of 2014 (Operational Land Imager—OLI), and Landsat 9 on April of 2022 (OLI-2) was used to fulfill the objectives of this study. Multidate image acquisition from near anniversary dates was chosen, as it allows the exploitation of change detection techniques regarding coastal changes. Cloud-free images were acquired to minimize errors regarding the coastal regions that were observed. The images were obtained at Surface Reflectance Level-1 processing level, and, for computational resources, efficiency purposes were subsequently cropped to the extent of the two examined river deltas before any further analysis. From the available spectral bands of each Landsat imagery, only the VNIR and SWIR spectral bands were used from the Landsat satellites whereas the thermal bands were excluded from the analysis.

5.3 Methods

To detect the morphological in the study area over the selected time, a GEE-based mapping approach was implemented with a direct photo-interpretation implemented for validation purposes. The methodological framework adopted to satisfy the study objectives is summarized in Fig. 5.2, whereas next is provided more detailed information concerning the overall methodology implementation to satisfy the study objectives.

5.3.1 Obtaining river deltas cartography

5.3.1.1 GEE-based approach

The GEE mapping approach was implemented by conducting a ML supervised pixel-based classification exploiting the capabilities of the GEE cloud platform to delineate water from land in two examined river deltas over the selected time period. GEE provides a variety of both supervised and unsupervised ML techniques of regression and classification that can be applied over geospatial data. The full list of the integrated ML techniques can be accessed freely. In addition to the already existing ML techniques in the toolset of GEE, users can exploit the synergistic use of GEE database with custom ML platforms (e.g.,TensorFlow) outside the Earth Engine environment. For the

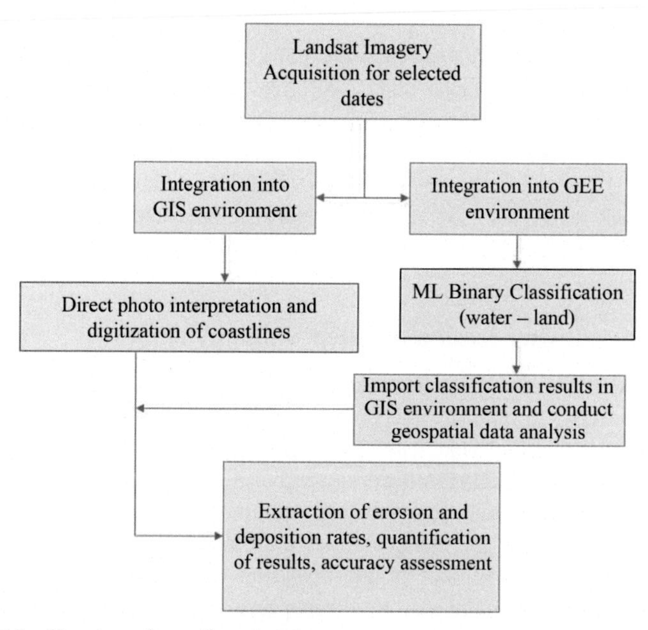

FIGURE 5.2 Flowchart of overall methodology.

purposes of our study, Random Forests (RFs) (Breiman, 2001) supervised pixel classifier was applied to the four scenes of Landsat satellite imagery inside the GEE cloud platform. RF is an ensemble pixel-based classification algorithm having shown promising results when applied to different types of EO imagery (Ireland & Petropoulos, 2015; Petropoulos et al., 2014). As RF classification algorithm randomly selects a part of the training samples assists in obtaining accurate predictions as well as avoids overfitting (Sheykhmousa et al., 2020).

The first step was to formulate the classification scheme, which consisted of the classes "Water" and "Land". Second, representative training sites for each class were collected from each of the selected Landsat imagery adopting a random sampling approach. In terms of training samples, an average of 200 hundred pixels were identified as training data in each image, representing the classes defined in the adopted classification scheme. The spectral separability between the target map classes was visually assessed by comparing the plotted mean spectra values of each classification class for each image. As far as the hyperparameters tuning of the RF classifier is concerned, emphasis was given on the number of the random trees that each classification was implemented. The optimal number of trees varied from 20 to 50, and it was decided after following trial and error procedures, a standard approach implemented also in other similar studies (Brown et al., 2018; Pandey et al., 2019; Srivastava et al., 2021).

5.3.1.2 Direct photo-interpretation approach

The direct photo-interpretation method included a direct vectorization of the deltaic surfaces of the chosen dates. As a result, the coastlines of the study areas were manually digitized using the satellite images with the use of appropriate pseudo-color composites to visibly separate the thin limits between the deltaic surface and water. Imagery from platforms such as Google Earth where also used to cross-reference abstruse spots on both study areas for the selective dates.

5.3.2 Quantifying erosion and deposition dynamics

Results obtained from the two approaches were subsequently imported in a GIS environment to quantify the erosion and deposition rates. The coastline from the first examined year, 1986, from both direct photo-interpretation and the GEE-based approach was used as the baseline. Then, the coastline from each year and mapping approach was compared to the defined baseline with the seaward shift of the coastline to be considered as deposition and the landward shift movement of the coastline to be considered as erosion. This approach was adopted also to similar case studies (Aguilar et al., 2019; Misra & Balaji, 2015; Valderrama-Landeros & Flores-de-Santiago, 2019).

The erosion and deposition rates were estimated using simple geospatial analysis tools (e.g., clip, erase, and intersect).

5.3.3 Validation approach

The GEE-based retrievals were compared against the photo-interpretation results, in terms of both spatial and temporal trends. To compare the deposition/erosion rates, a series of statistical metrics were estimated (Table 5.1). Those included the detected area efficiency, skipped area rate (omission error), and false area rate (commission error). This approach has also been used efficiently in other studies (e.g., Petropoulos et al., 2014). Furthermore, to enable overlay and facilitate efficiency in the erosion/deposition comparisons, the classified scenes from the Landsat archive were exported and subsequently transformed into ArcGIS Pro software platform (ESRI Inc., v. 9.3.1).

5.4 Results

5.4.1 GEE mapping

The present study aimed at exploring the use of GEE platform combined with Landsat archive to monitor the dynamics of erosion and deposition at two river deltas located in Greece for a period of 36 years. Table 5.2 summarizes the estimated, in km^2, total areas of erosion and deposition for Axios and Aliakmonas river deltas. The actual surface of the eroded and deposited land of the aforementioned river deltas for all chronical scales is presented in Fig. 5.3.

According to the GEE results (Fig. 5.3), Axios river delta exhibited its highest extent of erosion from 1986 to 2000. As it can be observed, erosion is the dominant process operating for the examined time period.

TABLE 5.1 Respectively DDA is the detected delta area (common area between the Google Earth Engine [GEE]-based and manual approaches), FDA is the false delta area (the area included in the GEE-based polygon but not in the reference manual polygon), and SDA is skipped delta area (the area included in the reference in situ polygon but not in the GEE-based polygon).

Detected area efficiency	DAE	$\frac{DDA}{DDA+SDA}$
Skipped area rate	SAR	$\frac{SDA}{DDA+SDA}$
False area rate	FAR	$\frac{FDA}{DDA+FDA}$

TABLE 5.2 Google Earth Engine results of total eroded and deposited areas.

Time period	Axios river delta		Aliakmonas river delta	
	Erosion (km^2)	Deposition (km^2)	Erosion (km^2)	Deposition (km^2)
1986–2000	1.52	1.84	0.73	0.76
2000–14	1.08	1.94	1.48	0.33
2014–22	0.99	1.05	0.57	0.48

More specifically, erosion in Axios river displays the most intense process from 1986 to 2000 with the eroded land reaching 1.52 km^2 (Table 5.2). During this time period, most of the eroded land in Axios river delta seemed to be mostly concentrated in the northern part of the delta. Since then, erosion rates of Axios displayed a reduction while the green-colored areas were just a few. Moreover, as illustrated in Fig. 5.3, the deposited land demonstrated its highest rates during the time period between 2000 and 2014, having reached 1.94 km^2 of deposition (Table 5.2).

Regarding Aliakmonas river delta, results demonstrated the highest erosion rates during the time period between 2000 and 2014, where the eroded land covered 1.48 km^2 (Table 5.2). More specifically, as displayed in Fig. 5.1, the eroded land for Aliakmonas river delta during this specific time period seemed to play a major role since the deposition rates, which are characterized by green color, seem to cover very small areas of the examined region in contrast with the red-colored areas which are attributed to erosion rates. As for the deposition rates, the highest erosion values correspond to the time period between 1986 and 2000, reaching 0.76 km^2 of deposited land mostly in the northern part of the examined delta.

Axios river deltas erosion and deposition values presented some interesting patterns since the most recent time period, 2014 to 2022, corresponded to the lowest rates for each process. This significant decrease is visible in Fig. 5.1, where green-colored and red-colored areas are almost hardly detected. On the other hand, Aliakmonas river deltas did not follow up to a specific pattern since erosion and deposition values corresponded to different time periods regarding their maximum and minimum values.

5.4.2 Erosion and deposition rates from GIS analysis

The final results from the GIS analysis of the erosion and deposition rates that were mapped and quantified for the consecutive time periods manifest significant discrepancies from those of the GEE approach. For the Axios

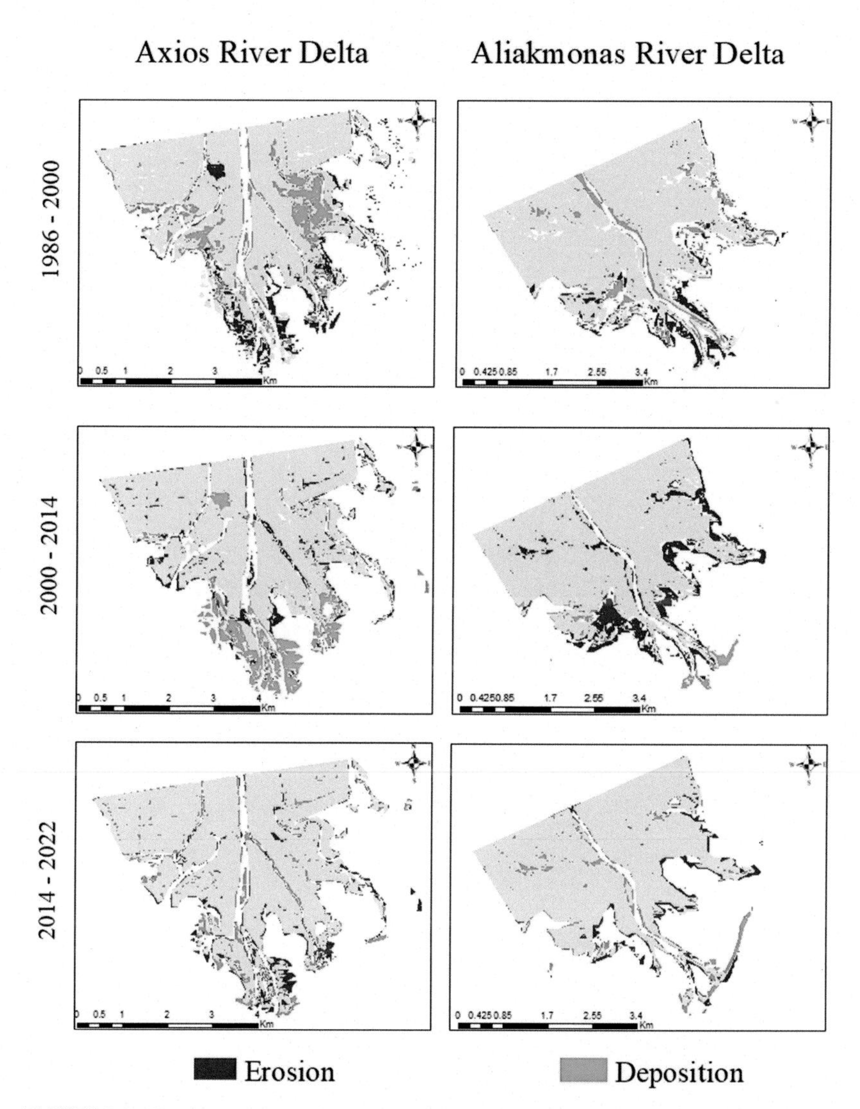

FIGURE 5.3 Erosion and deposition rates of the GEE approach. *GEE*, Google Earth Engine.

river delta, the deposition rates are higher than the erosion rates. The erosion and deposition rates for the respective time periods (Table 5.2) in Axios river delta were the highest during 2000–14 (Fig. 5.4), mainly at the southern part of the river delta, along the river streams and on the western and eastern parts. For that time period, the deposition rates where significantly lower and can be observed mostly along the river streams and the southern part of the study region. The opposite can be observed between 1986 and 2000, where deposition can be noticed mainly at the central part of the river delta, while

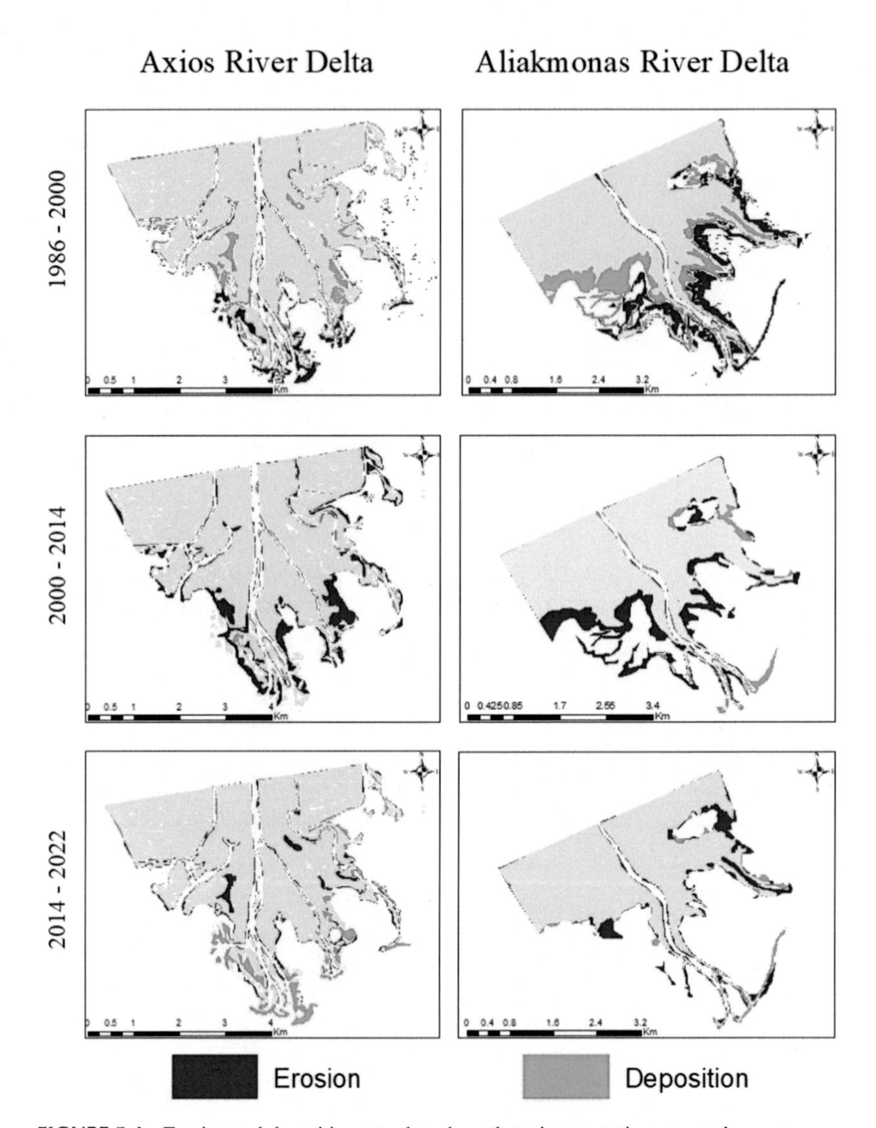

FIGURE 5.4 Erosion and deposition rates based on photo-interpretation approach.

erosion processes can be found on the southern region. Lastly, the most present temporal segment reveals equal results for the eroded and the newly formed land, approximately 0.88 km^2 (Table 5.2).

Regarding Aliakmonas river delta, erosion detected during 1986–2000 and 2000–14 are approximately of the same rate (approximately 1.69 km^2). For the last time period, deposition is the dominant process and can be noticed on the mid-western and northern-eastern parts of the river delta. On the other hand, erosion is mainly noticed on the southern and eastern parts of the studied region. During 2000–14, the main morphological change of the study area is erosion occuring primarily in the southern and western parts while deposition rates are significantly less, almost three times less than the erosion rates (Table 5.2). Finally, for the time period between 2014 and 2022 similar rates for the erosion and deposition are reported (Fig. 5.4; Table 5.3).

Comparing the final results of the two approaches implemented for the overall time period, the eroded and deposited surfaces differ. In the case of the Axios river delta during the photo-interpretation process, the final eroded surface was over 4 km^2 and the deposited surface was less than 3 km^2 (Table 5.4). On the other hand, as can be observed, the results of the GEE approach are the opposite, that is, the erosion rates are lower than 3 km^2 and the deposition rates are over 4 km^2 demonstrating significantly different results (Fig. 5.5). For the second study area, Aliakmonas river delta, results follow the same pattern in both erosion and deposition values (Fig. 5.5) where the total eroded surface is higher in both approaches than the total newly formed surface.

The overall analysis indicates that that erosion is the dominant process for both the Axios and Aliakmonas river deltas. On the contrary, the GEE approach results indicate a tendency for newly formed land for Axios rivel

TABLE 5.3 Results of total eroded and deposited areas from the direct photo-interpretation approach.

Time period	Axios river delta		Aliakmonas river delta	
	Erosion (km^2)	Deposition (km^2)	Erosion (km^2)	Deposition (km^2)
1986–2000	1.17	1.42	1.69	1.76
2000–14	2.18	0.34	1.69	0.42
2014–22	0.88	0.88	0.85	0.85

TABLE 5.4 Result comparison of both approaches from 1986 to 2022.

	Axios river delta		Aliakmonas river delta	
Approach	Erosion (km^2)	Deposition (km^2)	Erosion (km^2)	Deposition (km^2)
Photo-interpretation	4.23	2.64	4.23	3.03
GEE	3.59	4.83	2.78	1.57

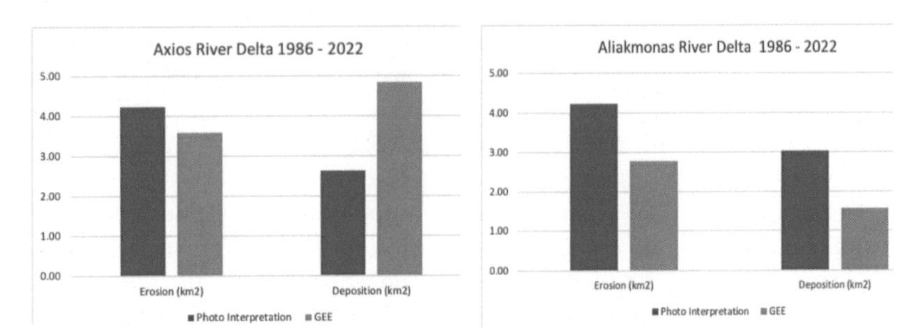

FIGURE 5.5 Result comparison for Axios river delta and Aliakmonas river delta.

delta, where for Aliakmonas river delta the results coincident with the photo-interpretation technique, meaning the erosion values surpass the deposition values.

The results from the statistical metrics for both river deltas (Axios river delta and Aliakmonas river delta) are summarized in Table 5.5. For the false delta area (FDA) in Axios river delta, the values vary from 0.65 to 1.31 km^2 from 1986 and 2022, respectively, while, for the Aliakmonas river delta, the values of FDA are higher, precisely from 2.24 km^2 during 1986 to 2.5 km^2 during 2022. On the contrary, the skipped delta area (SDA) values of Aliakmonas river delta lower than those of Axios river delta for both dates, where the minimum value of SDA is below than 0.5 km^2 for both dates, while in Axios river delta is above 0.5 km^2 for 1986 and 2022. The efficiency rates for both study areas are over 0.85 for all dates, with the highest rate detected on Aliakmonas river delta on 2022 at 0.99 and the lowest on Axios river delta at exactly 0.85. As for the commission errors, the maximum value is located again in Aliakmonas river delta during the 2022 and the lowest also in 2022 on Axios river delta. The omission error's minimum values are located during 2022 on both river deltas, and the maximum value is detected on Axios river delta on the year 1986, which is distinctive from the other three omission errors.

5.5 Discussion

The retreat of the deltaic coastlines, indicated by the photo-interpretation approach for both study areas and by the GEE approach only for Aliakmonas river delta, reveals the general tendency that has been described in other studies by Athanasiou (2009), Kapsimalis et al. (2005), and Petropoulos et al. (2015). The geomorphological processes of the two river catchments and the interaction between the deltaic surface and the sea of Thermaikos Gulf can both (meaning erosion and deposition processes) be observed. The causes for the coastlines retreat, which is noticed mainly across the studied regions, can be attributed to numerous complex factors such as channelization, human activity, and sea level rise. Furthermore, dams

TABLE 5.5 Axios and Aliakmonas river delta efficiency detection.

Year	Common area (km^2)		False delta area (km^2)		Skipped delta area (km^2)		Efficiency rate		Commission error		Omission error	
	Axios	Aliakmonas	Axios	Aliakmonas	Axios	Aliakmonas	Axios	Aliakmonas	Axios	Aliakmonas	Axios	Aliakmonas
1986	13.04	8.43	0.65	2.24	2.28	0.38	0.85	0.96	0.22	0.21	0.15	0.04
2022	13.62	6.97	1.31	2.50	0.51	0.08	0.96	0.99	0.09	0.26	0.01	0.01

in both river catchments have disrupted their natural discharge regime and sediment dynamics by decoupling the upper reaches from the deltaic regions, thus reducing the volume of sediment reaching those coastal areas (Petropoulos et al., 2015).

The final acquired spatial data by the two approaches implemented in this study exhibit evidently a diverse pattern. Specifically, the eroded and deposited surfaces for Axios river delta resulted from the execution of the GEE approach are not in compliance with those resulted from the photo-interpretation approach. The main error source of this discrepancy can be attributed to the spatial resolution of the Landsat imagery. This was particularly the case in areas where the width of the two rivers was lower than the imagery pixel size. For this reason, spectral unmixing techniques and object-based classification would be considered to deliver better results for studies such as this. Therefore the values of deposition for Axios river delta during 1986 up to 2022, provided from the GEE approach, can be called into question.

However, while the delineation of the river deltas through the digitization approach can be considered a more accurate technique for studies such as this, because of the user's ability to map the study area in detail and with careful precision, it is a time-consuming procedure, solely based on the digitizer's fluency, therefore prone to human mistakes, that cannot be automated and used in other study areas. Furthermore, during the digitization of the coastlines of the studied regions, there were parts of the deltas where maximum detail is needed, for example along the river streams and mainly to distinguish the thin, complex natural borders between land and water, and mistakes can easily occur. Once more, the main error source for this approach could be related to the pixel size of the images, which can be challenging for mapping river deltas and wetlands in general, which was also the case for the GEE-based approach.

All in all, direct digitization methodology developed for this project provides a user-friendly technique for studies such as this one. Semiautomatic classification, on the other hand, provides an advanced technique which requires specialized skills in delineating the river deltas through a GEE platform. In general, cloud-based platforms such as the GEE developed in this study enable cost-effective time series analysis without the need of any computational hardware or data management logistics. Regarding the semi-automated technique adapted in the current study, its improvement would be of vital meaning. Although, hand digitizing proved to be more efficient in providing highly accurate information toward the river deltas delineation.

5.6 Concluding remarks

In the present study, it was developed a cloud-based platform of GEE which was combined with Landsat archive to monitor the dynamics of erosion and deposition at Axios and Aliakmonas river deltas dynamics and the use of the method was demonstrated studying two of the largest river deltas in Greece for a period of 36 years. For this purpose, the supervised classification approach was developed and subsequently unimplemented in GEE environment to map river delta extent along with a direct photo-interpretation approach to validate the GEE-derived results.

Results obtained suggest that erosion of the deltaic coastline is dominant in both study areas, which is in accordance with previous studies conducted in these areas. A notable discrepancy was observed in absolute numbers between the GEE approach and the direct photo-interpretation approaches, where deposition is higher than degraded surface in Axios river delta. This specific discordance of the results extracted from the two approaches indicates that while cloud-based platforms can be useful for studies such as this, different approaches can be implemented, such as object-based classification or use of satellite imagery with higher spatial resolution, to ensure that efficient data can be delivered. Nonetheless, the results of the spatio-temporal analysis for Aliakmonas river delta agree with each other. Furthermore, our results depict the value of the Landsat archive when used for environmental applications and monitoring highly dynamic environments such as wetlands and river deltas. Moreover, our study once again highlights the contribution of Landsat data as a cornerstone in the rapid evolution of EO technology.

The significance of the present study is multifaceted. Our results indicate that the two river deltas are continuously shrinking in size. The latter is a tendency that is observed throughout the last 36 years, information that is vital for policy decision making and the regional activities and agriculture production. Last but not least, our study illustrates clearly the added value of cloud-based platforms, such as GEE, in the cost and time-efficient process for mapping river deltas dynamics in the context of EO big data usage. Possible future directions of this study are the use of other state-of-the-art techniques such as deep learning combined with the cloud-based platforms to fully automate the monitoring processes. This will enhance the limited transferability of the semisupervised classification approaches such as the one implemented in this study. The latter remains to be investigated.

References

Aguilar, F. J., Fernandez-Luque, I., Aguilar, M. A., García Lorca, A. M., & Viciana, A. R. (2019). The integration of multi-source remote sensing data for the modelling of shoreline change rates in a mediterranean coastal sector. *International Journal of Remote Sensing*, *40*(3), 1148–1174.

Athanasiou, H. (2009). Wetland Habitat Loss in Thessaloniki Plain (M.Sc. Dissertation 1990), Greece: University College, London.

Bishop-Taylor, R., Nanson, R., Sagar, S., & Lymburner, L. (2021). Mapping Australia's dynamic coastline at mean sea level using three decades of Landsat imagery. *Remote Sensing of Environment*, *267*, 112734.

Breiman, L. (2001). Random forests. *Machine Learning*, *45*(1), 5–32.

Brown, A. R., Petropoulos, G. P., & Ferentinos, K. P. (2018). Appraisal of the Sentinel-1 & 2 use in a large-scale wildfire assessment: A case study from Portugal's fires of 2017. *Applied Geography*, *100*, 78–89.

Cai, H., Li, C., Luan, X., Ai, B., Yan, L., & Wen, Z. (2022). Analysis of the spatiotemporal evolution of the coastline of Jiaozhou Bay and its driving factors. *Ocean & Coastal Management*, *226*, 106246.

Cao, W., Zhou, Y., Li, R., & Li, X. (2020). Mapping changes in coastlines and tidal flats in developing islands using the full time series of Landsat images. *Remote Sensing of Environment*, *239*, 111665.

Chen, S., Chen, L.-F., Liu, Q.-H., Li, X., & Tan, Q. (2005). Remote sensing and GIS-based integrated analysis of coastal changes and their impacts in Lingding Bay, Pearl River Estuary, South China. *Ocean & Coastal Management*, *48*, 65–83.

Dilalos, S., Alexopoulos, J. D., Vassilakis, E., & Poulos, S. E. (2022). Investigation of the structural control of a deltaic valley with geophysical methods. The case study of Pineios river delta (Thessaly, Greece). *Journal of Applied Geophysics*, *202*, 104652.

Fragou, S., Kalogeropoulos, K., Stathopoulos, N., Louka, P., Srivastava, P. K., Karpouzas, S., Kalivas, D. P., & Petropoulos, G. P. (2020a). Quantifying land cover changes in a Mediterranean environment using landsat TM and support vector machines. *Forests MDPI*, *11*, 750–769. Available from https://doi.org/10.3390/f11070750.

Fragou, S., Kalogeropoulos, K., Stathopoulos, N., Louka, P., Srivastava, P. K., Karpouzas, S., P Kalivas, D., & P Petropoulos, G. (2020b). Quantifying land cover changes in a mediterranean environment using Landsat TM and support vector machines. *Forests*, *11*(7), 750.

Google Earth Engine (GEE). Available at: https://code.earthengine.google.com/. Accessed 10 July 2022.

Gorelick, N., Hancher, M., Dixon, M., Ilyushchenko, S., Thau, D., & Moore, R. (2017). Google Earth Engine: Planetary-scale geospatial analysis for everyone. *Remote Sensing of Environment*, *202*, 18–27.

Ireland, G., & Petropoulos, G. P. (2015). Exploring the relationships between post-fire vegetation regeneration dynamics, topography and burn severity: A case study from the Montane Cordillera Ecozones of Western Canada. *Applied Geography*, *56*, 232–248.

Jiang, D., Hao, M., & Fu, J. (2016). Monitoring the coastal environment using remote sensing and GIS techniques. *Applied Studies of Coastal and Marine Environments*, 353–385.

Kapsimalis, V., Poulos, S. E., Karageorgis, A. P., Pavlakis, P., & Collins, M. (2005). Recent evolution of a Mediterranean deltaic coastal zone: Human impacts on the Inner Thermaikos Gulf, NW Aegean Sea. *Journal of the Geological Society*, *162*(6), 897–908.

Kazantzidis, S., & Goutner, V. (1996). Foraging ecology and conservation of feeding habitats of little egrets (Egretta garzetta) in the Axios River Delta, Macedonia, Greece. *Colonial Waterbirds*, 115–121.

Kuenzer, C., Bluemel, A., Gebhardt, S., Quoc, T. V., & Dech, S. (2011). Remote sensing of mangrove ecosystems: A review. *Remote Sensing*, *3*(5), 878–928.

Misra, A., & Balaji, R. A. (2015). A study on the shoreline changes and Land-use/land-cover along the South Gujarat coastline. *Procedia Engineering*, *116*, 381–389.

Pandey, P.C., N. Koutsias, G.P. Petropoulos, P.K. Srivastava & E.B. Dor (2019): Land use/land cover in view of Earth Observation: Data sources, input dimensions and classifiers—A review of the state of the art. Geocarto International, [in press]

Petropoulos, G. P., Griffiths, H. M., & Kalivas, D. P. (2014). Quantifying spatial and temporal vegetation recovery dynamics following a wildfire event in a Mediterranean landscape using EO data and GIS. *Applied Geography*, *50*, 120–131.

Petropoulos, G. P., Kalivas, D. P., Griffiths, H. M., & Dimou, P. P. (2015). Remote sensing and GIS analysis for mapping spatio-temporal changes of erosion and deposition of two Mediterranean river deltas: The case of the Axios and Aliakmonas rivers, Greece. *International Journal of Applied Earth Observation and Geoinformation, 35*, 217–228.

Petropoulos, G. P., Sandric, I., Hristopulos, D., & Carlson, T. N. (2020). Evaporative fluxes and surface soil moisture retrievals in a Mediterranean setting from sentinel-3 and the "simplified triangle". *Remote Sensing MDPI, 12*(19), 3192. Available from https://doi.org/10.3390/rs12193192.

Poulos, S. E., Ghionis, G., & Maroukian, H. (2009). The consequences of a future Eustaticsea-level rise on the deltaic coasts of Inner Thermaikos Gulg (Aegean Sea) and Kyparissiakos Gulf (Ionian Sea), Greece. *Geomorphology, 107*, 18–24.

Salimi, S., Almuktar, S. A., & Scholz, M. (2021). Impact of climate change on wetland ecosystems: A critical review of experimental wetlands. *Journal of Environmental Management, 286*, 112160.

Sandric, I., Irmia, R., Petropoulos, G. P., Anand, A., Srivastava, P. K., Pesolanu, A., Faraslis, I., Stateras, D., & Kalivas, D. (2022). Tree's detection and heath assessment from ultra-high resolution UAV imagery and deep learning. *Geocarto International*. Available from https://www.tandfonline.com/doi/full/10.1080/10106049.2022.2036824, in press.

Sheykhmousa, M., Mahdianpari, M., Ghanbari, H., Mohammadimanesh, F., Ghamisi, P., & Homayouni, S. (2020). Support vector machine versus random forest for remote sensing image classification: A meta-analysis and systematic review. *IEEE Journal of Selected Topics in Applied Earth Observations and Remote Sensing, 13*, 6308–6325.

Smardon, R. (2009). *The Axios river delta–Mediterranean wetland under siege. Sustaining the World's Wetlands* (pp. 57–92). New York, NY: Springer.

Srivastava, P. K., Gupta, M., Singh, U., Prasad, R., Pandey, P. C., Raghubanshi, A. S., & Petropoulos, G. P. (2021). Sensitivity analysis of artificial neural network for chlorophyll prediction using hyperspectral data. *Environment, Development and Sustainability, 23*, 5504–5519. Available from https://doi.org/10.1007/s10668-020-00827-6.

TensorFlow. Available at: https://www.tensorflow.org/. Accessed 12 July 2022.

Tsatsaris, A., Kalogeropoulos, K., Stathopoulos, N., Louka, P., Tsanakas, K., Tsesmelis, D. E., Krassanakis, V., Petropoulos, G. P., Pappas, V., & Chalkias, C. (2021). Geoinformation technologies in support of environmental hazards monitoring under climate change: An extensive review. *ISPRS International Journal of Geo-Information, 10*(2), 94.

Valderrama-Landeros, L., & Flores-de-Santiago, F. (2019). Assessing coastal erosion and accretion trends along two contrasting subtropical rivers based on remote sensing data. *Ocean & Coastal Management, 169*, 58–67.

Vörösmarty, C. J., Meybeck, M., Fekete, B., Sharma, K., Green, P., & Syvitski, J. P. (2003). Anthropogenic sediment retention: major global impact from registered river impoundments. *Global and Planetary Change, 39*(1–2), 169–190.

White, K., & El Asmar, H. M. (1999). Monitoring changing position of coastlines using Thematic Mapper imagery, an example from the Nile Delta. *Geomorphology, 29*(1-2), 93–105.

Yang, L., Driscol, J., Sarigai, S., Wu, Q., Chen, H., & Lippitt, C. D. (2022). Google Earth Engine and artificial intelligence (AI): A comprehensive review. *Remote Sensing, 14*(14), 3253.

Chapter 6

Crowdsourced geospatial data in human and Earth observations: opportunities and challenges

Xiao Huang[1], Xiao Li[2], Di Yang[3] and Lei Zou[4]
[1]Department of Geoscience, University of Arkansas, Fayetteville, AR, United States, [2]Texas A&M Transportation Institute, Texas A&M University, College Station, TX, United States, [3]Wyoming Geographic Information Science Center, University of Wyoming, Laramie, WY, United States, [4]Department of Geography, Texas A&M University, College Station, TX, United States

Crowdsourced geospatial data, referring to data sources gathered from networks of volunteers, has greatly transformed the paradigms of human and Earth observations in many ways, thanks to the emerging concepts of "Web 2.0," "Big Data," and "Citizen Sciences." This chapter (1) narrates popular crowdsourced geospatial data by describing the characteristics of their contributors, detailing the data collecting means and analytical frameworks, and showcasing their real-world applications; (2) discusses the challenges in crowdsourced geospatial data from the perspectives of data handling requirements, data quality problems, privacy concerns, governance, and repeatability/reproducibility-related issues. We illustrate the opportunities and challenges of crowdsourced geospatial data in human observations, Earth observations, and human–environment observations. By summarizing the crowdsourced geospatial data sources, analytical tools and frameworks, applications, and limitations, this chapter gives a big picture of the contemporary development of crowdsourced geospatial analytics as well as possible future directions where its potential can be better harnessed.

6.1 Introduction

For decades, authoritative data sources were dominant in both human observations (e.g., population census and surveys) and Earth observations (e.g.,

Geoinformatics for Geosciences. DOI: https://doi.org/10.1016/B978-0-323-98983-1.00007-7

satellite imagery and physical sensors). In recent years, however, we have seen a growing trend where citizens themselves have started to participate in the workflow of data mapping, monitoring, analyzing, and assessing, fueled by the emerging concepts of "Web 2.0" (O'reilly, 2007) and "Big Data" (McAfee et al., 2012).

Such a transition, that is, from the dominance of authoritative data sources to the popularity of user-contributed nonauthoritative ones, leads to many diverse definitions. For example, "crowdsourcing geospatial data" (used in this chapter) has been widely adopted to describe data acquisition by large and diverse groups of people who, in many cases, are not trained professionals (Heipke, 2010). The term "NeoGeography," coined by Turner (2006), describes location sharing that shapes context and conveys understanding through the knowledge of places using the increasing number of tools and resources that are freely available. Goodchild (2007) describes the trend of widespread engagement of large numbers of private citizens in the creation of geographic information as "Volunteered Geographic Information (VGI)". He defined VGI as "the harnessing of tools to create, assemble, and disseminate geographic data provided voluntarily by individuals". Another similar term is "Citizen Science" which describes public participation in scientific research, participatory monitoring, and participatory action research, whose outcomes are often advancements in scientific research by improving the scientific community's capacity, as well as increasing the public's understanding of science (Strasser, 2019; Doyle et al., 2019). Despite the varying focuses of these definitions, they all highlight the contribution of nonauthoritative sources.

The progress in rapid and accurate positioning technology, the popularity of digital devices, the availability of broadband communication links, and the advances in data managing techniques jointly fast-track the conceptual, methodological, and applicable development of crowdsourced geospatial data. Unlike traditional human and Earth observational means that are often exclusively coordinated and often carried out by governments and large organizations, the collecting process gradually involves regular users in the observing process, thus blurring the boundary that existed in traditional means between data producers and the data users. Such a novel and decentralized way of data collection approach, powered by a massive user base worldwide, potentially allows fine-grained spatiotemporal observations that would not be otherwise available.

This chapter illustrates the opportunities and challenges of crowdsourced geospatial data from three major domains (Fig. 6.1), that is, human observations (e.g., human mobility), Earth observations (e.g., land use/land cover mapping), and human–environment observations (e.g., disaster mitigation). We narrate the characteristics of data contributors for popular crowdsourced geospatial data sources and describe their data collecting means, analytical frameworks, and real-world applications. In addition, we further discuss the challenges in crowdsourced geospatial data from the perspectives of data handling requirements,

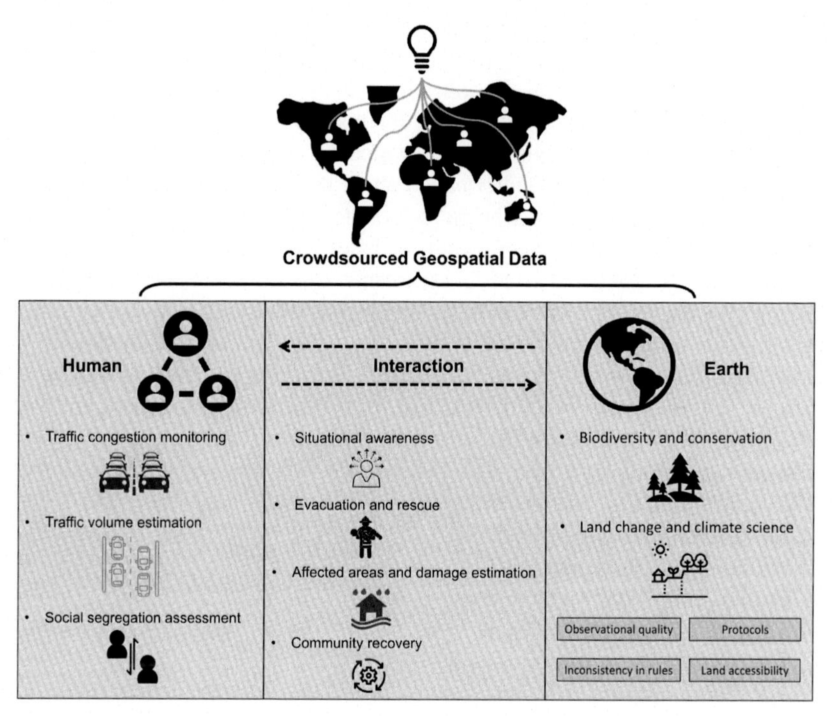

FIGURE 6.1 The structure of this chapter.

data quality problems, privacy concerns, governance, and repeatability/reproducibility-related issues. This chapter gives a big picture of the contemporary development of crowdsourced geospatial analytics as well as possible future directions where its potential can be better harnessed.

6.2 Crowdsourced geospatial data in human observations

Human mobility is an essential characteristic of human behavior, which directly reflects how individual humans move within a network or system. Effectively monitoring the dynamics of human mobility is of great importance in numerous research and application fields. This section briefly introduces three typical applications of crowdsourced data in transportation and urban studies, including traffic congestion monitoring, traffic volume estimation, and social segregation assessment.

6.2.1 Crowdsourced data in traffic congestion monitoring

Providing an accurate and practical traffic congestion estimation is of great importance to advance the understanding of urban flow dynamics and support intelligent transportation systems. Crowdsourced traffic data is promising in transportation monitoring and management. With the maturity and advancement of mobile sensing techniques and social media applications, smartphones have become a promising platform for collecting and sharing traffic information. For travel time data collection, crowdsourced methods are the most used private sector mechanism today. Mobile devices carried by drivers or installed in their vehicles can provide detailed information about their location, speed, and possibly additional information to a public or private entity, which could be used to generate traffic/travel time information for assessing the severity of congestion (The San Diego Association of Governments, 2017). For example, Waze, a leading crowdsourcing mobile app, has millions of registered active users sharing their real-time traffic and road information, which can be potentially used to monitor traffic conditions (Li et al., 2020). Meanwhile, different mobile apps, such as TBBW, Border Wait Times, and Best Time to Cross Border, are also developed to collect border wait times from their users (Lin et al., 2015). Many third-party companies provide high-quality traffic data by aggregating crowdsourced traffic information, including but not limited to:

Inrix—http://www.inrix.com/.
HERE—http://here.com/.
Cellint—http://www.cellint.com/.
Telenav—http://www.telenav.com/.
TrafficCast—http://trafficcast.com/.
TomTom—http://www.tomtom.com/en_gb/licensing/products/traffic/.
Cuebiq—https://www.cuebiq.com.
Waze—https://www.waze.com.

Driving data collected from connected vehicles (CV) is another emerging crowdsourced data source. CV include various communications technologies for vehicle-to-vehicle (V2V), vehicle-to-infrastructure (V2I), vehicle-to-cloud (V2C), and vehicle-to-everything (V2X) applications. Although CV-related technologies are still in the early stage of development, some leading manufacturers have already implemented V2C technologies in their vehicles and could directly collect driving data from millions of vehicles. Many automotive data companies also emerged, which partnered with these auto manufacturers to collect, standardize, and commercialize their driving data (Miles, 2019). For example, Wejo, as a leading CV data start-up, provides high sampling rates and multidimensional vehicle movement data (e.g., speed, location, and acceleration), which could collect data from millions of V2C vehicles with a sampling rate of 3 seconds per waypoint. This game-changing movement data could substantially improve the coverage and accuracy of existing solutions for traffic congestion assessment.

6.2.2 Crowdsourced data in traffic volume estimation

Traffic volume data is a fundamental ingredient in various transportation studies. Traditionally, the traffic volume was collected by implementing traffic counting devices or estimated from the roadside sensors' data, which are costly for deployment and maintenance and the resulting volume data is spatially limited. By leveraging crowdsourced probe/floating vehicle data, researchers could effectively estimate the traffic volume for extended areas not covered by existing traffic sensors. For example, Zhang and Chen (2020) utilized crowdsourced probe vehicle data acquired from a third-party data vendor to estimate statewide annual average daily traffic (AADT) volume. They generated an annual average daily probes variable using the crowdsourced probe data and the betweenness centrality variable based on the probe speed data. Their results indicated that both variables strongly correlated with AADT. By combining them with other commonly used predictor variables (e.g., functional class, area type, number of lanes, and population density), they can effectively model the AADT with an R2 of 0.92 for their best model. Zhang et al. (2020) innovatively integrated the crowdsourced floating car data with insufficient detector records to obtain a high-accurate network-level traffic volume estimation. They first generated a spatial affinity graph employing the correlation coefficients of speed data obtained from floating cars to quantify the similarities among road segments. Then, they performed a geometric matrix completion framework model based on the spatial affinity graph and temporal continuity constraint, which achieved an outstanding accuracy for estimating the network-level traffic flow.

Besides the vehicle volumes, the volume data for bicyclists is also of great importance for practitioners managing and planning multimodal transportation facilities. However, compared with the vehicle volume data, there is a greater lack of counts for nonmotorized modes that are not feasible to collect with field equipment. To fill this data gap, researchers started leveraging crowdsourced biking activity data tracked by smartphone apps to estimate their volumes. For example, as one of the most popular exercise tracking mobile apps, Strava Metro has become a leading source of crowdsourced active transportation activities, including volumes for walking, running, hiking, and biking obtained from their registered users (Dadashova & Griffin, 2020). Dadashova et al. (2020) developed a direct-demand model based on the crowdsourced bike volumes obtained from Strava along with the road inventory and sociodemographic data to estimate annual average daily bicycle counts. Their results indicated that the Strava bike count is a valued predictor for estimating bike volumes. Lee and Sener (2021) systematically reviewed existing applications of Strava data for bicycle monitoring. Their review concluded that Strava is a valuable data source for estimating bicyclist volumes, which also shows potential in other applications, such as travel demand estimation, route choice analysis, and air pollution exposure assessment, among others.

6.2.3 Crowdsourced data in social segregation assessment

Segregation, as a long-standing social phenomenon, is broadly defined as the severity of isolation/separation between two or more population groups, limiting their contacts, communications, and social relations. Effectively measuring social segregation facilitates the understanding of social inequity issues and benefits urban planning and policymaking (Li et al., 2022). With the advancement of transportation technologies, people's mobility patterns and social interactions are commonly taken place at varying locations, resulting in a growing research trend to measure social segregation from a mobility perspective by leveraging crowdsourced mobility data (Park & Kwan, 2018). There are three types of crowdsourced mobility data sources that have been intensively utilized in existing studies, that is, social media, call details records (CDRs), and activity-tracking mobile apps.

Social media data (e.g., geotagged Twitter posts [tweets]) also have been intensively utilized in recent studies to quantify human–place interactions (Li, Huang, Hu, et al., 2021; Li, Huang, Ye, et al., 2021). Wang et al. (2018) utilized geocoded Twitter messages to quantify the neighborhood isolation in the 50 largest US cities. They utilized 650 million geotagged tweets to estimate the mobility patterns and home locations for over 400,000 residents. They evaluated the average travel distance and number of visited neighborhoods for different population groups in terms of race and income level. Their results indicated that the residents from the primarily black and Hispanic neighborhoods expressed relatively higher isolation than those from primarily white neighborhoods. Liu (2021) systematically discussed how geotagged tweets could be used as mobility records to study spatial-social segregation. The sequenced geotagged tweets were used to generate origin neighborhoods and destination neighborhoods of population flows. Meanwhile, the neighborhoods were classified into four racial types. This study found that the population flows between the same type of neighborhoods were significantly higher than those between different neighborhood types.

Another commonly used phone-based data is CDRs. For example, Amini et al. (2014) utilized the CDRs to assess to influence of social segregation on the human mobility patterns between Ivory Coast and Portugal. Each CDR contains the I.D. of the antenna that the phone was connected with when making the call, which can be used to estimate phone users' locations. Their study demonstrated that CDRs could successfully depict the differences in mobility patterns influenced by different social factors such as tribal, cultural, and lingual. Xu et al. (2019) proposed communication and physical segregation indices to assess individuals' social and physical segregation levels based on the CDRs collected from a major mobile operator in Singapore to depict individuals' communication patterns and location footprints. Meanwhile, they utilized the housing prices to represent and quantify the socioeconomic status of individuals. This study found that the richer people more closely interact with the same socioeconomic class of people, indicating higher segregation both socially and physically.

There are also a group of studies that took advantage of activity-tracking apps to collect mobility data from their research participants. For example,

Yip et al. (2016) made use of an Android-based mobile app to record the locations of 71 participants every 5 minutes for 7 days, which were used to track each participant's trips that occurred between two neighborhoods (census tracts) during the study period. They divided the neighborhoods into six levels based on the aggregated households' income levels within the neighborhood. Among the 1073 recorded trips, only 29% of the trips took place between neighborhoods with similar income levels. Most of the trips were either from rich areas to less rich places or from poor places to richer places. Meanwhile, their study also identified that participants from poor neighborhoods are more likely to visit other poor neighborhoods, suggesting that although the social segregation at the city level is low, it may vary across different social groups.

6.3 Crowdsourced geospatial data in Earth observations

Recent developments in cloud computing, collaborative mapping, and user-generated content platforms have spawned a new era in geographic visualization (geo-visualization or mapping and visualizing the world). The combination of crowdsourced mapping and cloud-based computing may bridge many research gaps in biology, ecology, urban planning, and environmental sciences (Fritz et al., 2017).

The crowdsourcing geospatial data platform inaugurates a new era in mapping and visualizing land systems (Dong et al., 2019; Haklay, 2010, 2013; Haklay et al., 2018). OpenStreetMap (OSM) is an excellent example of a geospatial crowdsourcing platform that permits advancement with the rapid expansion of big data and cloud-based computing (Neis & Zielstra, 2014). Moreover, OSM possesses the unique advantage of providing a wider range of applications than official geographic road databases. OSM has the capability to create superior maps when considering temporal change by providing "up-to-date" data (Estima & Painho, 2013; Girres & Touya, 2010).

6.3.1 Leverage emerging crowdsourcing data sources and remote sensing data

Advances in Earth monitoring and observation can benefit from a wide range of new and emerging crowdsourcing data sources and geospatial analytic techniques. Ground-based geotagged cameras provide wide geographic coverage and a huge amount of information for Earth observations but require improved analytics techniques of data processing. Machine learning and artificial intelligence show great promise for large-scale analysis of crowdsourcing data across a range of platforms (Yang et al., 2017; Vaughan, 2017; Zhou et al., 2018). Given the diversity of the efforts, platforms, and tools to support Earth observations in recent decades, it is paramount to learn from the applications and past trails of crowdsourcing data in different fields in Earth observation.

6.3.1.1 Biodiversity and conservation

Volunteers have been actively contributing valuable data in biodiversity monitoring and conservation. There are hundreds of different citizen science projects in this area to improve people's awareness of protecting the surrounding environments and promoting data literacy. Many citizen science projects have geospatial components (e.g., the location of observatory data) and related images. It is hard to comprehensively list all of them or even touch upon all domains. From a geographical scale perspective, some of the biodiversity monitoring and conservation programs are local projects, where the citizen scientists and local residents collect data on a relatively small scale, while the others are on a continental and global scale, such as eBird and Zooniverse (Simpson et al., 2014; Sullivan et al., 2014). Notably, some global-scale platforms can include small-scale projects, such as special events in iNaturalist. In this way, citizen scientists can have more clear targets on what biological and environmental data to collect, which has a great impact on data quality.

6.3.1.2 Land change and climate science

Geospatial crowdsourcing data is an important resource for mapping land cover and land use. Some of the current products have been created globally, for example, Globe-land30 (Chen et al., 2015), OpenStreetMap (Haklay & Weber, 2008), and NASA GLOBE Observer (Amos et al., 2020). There are also continental geospatial crowdsourcing sources for mapping land cover and land use, such as CORINE land cover and AFRICOVER for some African Countries (Kalensky, 1998; Bossard et al., 2000). Wikimapia was founded in 2006 as a for-profit crowdsourcing geospatial data project, allowing users to draw bounding boxes around geographic features and attach media-rich content, such as texts or images, to the geometry. Because of its functionality, the project was considered a crowdsourced gazetteer with the potential to supplement the official efforts (Goodchild, 2007). At its peak time, Wikimapia had a user community whose size was comparable to OSM's; and these two projects had been regularly cited as successful volunteer-based geospatial data collection projects. A wide variety of place-based information has been shared through the platform, including many belonging to the land use and land cover categories. The contribution of these Earth observations includes the development in the interest of people in the field and facilitates tasks such as geographic object-based image analysis. However, OSM has seen a constantly growing community with a broad usage in many GIS projects, whereas Wikimapia suffers from gradually dwindling user attention and interest. In addition, the project's for-profit nature and its partnership relationship with service providers have complicated the legal status of the crowdsourced data (Ballatore and Jokar Arsanjani, 2019). Nonetheless, Wikimapia still contributes valuable geographic information, especially to the countries and regions with relatively scarce geospatial resources.

6.3.2 Notable issues in geospatial crowdsourced Earth observations

6.3.2.1 Observational quality

One of the main challenges that have always been raised with crowdsourcing geospatial data, which is also considered a barrier to the future usage of crowdsourcing geospatial data in research and management practices, is the quality of data. For this reason, researchers have spent a considerable effort on the data quality of crowdsourcing geospatial data Earth observation data (Meek et al., 2014; Leibovici et al., 2017; Hillen & Hofle, 2015; Wittmann et al., 2019). There is a standard for spatial quality that can be applied to crowdsourcing data, but additional quality indicators are required due to the characteristics of different databases and platforms. For example, OSM has white papers about open data quality (Hagen, 2020), which focus on metadata, data quality, and related issues.

6.3.2.2 Inconsistency in classification rules

A number of land cover-related crowdsourcing geospatial data product uses satellite and aerial imagery in combination with their mapping strategies and classification rules. One of the problems that have been highlighted by researchers is that there is a high potential for disagreement when comparing different land cover maps from different organizations. Researchers and citizen scientists are undertaking activities with a promising contribution from crowdsourcing geospatial data. Additionally, in situ data have great value for calibration and validation of products from Earth observation, especially in terms of data distributed density and the updating frequency (Fritz et al., 2017).

6.3.2.3 Policy and ethics

Crowdsourcing geospatial data contributed by citizen scientists are in the public interest, and crowdsourcing geospatial data collectors should be allowed the option to opt-in/out for correspondence, such as news and events. There are a number of organizations incorporating policy, ethics, and legal issues into their white papers, such as OSM (Hagen, 2020). Crowdsourcing data ownership is another challenge in terms of ownership of individual observations (e.g., photos) and data ownership of the full dataset (Newell et al., 2012).

6.3.2.4 *Measurement protocols*

Collecting relevant geospatial crowdsourcing observations largely depends on the measurement protocols, which are the methods used to acquire observations. The goal of the geospatial crowdsourcing measurement protocols is to minimize individual bias and error while obtaining and measuring the observations. The measurement protocol should be described clearly and concisely. For example, NASA Globe Land Cover has a detailed measurement protocol (Amos et al., 2020). The details might include: (1) the precise geographic information, (2) sampling strategy and repeatability, (3) quality assurance, and (4) data format and metadata.

6.3.2.5 *Land accessibility*

In some countries and regions, private land ownerships take a large proportion of the land, and this will have a potential impact on the data distribution of many geospatial crowdsourcing platforms (Fennell, 2012).

6.4 Crowdsourced geospatial data in gauging human–environment interactions

Human–environment interactions describe the connections and feedback between social and natural systems. For example, human beings explore the environment to obtain essential resources (e.g., food, water, and energy), modify the natural environment for economic development (e.g., urbanization, river diversion, and greenhouse gas emission), while responding and adapting to environmental threats (e.g., natural hazards, extreme climate conditions, and sea-level rise) (Cai et al., 2016; Lam et al., 2018). The population boom in recent decades has led to increased residents in hazard-prone areas worldwide. Meanwhile, climate change, partially because of human activities in exploring and modifying the environment, has caused more capricious natural hazards, resulting in recurrent, devastating impacts on human communities (Lam et al., 2016). Therefore gauging the interactions between social and environmental systems, especially during natural hazards, is crucial to understanding the effect of environmental threats on local communities, informing governments and residents to plan early for risk mitigation, and achieving long-term human–nature sustainability.

Quantitatively analyzing human–environment interactions is vital for disaster management and risk mitigation but remains challenging, as data describing fine-grained human behaviors, social impacts, and environmental conditions during hazardous events are generally unavailable in traditional data. Previous studies have leveraged remote sensing images to monitor disaster-affected areas and estimate property damages (Dong & Guo, 2012), but remote sensing data fall short in capturing ground-truthing information such as social behaviors and impacts, for example, evacuation, rescue, and malfunctioned transportation systems. Surveys were conducted to supplement disaster-triggered behaviors and impacts on the social system, for example, postdisaster economic recovery (Lam et al., 2009) and human migration decision-making (Correll et al., 2021) but collecting a large amount of representative and real-time survey data is difficult.

The emerging crowdsourced geospatial data generated by users of social media, mobile devices, and webGIS applications offer a unique lens to observe real-time, multidimensional human behaviors and impacts under environmental threats at an unprecedented scale. During disasters, many citizens turn to social media, cellphone apps or services (e.g., 311 Call System in the United States), and webGIS to access disaster-related information, communicate with others, request help, and report local conditions and impacts (Zou et al., 2019). This actively generated information reflects how residents in different communities prepare for, respond to, and recover from disasters (Zou et al., 2018). People also unintentionally create geospatial data by carrying GPS-enabled mobile devices or using those platforms for other purposes (Poblet et al., 2018). This passively crowdsourced information indicates the general human behaviors, for example, mobility and visitations, which are also shaped by natural hazards and reflect disaster social impacts. As a result, the popularity of incorporating crowdsourced geospatial data into the disaster management cycle continues to grow (Wang et al., 2021). This section will briefly introduce the uses of crowdsourced geospatial data in gauging two human–environment interactions: human responses to natural hazards and disaster impacts on the human community.

6.4.1 Human responses to natural hazards

During natural hazards, governments, organizations, communities, and individuals undertake various measures to better cope with and recover from the immediate aftermath of natural hazards. This section elaborates on the assessment of three human responses to natural hazards, including situational awareness (SA), evacuation, and rescue.

6.4.1.1 Situational awareness

Whether a community or an individual is sufficiently prepared for an impending disaster depends on if they can access the needed information and perceive the risk, broadly defined as SA. The popularity of using social media to share disaster information enables observing the diffusion of disaster-related information among social networks and, thus, estimating SA (Huang & Xiao, 2015). One common approach to measure SA from social media is to identify messages relevant to disaster and explore their spatial-temporal patterns under the consideration of disparate social media use in different communities. For instance, Zou et al. (2018) proposed a Twitter-based normalized ratio index

(NRI) to evaluate county-level SA before, during, and after Hurricanes Sandy in 2012. NRI measures the number of disaster-related tweets in an area divided by the total number of background tweets in the same region and normalized by the disaster threat level. The results show significant geographical disparities in SA in the preparedness, response, and recovery phases of the emergency management cycle. Under the same disaster threat level, communities with better socioeconomic conditions showed significantly larger SA scores in all three phases, demonstrating that those communities had higher perceived risks and were more resilient to Hurricane Sandy. This phenomenon was further validated in Hurricanes Isaac in 2012 (Wang et al., 2021) and Hurricane Harvey in 2017 (Zou et al., 2019).

6.4.1.2 Evacuation

Evacuation orders are demonstrated as an efficient disaster response strategy to minimize fatalities, but the compliance rates with evacuation orders are often significantly lower than 100% (Martin et al., 2017). Monitoring evacuation compliance and movements under disasters in a timely fashion offers valuable information in locating communities that failed to evacuate and sending help. Scholars have analyzed individual/group-level mobility patterns collected from different crowdsourced geospatial data and investigated their changes before, during, and after disasters to uncover evacuation behaviors. For example, Martin et al. (2017) and Jiang et al. (2019) tracked a group of Twitter users who lived in the affected areas and posted geotagged messages before, during, and after 2016 Hurricane Matthew. By analyzing the spatial-temporal patterns of when and where those users evacuated and returned, they confirmed the utility of social media in monitoring evacuation behaviors in response to Hurricane Matthew. They also find that people in the same social networks tend to make the same evacuation decisions, and evacuated people tend to have larger long-term activity space and smaller long-term sentiment variances. Cellphone-based human mobility data provided by companies like Cuebiq and SafeGraph are another type of crowdsourced geospatial data widely applied in evacuation monitoring. For example, Deng et al. (2021) obtained 30 million GPS data points corresponding to roughly 150,000 unique, anonymous users from the Cuebiq company to analyze Houston residents' evacuation behaviors during Hurricane Harvey. By computing the evacuation rates and distances, they revealed that disadvantaged minority populations were less likely to evacuate than wealthier residents. Solid social cohesion exists among evacuees from advantaged neighborhoods in their destination choices.

6.4.1.3 Rescue

When disaster victims fail to evacuate on time, surveilling neighborhoods needing rescue and sending first responders becomes the final defensive line for saving people's lives. Multiple crowdsourced geospatial data have been utilized to observe the emergency rescue operation under hazard threats. For example, disaster victims have resorted to Twitter and Facebook to request help during natural hazards since Hurricane Harvey in 2017, making it possible to observe emergency rescue operations via social media. Monitoring rescue needs during disasters through social media takes three steps: (1) searching rescue request messages, (2) extracting geographical information, and (3) converting locations into pairs of coordinates. Previous work has incorporated advanced natural language processing models into three steps to improve the understanding of contents in social media messages and enhance the performance of observing disaster rescue. For instance, Zhou et al. (2022) took advantage of the Bidirectional Encoder Representations from Transformers model in the first step to identify rescue request tweets during Hurricane Harvey and rendered state-of-the-art performance with an F1-score of 0.92. Wang et al. (2020) developed a novel toponym recognition model (NeuroTPR) for the second step based on pretrained language models and the Long Short-Term Memory model and achieved an F1-score of 0.73. Zou et al. (2021) further analyzed rescue requests on social media and found that communities needing additional assistance in disaster rescue were environmentally and socially vulnerable to natural hazards. Crowdsourcing websites are also powerful tools for gauging rescue needs and operations under disasters. One example is the Crowdsource Rescue (CSR) website https://crowdsourcerescue.com/. Disaster-affected residents can submit a help request (e.g., need for rescue) along with their addresses to CSR. At the same time, volunteers can sign up through the CSR platform and respond to the nearest requests. CSR was established in 2017 during Hurricane Harvey and has supported 28 incidents and helped 94,036 survivors during disasters, including natural hazards and the ongoing health crisis COVID-19.

6.4.2 Disaster impacts on human communities

Crowdsourced geospatial data also play increasingly important roles in assessing disaster impacts on human communities. This section covers three intensively investigated disaster impacts through the analysis of crowdsourced geospatial data.

6.4.2.1 Affected areas

Attributing to crowdsourced geospatial data's agility in reporting ground-true information, they have been applied in detecting disaster-affected areas since the emergence of crowdsourcing platforms. At the early stage, this task was achieved by detecting messages or updates posted on social media or web applications and geographically visualizing those messages to reveal the affected areas. For instance, Earle and others (2010) successfully mapped areas affected by the 2009 Morgan Hill earthquake in California through geocoded counts of tweets relevant to this event. Recently, researchers have attempted to incorporate information obtained from crowdsourcing platforms with physical data and models to resolve the user biases in social media and accurately estimate affected areas during natural hazards. Using a severe flood in South Carolina in 2015 as an example, a study fuses Twitter data with field gauge and elevation data to visualize the flood extending in near real-time (Li et al., 2018). Huang further improved the model by adding postflood soil moisture information derived from satellite observations (Huang et al., 2018).

6.4.2.2 Damage estimation

Monitoring real-time natural hazard damages is critical for allocating resources for immediate repairing and rebuilding and can be achieved through crowdsourced geospatial data. For example, social media data have been validated as a potential data source for observing onsite disaster damages. Several studies have examined the value of social media data in postdisaster damage estimation (Kryvasheyeu et al., 2016; Zou et al., 2018). Using Hurricane Sandy as an example, these studies found significant correlations between disaster-related Twitter activities and economic losses at multiple spatial and temporal scales. Information crowdsourcing from public service applications is also viable for estimating disaster damages. For example, many Houston residents called the 311 services to report local impacts caused by Hurricane Harvey, which the City of Houston adopted in the Harvey damage estimation report.

6.4.2.3 Community recovery

Finally, crowdsourced geospatial data are also valuable for monitoring communities' recovery from natural hazards. For example, Guan and Chen (2014) examined the temporal trends and relationships between disaster-related tweets and the power outage caused by Hurricane Sandy in the affected regions. They found a significant correlation between disaster-related Twitter use and the power outage recovery, indicating the potential of monitoring power system recovery through crowdsourced datasets. Another recent study developed a machine learning-based algorithm to identify people who experienced Hurricane Sandy and assess their concerns by analyzing their Twitter data during the postdisaster recovery phase of Hurricane Sandy (Jamali et al., 2019). Their results reveal that information from mining social media could improve the understanding of the priorities of people impacted by natural disasters, which is vital for effective recovery policymaking.

6.5 Challenges and opportunities

Despite the advances in crowdsourced geospatial data in human and Earth observations, many challenges, as well as opportunities, need to be mentioned.

Crowdsourced geospatial data are always biased in one or more ways due to their voluntary nature, as only people who are willing to and have chosen to contribute are included. Thus inferences from a voluntary response sample are not as credible as conclusions are based on a random sample of the entire population (Basiri et al., 2019; Huang, Wang et al., 2021). Twitter, a popular social media platform favored by many crowdsourcing studies, has been criticized for its data representativeness, as many pieces of strong evidence suggest that Twitter data is undersampled toward the elderly, the poor, and those who do not have access to digital devices and are not willing to share information online (Huang, Wang et al., 2019; Huang, Li et al. 2019; Huang et al., 2022). OSM, a popular collaborative project for land use and land cover mapping, also owns severe data distribution issues, evidenced by its more comprehensiveness in urban than in rural areas and in developed countries than in developing countries (Huang, Li et al., 2021; Haklay, 2010). Similar data distribution biases have been noted in other crowdsourcing projects that include eBird (https://ebird.org/home), Geo-Wiki (https://www.geo-wiki.org/), iNaturalist (https://www.inaturalist.org/), Mapillary (https://www.mapillary.com/), to list a few. In addition, the volunteers, as individuals, can have different aspects and levels of quality of judgment and decision-making, and their decisions, opinions, and preferences could be significantly represented and/or influence their data contribution (Basiri et al., 2019). To mitigate or address the biases in crowdsourced geospatial data, additional efforts are encouraged to better understand contributors' profiles, design data fusion paradigms, and establish standardized quality control workflows.

Issues in data ownership, user privacy, and product licensing will continue to draw attention in the future, especially when citizen-contributed data are integrated with other products or used by third parties. Who has or should have the data ownership in crowdsourcing projects has also been a heated debate. We have to acknowledge that traditional concepts of copyright and intellectual property have been challenged by the attitudes toward ownership and copyright under the context of crowdsourcing, and such transition in data collecting paradigms is expected to continue challenging industrial and academic fields for years to come. The privacy concerns, especially over crowdsourcing projects built upon location-based services, aroused much attention during the COVID-19 pandemic when many commercial companies made human moving patterns available to assist targeted mitigation strategies (Huang et al., 2020). Although data privacy laws generally require the protection of personal information to prevent data providers of crowdsourcing projects from being identified, great risks can not be ignored, as users' locational information as well as other identities can be easily disclosed without consent if the data providers are unaware of the data collecting process and/or data users are not careful in how the data are handled and subsequently employed. We thus encourage future crowdsourcing projects to thoroughly consider the issues regarding data ownership, user privacy, and product licensing and exert every effort to ensure data providers' privacy is protected.

Crowdsourcing platforms greatly differ in terms of their contributors' engagement. Certain platforms receive a massive number of consistent contributions from a variety of users across the world, while others struggle to acquire sufficient data for analysis. The difference in user base can be translated to data biases, ultimately influencing the quality of the crowdsourcing project. Thus a better understanding of the motivation of data contributors is expected to assist in boosting users' participation in crowdsourcing platforms. Hossain (2012) classified contributors' motivation into intrinsic motivation (task-driven and not relying on external pressure) and extrinsic motivations (e.g., reputation, status, peer pressure, fame, community identification). Unfortunately, our understanding of the mechanism of participation motivation falls behind the booming development of crowdsourcing platforms for human and Earth observations. Thus underexploited domain deserves additional attention.

By summarizing the crowdsourced geospatial data sources, analytical tools and frameworks, applications, and limitations, this chapter recaps the contemporary development of crowdsourced geospatial analytics, presents existing challenges, and discusses possible future directions. Despite the remaining challenges, the future is bright for crowdsourcing geospatial data in addressing human and Earth observational challenges.

References

Amini, A., Kung, K., Kang, C., Sobolevsky, S., & Ratti, C. (2014). The impact of social segregation on human mobility in developing and industrialized regions. *EPJ Data Science*, *3*(1), 1–20. Available from https://doi.org/10.1140/epjds31.

Amos, H. M., Starke, M. J., Rogerson, T. M., Colón Robles, M., Andersen, T., Boger, R., & Schwerin, T. G. (2020). GLOBE observer data. *Earth and Space Science*, *7*(8), e2020EA001175.

Ballatore, A., & Jokar Arsanjani, J. (2019). Placing Wikimapia: An exploratory analysis. *International Journal of Geographical Information Science*, *33*(8), 1633–1650.

Basiri, A., Haklay, M., Foody, G., & Mooney, P. (2019). Crowdsourced geospatial data quality: Challenges and future directions. *International Journal of Geographical Information Science*, *33*(8), 1588–1593.

Bossard, M., Feranec, J., & Otahel, J. (2000). CORINE land cover technical guide: Addendum 2000 . (Vol. 40). Copenhagen: European Environment Agency.

Cai, H., Lam, N. S.-N., Zou, L., Qiang, Y., & Li, K. (2016). Assessing community resilience to coastal hazards in the lower Mississippi River Basin. *Water*, *8*(2), 46. Available from https://doi.org/10.3390/w8020046.

Chen, J., Chen, J., Liao, A., Cao, X., Chen, L., Chen, X., & Mills, J. (2015). Global land cover mapping at 30 m resolution: A POK-based operational approach. *ISPRS Journal of Photogrammetry and Remote Sensing, 103*, 7–27.

Correll, R. M., Lam, N. S. N., Mihunov, V. V., Zou, L., & Cai, H. (2021). Economics over risk: Flooding is not the only driving factor of migration considerations on a vulnerable coast. *Annals of the American Association of Geographers, 111*(1), 300–315. Available from https://doi.org/10.1080/24694452.2020.1766409.

Dadashova, B., & Griffin, G. P. (2020). Random parameter models for estimating statewide daily bicycle counts using crowdsourced data. *Transportation Research Part D: Transport and Environment, 84*. Available from https://doi.org/10.1016/j.trd.2020.102368.

Dadashova, B., Griffin, G. P., Das, S., Turner, S., & Sherman, B. (2020). Estimation of average annual daily bicycle counts using crowdsourced strava data. *Transportation Research Record, 2674*(11), 390–402. Available from https://doi.org/10.1177/0361198120946016.

Deng, H., Aldrich, D. P., Danziger, M. M., Gao, J., Phillips, N. E., Cornelius, S. P., & Wang, Q. R. (2021). High-resolution human mobility data reveal race and wealth disparities in disaster evacuation patterns. *Humanities and Social Sciences Communications, 8*(1), 1–8. Available from https://doi.org/10.1057/s41599-021-00824-8.

Dong, J., Metternicht, G., Hostert, P., Fensholt, R., & Chowdhury, R. R. (2019). Remote sensing and geospatial technologies in support of a normative land system science: Status and prospects. *Current Opinion in Environmental Sustainability, 38*, 44–52.

Dong, P., & Guo, H. (2012). A framework for automated assessment of post-earthquake building damage using geospatial data. *International Journal of Remote Sensing, 33*(1), 81–100. Available from https://doi.org/10.1080/01431161.2011.582188.

Doyle, C., David, R. Li, Y., Luczak-Roesch, M., Anderson, D., Pierson, C.M. (2019). Using the web for science in the classroom: Online citizen science participation in teaching and learning. In Proceedings of the 10th ACM conference on web science, June 2019, p. 7180

Earle, P., Guy, M., Buckmaster, R., Ostrum, C., Horvath, S., & Vaughan, A. (2010). OMG earthquake! Can Twitter improve earthquake response? *Seismological Research Letters, 81*(2), 246–251. Available from https://doi.org/10.1785/gssrl.81.2.246.

Estima, J., Painho, M. (2013). Exploratory analysis of OpenStreetMap for land use classification. In Proceedings of the second ACM SIGSPATIAL international workshop on crowdsourced and volunteered geographic information, p. 3946

Fennell, L. A. (2012). Crowdsourcing land use. *Brooklyn Law Review, 78*, 385.

Fritz, S., Fonte, C. C., & See, L. (2017). The role of citizen science in earth observation. *Remote Sensing, 9*(4), 357.

Girres, J. F., & Touya, G. (2010). Quality assessment of the French OpenStreetMap dataset. *Transactions in GIS, 14*(4), 435–459.

Goodchild, M. F. (2007). Citizens as sensors: The world of volunteered geography. *GeoJournal, 69*(4), 211–221.

Guan, X., & Chen, C. (2014). Using social media data to understand and assess disasters. *Natural Hazards, 74*(2), 837–850. Available from https://doi.org/10.1007/s11069-014-1217-1.

Hagen, E. (2020). *Sustainability in OpenStreetMap building a more stable ecosystem in OSM for development and humanitarianism.*

Haklay, M. (2010). How good is volunteered geographical information? A comparative study of OpenStreetMap and Ordnance Survey datasets. *Environment and Planning B: Planning and Design, 37*(4), 682–703.

Haklay, M. (2013). Citizen science and volunteered geographic information: Overview and typology of participation. *Crowdsourcing Geographic Knowledge*, 105–122.

Haklay, M., & Weber, P. (2008). Openstreetmap: User-generated street maps. *IEEE Pervasive Computing*, *7*(4), 12–18.

Haklay, M., Mazumdar, S., & Wardlaw, J. (2018). *Citizen science for observing and understanding the Earth*. Springer.

Heipke, C. (2010). Crowdsourcing geospatial data. *ISPRS Journal of Photogrammetry and Remote Sensing*, *65*(6), 550–557.

Hillen, F., & Höfle, B. (2015). Geo-reCAPTCHA: Crowdsourcing large amounts of geographic information from earth observation data. *International Journal of Applied Earth Observation and Geoinformation*, *40*, 29–38.

Hossain, M. (2012). *Crowdsourcing: Activities, incentives and users' motivations to participate. International Conference on Innovation Management and Technology Research, May 2012* (pp. 501–506). IEEE.

Huang, Q., & Xiao, Y. (2015). Geographic situational awareness: Mining Tweets for disaster preparedness, emergency response, impact, and recovery. *ISPRS International Journal of Geo-Information*, *4*(3), 1549–1568. Available from https://doi.org/10.3390/ijgi4031549.

Huang, X., Wang, C., & Li, Z. (2018). A near real-time flood-mapping approach by integrating social media and post-event satellite imagery. *Annals of GIS*, *24*(2), 113–123. Available from https://doi.org/10.1080/19475683.2018.1450787.

Huang, X., Wang, C., Li, Z., & Ning, H. (2019a). A visual–textual fused approach to automated tagging of flood-related tweets during a flood event. *International Journal of Digital Earth*, *12*(11), 1248–1264.

Huang, X., Li, Z., Wang, C., & Ning, H. (2019b). Identifying disaster related social media for rapid response: A visual-textual fused CNN architecture. *International Journal of Digital Earth*, *13*(9), 1017–1039.

Huang, X., Li, Z., Jiang, Y., Li, X., & Porter, D. (2020). Twitter reveals human mobility dynamics during the COVID-19 pandemic. *PLoS One*, *15*(11), e0241957.

Huang, X., Li, Z., Jiang, Y., Ye, X., Deng, C., Zhang, J., & Li, X. (2021a). The characteristics of multi-source mobility datasets and how they reveal the luxury nature of social distancing in the US during the COVID-19 pandemic. *International Journal of Digital Earth*, *14*(4), 424–442.

Huang, X., Wang, C., Li, Z., & Ning, H. (2021b). A 100 m population grid in the CONUS by disaggregating census data with open-source Microsoft building footprints. *Big Earth Data*, *5*(1), 112–133.

Huang, X., Martin, Y., Wang, S., Zhang, M., Gong, X., Ge, Y., & Li, Z. (2022). The promise of excess mobility analysis: Measuring episodic-mobility with geotagged social media data. *Cartography and Geographic Information Science*. Available from https://doi.org/10.1080/15230406.2021.2023366.

Jamali, M., Nejat, A., Ghosh, S., Jin, F., & Cao, G. (2019). Social media data and post-disaster recovery. *International Journal of Information Management*, *44*, 25–37. Available from https://doi.org/10.1016/j.ijinfomgt.2018.09.005.

Jiang, Y., Li, Z., & Cutter, S. L. (2019). Social network, activity space, sentiment, and evacuation: What can social media tell us? *Annals of the American Association of Geographers*, *109*(6), 1795–1810. Available from https://doi.org/10.1080/24694452.2019.1592660.

Kalensky, Z. D. (1998). AFRICOVER land cover database and map of Africa. *Canadian Journal of Remote Sensing*, *24*(3), 292–297.

Kryvasheyeu, Y., Chen, H., Obradovich, N., Moro, E., Van Hentenryck, P., Fowler, J., & Cebrian, M. (2016). Rapid assessment of disaster damage using social media activity. *Science Advances*, *2*(3), e1500779. Available from https://doi.org/10.1126/sciadv.1500779.

Lam, N. S. N., Pace, K., Campanella, R., LeSage, J., & Arenas, H. (2009). Business return in New Orleans: Decision making amid post-Katrina uncertainty. *PLoS ONE, 4*(8), e6765. Available from https://doi.org/10.1371/journal.pone.0006765.

Lam, N. S. N., Reams, M., Li, K., Li, C., & Mata, L. P (2016). Measuring community resilience to coastal hazards along the Northern Gulf of Mexico. *Natural Hazards Review, 17*(1). Available from https://doi.org/10.1061/(ASCE)NH.1527-6996.0000193, 04015013.

Lam, N. S.-N., Xu, Y. J., Liu, K., Dismukes, D. E., Reams, M., Pace, R. K., Qiang, Y., Narra, S., Li, K., Bianchette, T. A., Cai, H., Zou, L., & Mihunov, V. (2018). Understanding the Mississippi River Delta as a coupled natural-human system: Research methods, challenges, and prospects. *Water, 10*(8), 1054. Available from https://doi.org/10.3390/w10081054.

Lee, K., & Sener, I. N. (2021). Strava Metro data for bicycle monitoring: A literature review. *Transport Reviews, 41*(1). Available from https://doi.org/10.1080/01441647.2020.1798558.

Leibovici, D. G., Rosser, J. F., Hodges, C., Evans, B., Jackson, M. J., & Higgins, C. I. (2017). On data quality assurance and conflation entanglement in crowdsourcing for environmental studies. *ISPRS International Journal of Geo-Information, 6*(3), 78.

Li, X., Dadashova, B., Yu, S., & Zhang, Z. (2020). Rethinking highway safety analysis by leveraging crowdsourced waze data. *Sustainability (Switzerland), 12*(23). Available from https://doi.org/10.3390/su122310127.

Li, X., Huang, X., Li, D., & Xu, Y. (2022). Aggravated social segregation during the COVID-19 pandemic: Evidence from crowdsourced mobility data in twelve most populated U.S. metropolitan areas. *Sustainable Cities and Society, 81*(March), 103869. Available from https://doi.org/10.1016/j.scs.2022.103869.

Li, Z., Wang, C., Emrich, C. T., & Guo, D. (2018). A novel approach to leveraging social media for rapid flood mapping: A case study of the 2015 South Carolina floods. *Cartography and Geographic Information Science, 45*(2), 97–110. Available from https://doi.org/10.1080/15230406.2016.1271356.

Li, Z., Huang, X., Ye, X., Jiang, Y., Martin, Y., Ning, H., & Li, X. (2021a). Measuring global multi-scale place connectivity using geotagged social media data. *Scientific Reports, 11*(1), 1–19.

Li, Z., Huang, X., Hu, T., Ning, H., Ye, X., Huang, B., & Li, X. (2021b). ODT FLOW: Extracting, analyzing, and sharing multi-source multi-scale human mobility. *PLoS One, 16* (8), e0255259.

Lin, L., Wang, Q., Sadek, A., Kott, G. (2019). An Android smartphone application for collecting, sharing, and predicting border crossing wait time. In Proc. of the Transportation Research Board Annual Meeting (TRB'14), February 2019

Liu, Q. (2021). *The role of mobility in the socio-spatial segregation assessment with social media data*. Kent State University. Available from http://rave.ohiolink.edu/etdc/view?acc_num = kent1618913543377221.

Martín, Y., Li, Z., & Cutter, S. L. (2017). Leveraging Twitter to gauge evacuation compliance: Spatiotemporal analysis of Hurricane Matthew. *PLoS One, 12*(7), e0181701. Available from https://doi.org/10.1371/journal.pone.0181701.

McAfee, A., Brynjolfsson, E., Davenport, T. H., Patil, D. J., & Barton, D. (2012). Big data: The management revolution. *Harvard Business Review, 90*(10), 60–68.

Meek, S., Jackson, M. J., & Leibovici, D. G. (2014). *A flexible framework for assessing the quality of crowdsourced data*.

Miles, S. (2019). Automotive Data Services Platforms. Available at: https://streetfightmag.com/2019/08/23/6-automotive-data-services-platforms/#.YMlWW_lKguU6.

Neis, P., & Zielstra, D. (2014). Recent developments and future trends in volunteered geographic information research: The case of OpenStreetMap. *Future Internet*, *6*(1), 76–106.

Newell, D. A., Pembroke, M. M., & Boyd, W. E. (2012). Crowd sourcing for conservation: Web 2.0 a powerful tool for biologists. *Future Internet*, *4*(2), 551–562.

O'reilly, T. (2007). What is Web 2.0: Design patterns and business models for the next generation of software. *Communications & Strategies* (1), 17.

Park, Y. M., & Kwan, M. P. (2018). Beyond residential segregation: A spatiotemporal approach to examining multi-contextual segregation. *Computers, Environment and Urban Systems*, *71* (September 2017), 98–108. Available from https://doi.org/10.1016/j.compenvurbsys.2018.05.001.

Poblet, M., García-Cuesta, E., & Casanovas, P. (2018). Crowdsourcing roles, methods and tools for data-intensive disaster management. *Information Systems Frontiers*, *20*(6), 1363–1379. Available from https://doi.org/10.1007/s10796-017-9734-6.

Simpson, R., Page, K. R., & De Roure, D. (2014). Zooniverse: Observing the world's largest citizen science platform. *In Proceedings of the 23rd international conference on world wide web*, 1049–1054.

Strasser, B., Baudry, J., Mahr, D., Sanchez, G., & Tancoigne, E. (2019). "Citizen science"? Rethinking science and public participation. *Science & Technology Studies*, *32*, 52–76.

Sullivan, B. L., Aycrigg, J. L., Barry, J. H., Bonney, R. E., Bruns, N., Cooper, C. B., & Kelling, S. (2014). The eBird enterprise: An integrated approach to development and application of citizen science. *Biological Conservation*, *169*, 31–40.

The San Diego Association of Governments (2017). Border Wait Time Technologies and Information Systems White Paper. Available from: https://www.sandag.org/uploads/publicationid/publicationid_4469_23227.pdf.

Turner, A. (2006). *Introduction to neogeography* (p. 54) O'Reilly Media Publisher.

Vaughan, J. W. (2017). Making better use of the crowd: How crowdsourcing can advance machine learning research. *Journal of Machine Learning Research*, *18*(1), 7026–7071.

Wang, J., Hu, Y., & Joseph, K. (2020). NeuroTPR: A neuro-net toponym recognition model for extracting locations from social media messages. *Transactions in GIS*, *24*(3), 719–735. Available from https://doi.org/10.1111/tgis.12627.

Wang, K., Lam, N. S. N., Zou, L., & Mihunov, V. (2021). Twitter use in Hurricane Isaac and its implications for disaster resilience. *ISPRS International Journal of Geo-Information*, *10*(3), 116. Available from https://doi.org/10.3390/ijgi10030116.

Wang, Q., Phillips, N. E., Small, M. L., & Sampson, R. J. (2018). Urban mobility and neighborhood isolation in America's 50 largest cities. *Proceedings of the National Academy of Sciences of the United States of America*, *115*(30), 7735–7740. Available from https://doi.org/10.1073/pnas.1802537115.

Wittmann, J., Girman, D., & Crocker, D. (2019). Using iNaturalist in a coverboard protocol to measure data quality: Suggestions for project design. *Citizen Science: Theory and Practice*, *4*(1).

Xu, Y., Belyi, A., Santi, P., & Ratti, C. (2019). Quantifying segregation in an integrated urban physical-social space. *Journal of the Royal Society Interface*, *16*(160). Available from https://doi.org/10.1098/rsif.2019.0536.

Yang, D., Fu, C. S., Smith, A. C., & Yu, Q. (2017). Open land-use map: a regional land-use mapping strategy for incorporating OpenStreetMap with earth observations. *Geospatial Information Science*, *20*(3), 269–281.

Yip, N. M., Forrest, R., & Xian, S. (2016). Exploring segregation and mobilities: Application of an activity tracking app on mobile phone. *Cities*, *59*, 156–163. Available from https://doi.org/10.1016/j.cities.2016.02.003.

Zhang, X., & Chen, M. (2020). Enhancing statewide annual average daily traffic estimation with ubiquitous probe vehicle data. *Transportation Research Record, 2674*(9). Available from https://doi.org/10.1177/0361198120931100.

Zhang, Z., Li, M., Lin, X., & Wang, Y. (2020). Network-wide traffic flow estimation with insufficient volume detection and crowdsourcing data. *Transportation Research Part C: Emerging Technologies, 121*. Available from https://doi.org/10.1016/j.trc.2020.102870.

Zhou, B., Zou, L., Mostafavi, A., Lin, B., Yang, M., Gharaibeh, N., Cai, H., Abedin, J., & Mandal, D. (2022). VictimFinder: Harvesting rescue requests in disaster response from social media with BERT. *Computers, Environment and Urban Systems, 95*, 101824. Available from https://doi.org/10.1016/j.compenvurbsys.2022.101824.

Zhou, N., Siegel, Z. D., Zarecor, S., Lee, N., Campbell, D. A., Andorf, C. M., & Friedberg, I. (2018). Crowdsourcing image analysis for plant phenomics to generate ground truth data for machine learning. *PLoS Computational Biology, 14*(7), e1006337.

Zou, L., Lam, N. S. N., Cai, H., & Qiang, Y. (2018). Mining Twitter data for improved understanding of disaster resilience. *Annals of the American Association of Geographers, 108*(5), 1422–1441. Available from https://doi.org/10.1080/24694452.2017.1421897.

Zou, L., Liao, D., Lam, N. S. N., Meyer, M., Gharaibeh, N. G., Cai, H., Zhou, B., & Li, D. (2021). Social media for emergency rescue: An analysis of rescue requests on Twitter during Hurricane Harvey. *arXiv*. Available from https://doi.org/10.48550/arXiv.2111.07187arXiv: 2111.07187.

Zou, L., Lam, N. S. N., Shams, S., Cai, H., Meyer, M. A., Yang, S., Lee, K., Park, S.-J., & Reams, M. A. (2019). Social and geographical disparities in Twitter use during Hurricane Harvey. *International Journal of Digital Earth, 12*(11), 1300–1318. Available from https://doi.org/10.1080/17538947.2018.1545878.

Chapter 7

Google Earth Engine and machine learning classifiers for obtaining burnt area cartography: a case study from a Mediterranean setting

Ioanna Tselka[1,2], Spyridon E. Detsikas[1], George P. Petropoulos[1] and Isidora Isis Demertzi[1,2]

[1]Department of Geography, Harokopio University of Athens, Athens, Greece, [2]School of Rural and Surveying Engineering, National Technical University of Athens, Zografou Campus, Athens, Greece

7.1 Introduction

Wildfire is a natural component of many ecosystems and traces evidence fire existence even 400 million years ago, as a tool of ecosystems regeneration (Amos et al., 2018; Karamesouti et al., 2016). Although wildfire constitutes a natural phenomenon, climate change has resulted to a dramatic raise of both fire events occurrence and intensity. This is heavily connected with human activity that causes ecosystem interference (Brown et al., 2018). According to IPCC, anthropogenic-related activities have resulted to a global mean land surface air temperature raise by 1.53°C since that start of industrial revolution. Climate change scenarios indicate that the global temperatures will very likely continue to grow. Predictions also show that wildfires will very likely continue to be more devastating and frequent, particularly in regions such as North and South America, central Asia, southern Europe and Africa, and Australia (Tsatsaris et al., 2021). In Europe in particular, about 90% of wildfires occur in Mediterranean countries, and this region is expected to be a prime hotspot of wildfires occurrence in the near feature (Evans et al., 2018; Tselka et al., 2021).

Due to their catastrophic impact, wildfires have been receiving a growing attention. In this framework, it is of paramount importance to acquire

Geoinformatics for Geosciences. DOI: https://doi.org/10.1016/B978-0-323-98983-1.00008-9

information on the location of fire incidents. Precise burnt area mapping enables the accurate estimate of the fire consequences so as to allow rapid establishing of restoration guidelines within the affected regions (Petropoulos et al., 2014). Furthermore, postfire assessment provides an important tool in preparation activities conducted by forestry services (Kontoes et al., 2009). Moreover, burnt area delineation is required in estimating postfire impacts of vegetation burning and calculating related emissions. Consequently, it is of key essence to develop innovative techniques of burnt area mapping (Colson et al., 2018; Kalivas et al., 2013).

Earth Observation (EO) and its recent technological advancements particularly so during the past three decades have allowed scientists to map wildfires destruction. The important role of EO in fire analysis, including the mapping of burnt areas from wildfires, was recognized even from the early 1970s (Jayaweera & Ahlnas, 1974). In comparison to traditional mapping approaches such as field surveys, EO data usage provides many advantages regarding wildfire monitoring, such as rapid and cost-efficient mapping (Petropoulos et al., 2014; Tsatsaris et al., 2021). Also, EO data provide access to unapproachable regions at regular time intervals. Moreover, Geographic Information Systems (GIS) use EO data to enable further processing through a wide variety of tools for spatial information extraction and analysis (Pandey et al., 2019). In recent years, the development of more relevant operational products is increasing, offering a wide range of spatial resolutions (Kalivas et al., 2013).

Contemporary satellite sensors that have been launched during this decade, coupled with highly sophisticated image analysis techniques, consist of an active research topic of critical importance and priority (Ireland & Petropoulos, 2015). In addition, the plethora of the available EO datasets has created new challenges in EO data processing. Consequently, cloud-based platforms such as Google Earth Engine (GEE) and Microsoft Planetary Computer have emerged as a very useful tool to handle the abundance of big EO datasets. These cloud-based platforms allow a cost and computational efficient way to process big EO data on the cloud while also supporting the deployment of sophisticate algorithms such as machine learning (ML) techniques useful in pattern recognition and classification. (Yang et al., 2022).

In purview of the above, this study objective has been to explore the capability of ML techniques into a GEE environment for automating burnt areas mapping from ESA's Sentinel-2 imagery. The work presented herein is structured as follows: Section 7.2 provides a description of the study site and the geospatial datasets used in the study; Section 7.3 outlines the methodological framework implemented to satisfy the study objectives; Sections 7.4 and 7.5 present and discuss the main study results. The last section, Section 7.6, summarizes the key study findings and provides recommendations for continuation of the present work.

7.2 Experimental setup

7.2.1 Study area description

During the summer of 2021, the scorching temperatures in the Mediterranean basin ignited large wildfires causing the destruction of hundreds of thousands hectares of forests and agricultural areas (Giannaros et al., 2022). This present study, examines a large-scale wildfire taking place in a suburban area of Athens approximately 20 km away from the center of Greece's capital city extending relatively around 23°47′ East and 38°8′ North (Fig. 7.1). The studied fire event broke out on August 3, 2021, lasted 4 days resulting in destroying 6747 Ha of forests as well as urban areas within the studied region (National's Observatory of Athens BEYOND, 2021). With regard to land cover, the largest part of the burnt land was covered by forest vegetation (mainly broad-leaved, coniferous, and mixed forests) and transitional zones between forest and shrub vegetation, whereas a smaller part was covered by cultivable and/or cultivated areas and urban fabric (Evelpidou et al., 2021).

7.2.2 Datasets and methodology

7.2.2.1 Datasets and preprocessing

To map the burned area at the selected study site, multispectral Sentinel-2 (S2) satellite imagery was obtained from August 8, 2021, that being one day after the wildfire suppression. S2 is an optical EO orbital system within the fleet of the European Union-owned Copernicus Sentinel program jointly coordinated by Space Agencies ESA and EUMETSTAT (Borgeaud et al., 2015). S2 is consisted from two twin polar-orbiting sun-synchronous

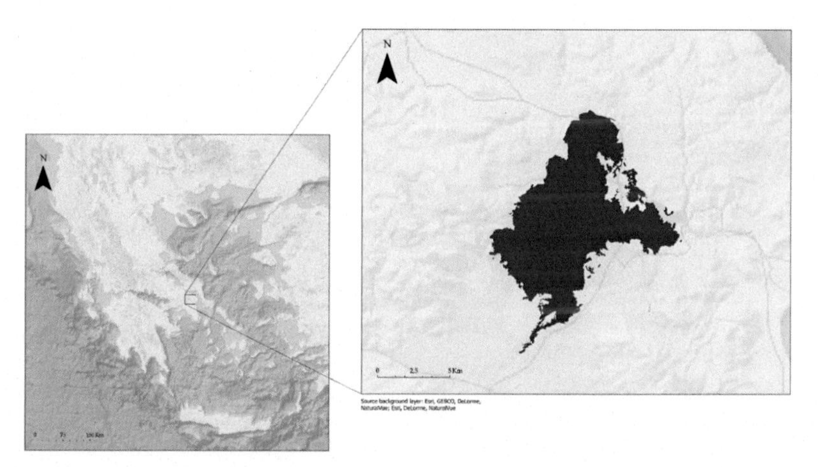

FIGURE 7.1 Study area location.

satellites, with S2-A and S2-B launched on June 23, 2015 and March 07, 2017 accordingly, operating as a constellation and thus enabling the frequent revisit at approximately 5 days. The Multispectral Instrument (MSI) aboard the S2 constellation has 13 spectral bands covering from UVIS to SWIR spectral regions. The high spatial resolution and frequent revisit time of S2 constellation is suitable for the time-sensitive research topics such as burnt area mapping (Colson et al., 2018; Quintano et al., 2018). In Table 7.1, a detailed description of S2 MSI sensor's spectral bands, their corresponding wavelength and spatial resolution, is provided.

The S2 imagery used in this study was obtained through the interactive, web-based GEE code editor using JavaScript programming language. GEE is a publicly available, at no cost for research and education purposes, cloud-based platform that enables labor and time-efficient geospatial analysis at planetary scale. GEE exploits Google Cloud computational resources coupled with a constantly growing 50 Petabyte database integrating data from different EO missions (Gorelick et al., 2017). It provides an array of geospatial analysis tools available through a simple, yet powerful application programming interface accessible both in JavaScript and Python programming

TABLE 7.1 Sentinel-2 bands with resolution, central wavelengths, and bandwidths.

Band number and name	Spatial resolution (m)	Central wavelength (nm)	Bandwidth (nm)
B1 Coastal Aerosol	60	443	20
B2 Blue	10	490	65
B3 Green	10	560	35
B4 Red	10	665	30
B5 Red Edge 1	20	705	15
B6 Red Edge 2	20	740	15
B7 Red Edge 3	20	783	20
B8 NIR	10	842	115
B8A NIR Narrow	20	865	20
B9 Water Vapor	60	945	20
B10 Cirus	60	1380	30
B11 SWIR1	20	1610	90
B12 SWIR2	20	2190	180

languages. S2 data from different levels of processing is being fully integrated to GEE database within days from ingestion.

To find suitable imagery for this case study, the GEE's S2 Surface Reflectance (SR) Level 2A Image Collection, which contains atmospherically and topographically corrected data, was filtered to dates predating (August 2, 2021) and postdating (August 9, 2021) the fire event. Additionally, a cloud cover threshold lower than 5% was set. Following the filtering selection, its results were printed in the GEE code editor console evidencing that only a S2 single scene were satisfying the defined criteria. After finding the closest to the fire event and most suitable for the purposes of this analysis imagery, this S2 scene was imported in the code editor environment. As a next step, a scaling factor of 0.0001, as suggested from S2 User Guide, was applied to all available S2 spectral bands. Continuing the preprocessing steps to the S2 scene, a polygon was drawn using the interactive geometry drawing tools that GEE platform offers, to clip the acquired imagery to the present study's area. After these preprocessing steps were concluded, all the spectral bands of the S2 imagery were resampled to 10 m spatial resolution using the bilinear resampling technique.

In addition to Sentinel-2, a reference dataset from the Copernicus Emergency Management Service (CEMS) was acquired to conduct comparisons between the service's burnt scar estimates and the ones resulting from the techniques applied to this case study. CEMS is activated when large-scale natural or anthropogenic catastrophes, such as wildfires, floods, explosion etc., are being placed enabling accurate, time, and labor efficient information to decision makers and national authorities. This particular service produces geospatial data in different data formats (vector, raster, and thematic maps) and in different categories. The first product that CEMS procures is the first estimate product with the delineation product being next, that assess the extent of the event. Last but not least the final product of the CEMS for a specific fire event is the grading product assessing the produced from the wildfire damage. Herein, the CEMSR527 Grading Product was used, produced specifically for the examined study area and fire event. More specifically, the grading product, version 1 release 1, which was published on August 9, 2021 at 22:17 (UTC), was downloaded on a vector format from the official Copernicus website and then implemented on a GIS environment where the Web Mercator projecting system was attributed to the shapefile.

7.3 Methods for burnt area mapping

Fig. 7.2 presents an overview of the adopted herein methodology, including the data preprocessing covered earlier, whereas next is provided more detailed information concerning the overall methodology implementation to satisfy the study objectives.

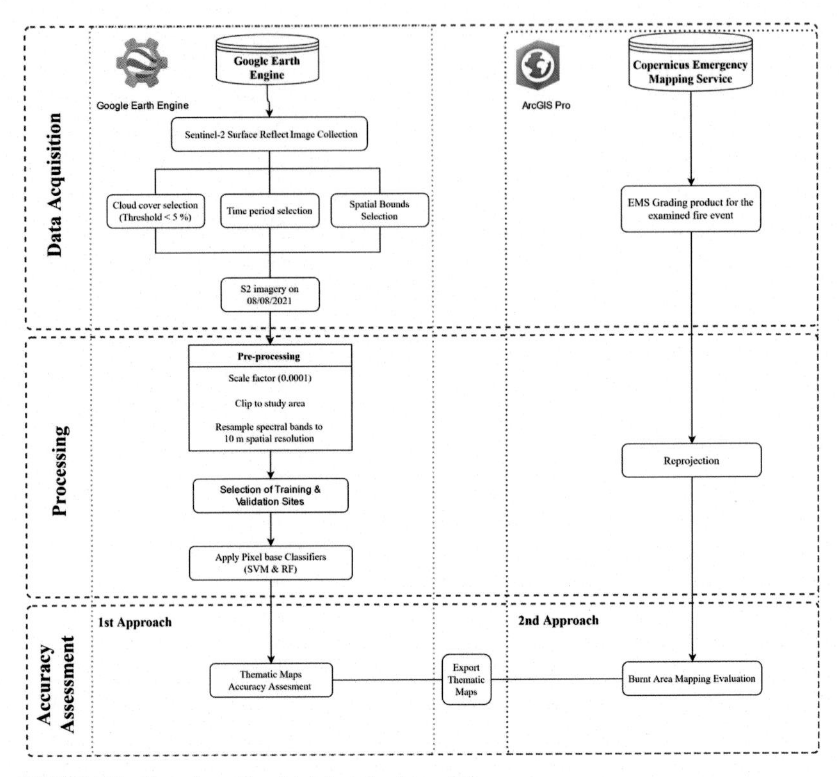

FIGURE 7.2 Overview of methodology implementation.

7.3.1 Google Earth Engine-based burnt area mapping

Support vector machines (SVM) and random forests (RF) (Breiman, 2001) supervised pixel-based classifiers were applied to the S2 satellite imagery inside the GEE cloud platform to delineate the burnt area for the studied region. GEE provides a variety of both supervised and unsupervised ML techniques of regression and classification that can be applied over geospatial data. The full list of the integrated ML techniques can be accessed freely. In addition to the already existing ML techniques in the toolset of GEE, users can exploit the synergistic use of GEE database with custom ML platforms (e.g., Tensor Flow) outside the Earth Engine environment.

These techniques are two of the most frequently used ML classifiers, and they were chosen as they are characterized by different assumptions in their implementation. SVM and RF are well-developed and more sophisticated variations of pixel-based classification techniques, having generally shown promising results when applied to different types of remote sensing data (Amos et al., 2018; Ireland & Petropoulos, 2015; Petropoulos et al., 2014). SVM ML method performs classification based on statistical learning theory

(Cortes & Vapnik, 1995). The SVM algorithm classification algorithm creates decision surfaces, so called hyperplane, on which the optimal class separation takes place. To represent more complex shapes than linear hyperplanes, the classifier may use kernel functions, with the most of them being include the polynomial, the radial basis function (RBF), and the sigmoid kernels, whereas new kernels can also be developed from mathematical operations between kernels. On the other hand, RF classifier is a multiple decision trees ensemble ML technique choosing randomly a part of the training samples and variable enabling accurate. RF classification its randomness in choosing a part of the training samples results in accurate predictions and high accurate classifications (Sheykhmousa et al., 2020).

The first step was to formulate the classification scheme, which was consisted of the classes "Burnt Area," "Artificial Surfaces," "Water Bodies," "Forests," and "grasslands/shrublands/mixed vegetation." Second, representative training sites for the previously mentioned classes were collected from the Sentinel-2 imagery adopting a random sampling approach. A total of 1400 pixels were identified as training data representing the classes defined in the adopted classification scheme. The spectral separability between the target map classes was explored through the generation of a scatter plot of the mean values of each classification class, presenting a good spectral separability of each class.

Parameterization of each classifier was decided following trial and error procedures, which is a standard approach implemented also in other similar studies (e.g., Brown et al., 2018; Petropoulos et al., 2010; 2014). For the RF classifier, the number of the random trees that were producing the best OA, and hence that number of random trees was used. Following similar procedures, for the SVM classifier, the best OA was perform by using a linear kernel; hence, the linear kernel was used for the classification. To facilitate a direct comparison between the burnt area estimates from the two methodologies, the same set of training points was used in performing the classification to the selected S2 scene.

7.3.2 Validation approach

In the present study, validation approach was twofold including (1) estimation of the confusion matrix using a collected validation dataset and (2) comparison of the derived burnt areas estimates from the SVM and RF classifiers with Copernicus EMS Grading derived product for this specific fire event.

7.3.2.1 Classification accuracy assessment

Accuracy assessment based on the confusion matrix was conducted using a validation dataset that was collected using the same principles used for the training dataset exploiting GEE's interactive drawing capabilities. The evaluation of the classification maps produced by the SVM and RF classifiers was performed using the same set of validation points to enable a direct comparison of the results obtained by the different ML classifiers. Statistical metrics, estimated from the confusion matrix, used in this study included the user's (UA) and producer's (PA) accuracies. Those correspond to error of commission or exclusion and to errors of omission or inclusion from the map user and producer perspective respectively. In addition, the overall accuracy (OA) and the Kappa index (K_c) were used to address the overall classification accuracy (Congalton & Green, 1999). OA is expressed as a percentage of the number correct classified pixels divided by the pixels total number. To evaluate the discrepancies between the expected and the actual classification accuracy, the Kappa statistical index was used. The mathematical equations used to estimate the previously mention statistics are described in Table 7.2 (Equation numbers 1–4). The same set of validation points was also used in both classifications, allowing a direct comparison between the derived classification results.

TABLE 7.2 Statistical metrics used to evaluate the overall classification accuracy and the derived burnt area estimates.

Equation number	Index	Symbol	Equation
(1)	Overall accuracy	OA	$\frac{1}{N}\sum_{\iota=1}^{r} n_{ii}$
(2)	Producers accuracy	PA	$\frac{n_{ii}}{n_{icol}}$
(3)	User's accuracy	UA	$\frac{n_{ii}}{n_{irow}}$
(4)	Kappa index	K_c	$N\sum_{\iota=1}^{r} n_{ii} - \sum_{\iota=1}^{r} \frac{n_{icol}n_{irow}}{N^2} - \sum_{\iota=1}^{r} n_{icol}n_{irow}$
(5)	Detected area efficiency	DAE	$\frac{DBA}{DBA+SBA}$
(6)	Skipped area rate	SAR	$\frac{SBA}{DBA+SBA}$
(7)	False area rate	FAR	$\frac{FBA}{DBA+FBA}$

In equation (1)–(4), n_{ii} is the number of pixels correctly classified in a category; N is the total number of pixels in the confusion matrix; r is the number of rows; and n_{icol} and n_{irow} are the column (reference data) and row (predicted classes) total. Respectively DBA is the detected burnt area (common area between the generated burn scar polygon and the reference in situ polygon), FBA is the false burnt area (the area included in the generated burn scar polygon but not in the reference in situ polygon), and SBA is skipped burnt area (the area included in the reference in situ polygon but not in the generated burn scar polygon).

7.3.2.2 Burnt area detection accuracy

The second part of the validation approach was to estimate the burnt scars-derived SVM and RF classifiers with the reference estimate of the burnt area acquired from Copernicus EMS. This approach has also been used efficiently in other studies (e.g., Petropoulos et al., 2014). Subsequently, the burnt area detection accuracy was expressed in terms of detected area efficiency (DAE), skipped area rate (SBA, omission error), and false area rate (FBA, commission error), adopting the equations shown in Table 7.2. To enable overlay and facilitate efficiency in the burnt area comparisons, the burnt area estimates from Sentinel-2 and the EMSR132 Copernicus product were extracted and were subsequently transformed into ArcGIS Pro software platform (ESRI Inc., v. 9.3.1).

7.4 Results

7.4.1 Classification accuracy assessment

The derived classification maps from the implementation of the different ML classifiers are illustrated in Fig. 7.3, and the corresponding classification accuracy results are summarized in Table 7.3. As can be observed, in terms of OAs, the RF classifier outperformed the SVM having scored 93.61% of accuracy while SVM scored 90.63%. The UA and PAs of RF and SVM were estimated for five different land cover categories, burnt areas, artificial surfaces, water bodies, forests, and grasslands/shrublands/mixed vegetation. The UA corresponded to 85.70%–100.00% for RF classifier and 83.00%–100.00% for SVM accordingly, whereas the PAs that occurred were 87.38%–97.37% for RF and 74.76%–96.09% for

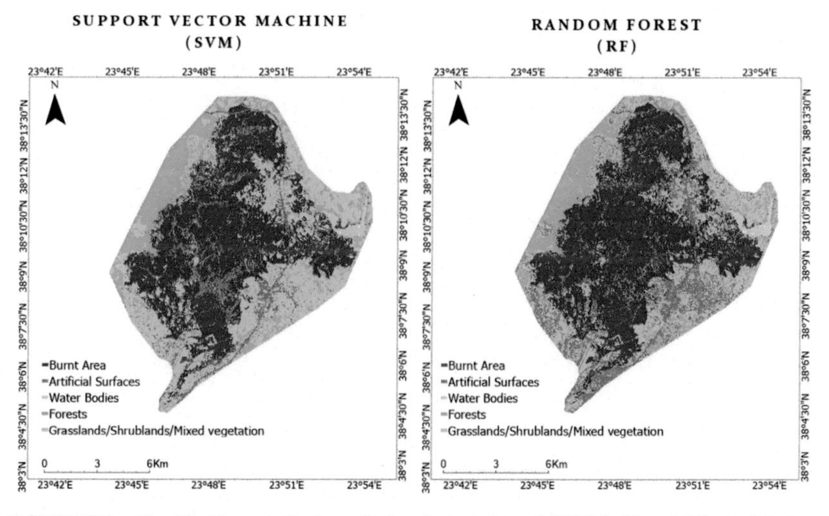

FIGURE 7.3 Classification results from the implementation of SVM (left) and RFs (right) classifiers. *SVM*, support vector machines; *RF*, random forest.

TABLE 7.3 Classifications results obtained from the support vector machines (SVM) and the random forests (RF) implementation.

Classification method	SVM		RF	
Land cover classes	**Producer's accuracy (%)**	**User's accuracy (%)**	**Producer's accuracy (%)**	**User's accuracy (%)**
Burnt areas	74.76	91.67	87.38	93.75
Artificial surfaces	93.63	84.00	95.54	90.36
Water bodies	92.11	100.00	97.37	100.00
Forests	96.09	100.00	94.53	99.20
Grasslands/ shrublands/ mixed vegetation	100	83.00	95.45	85.70
Overall accuracy (%)	90.63		93.61	
Kappa coefficient	0.875		0.915	

SVM. Specifically, the UA scored 93.75% of accuracy of the RF classifier, outperforming the SVM one, which scored 91.67%, whereas the PAs for burnt area classification was 74.76% for the SVM and 87.38% for the RF accordingly. The UA artificial surfaces scored 84.00% of accuracy of the SVM classifier, whereas the RF classifier outperformed SVM which scored 90.36% of accuracy. As for the PAs corresponding to artificial surfaces, RF reached an accuracy of 95.54%, while SVM reached 93.63%. Moreover, the UA water bodies demonstrated high accuracies for both ML classification techniques with the score of 100%. Nevertheless, the PAs of the same category corresponded to 94.53% of the RF classifier and to 92.11% of the SVM. Forests' classification using SVM classifier scored 100% of accuracy regarding CA and 96.09% for PA, while RF scored 99.20% and 94.53% accordingly. Out of the two ML algorithms, RF's performance has been operated in slightly higher accuracy regarding the overall accuracy assessment. Furthermore, the Kappa coefficient distributed the values of 0.91 for RF algorithm and 0.87 for SVM.

7.4.2 Burnt area mapping comparisons

Results of the burnt area mapping derived from the two implemented techniques were compared against the reference product of Copernicus rapid mapping service. For this purpose, the Copernicus Grading Product was utilized in validating the outputs of the two ML algorithms. Fig. 7.4 illustrates the outputs from each ML technique as well as the Copernicus product. More specifically, RF and SVM algorithms detected 63.08 km^2 and 56.55 km^2 of burnt land accordingly, while the Copernicus product estimated 80.09 km^2 (Table 7.4). As can be observed, detected burnt area (DBA) is resulted to RF's DBA being higher than SVM's DBA, where 56.04 km^2 corresponded to the SVM algorithm while 62.62 km^2 corresponded to the RF. Further indexes were also exploited, such as the false burned area (FBA) and the skipped burnt area (SBA). FBA demonstrated low values for each algorithm, 0.5 km^2 for SVM classifier and 0.46 km^2 for RF, while SBA resulted in 27.10 km^2 for SVM and 20.53 km^2 for RF. The detection efficiency rate is higher for RF classifier (0.75%) than SVM (0.67%). Also, as can be observed in Table 7.4, commission error is low giving the values 0.009 for SVM classifier and 0.007 for RF, while omission error varies from 0.25 for RF to 0.33 for SVM. Overall, omission error is higher than omission for both ML classifiers.

7.5 Discussion

ML algorithms are widely used as one of the most efficient techniques in mapping burnt areas (Bar et al., 2020; Qiu et al., 2022; Seydi et al., 2022). Burnt area mapping using EO datasets constitutes a challenge since there is a wide variety of methods that can generate different results and accuracies (Arjasakusuma et al., 2022; Petropoulos et al., 2011). The present study explored the potential use of high-resolution satellite imagery of Sentinel-2 in burnt area delineation over a fire event occurred in the outskirts of Athens exploiting the advantages offered by the GEE cloud-based platform, which provided the training and classification processes using ML-based algorithms. The land use/cover and burnt area mapping assessment of the specific

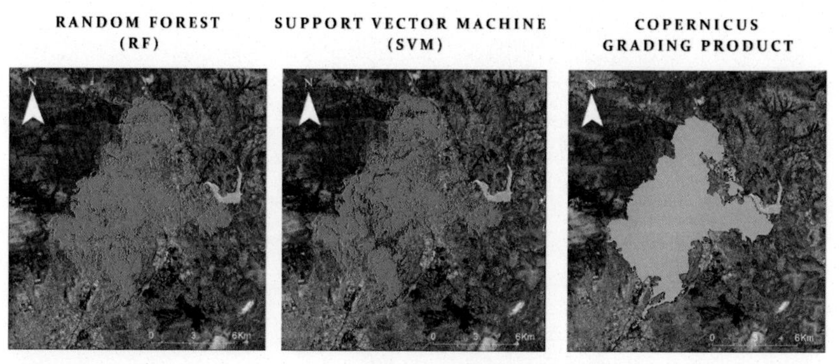

FIGURE 7.4 Burnt area detected by RFs (left), SVMs (middle), and the Copernicus product (right). *SVM*, support vector machines; *RF*, random forest.

TABLE 7.4 Detection efficiency rate results concerning the burnt area estimates.

Classification method	Detected burnt area (DBA) (km^2)	False burnt area (FBA) (km^2)	Skipped burnt area (SBA) (km^2)	Detection efficiency rate	Commission error	Omission error
SVM	56.04	0.50	27.10	0.67	0.009	0.33
RF	62.61	0.46	20.53	0.75	0.007	0.25

SVM, support vector machines; *RF*, random forest.

study area was generated from the application of the stand-alone SVM classifier and the tree-based RF classifier into a cloud environment, specifically GEE. The results of our study exhibited the potential of two ML-based algorithms in accurate burnt area mapping. Specifically, the examined ML approaches classified burnt and unburnt surfaces based on Sentinel-2 imagery and produced generally comparable outputs. Next are discussed the main study findings, concerning first the overall classification and subsequently the delineation of the burnt areas.

The performance of the two classifiers and their accuracy was thoroughly assessed herein. Regarding the UA and PA statistics, land cover classification results were reported over 70% for each class. Generally, this provides adequate information in accurate classification and mapping different land cove types (Petropoulos et al., 2012). The overall accuracy of the classifiers was reported always higher than 90%, indicating comparable and highly accurate classification maps (Fig. 7.5). Also the high values of Kappa statistic further support this finding. Concerning the quantitative results of the ML techniques applied, RF outperformed the SVM classifier in terms of OA results. This was expected since RF is frequently used in several studies as a powerful and accurate tool in burnt area mapping (Bar et al., 2020). In terms of classification performance assessment, the RF classifier showed a better value distribution of burnt areas than the SVM. Nonetheless, the land cover category corresponding to forestry area classification, distributed some different values in terms of accuracy, since the SVM performed better than RF. Similar findings were revealed in Uttarakhand, Western Himalaya by

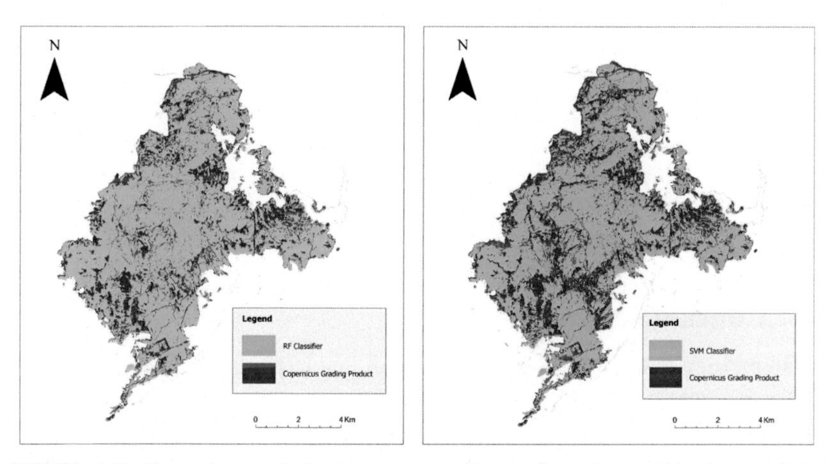

FIGURE 7.5 Comparisons of the burnt area estimates from the combined use of the Copernicus Grading product with RF and SVM classifiers accordingly. *SVM*, support vector machines; *RF*, random forest.

Bar et al. (2020), where Landsat-8 and Sentinel-2 imageries are classified using the SVM and RF ML-based algorithms. Generally, RF's behavior in the current study demonstrates some similarities to other researches regarding its highly accurate performance in comparison to multiple ML classifiers (Bui et al., 2017; Iban & Sekertekin, 2022; Mohajane et al., 2021; Pourtaghi et al., 2016).

Regarding the burnt area detection, some indexes were created so as to assess the accuracy of the burnt area delineation, namely the DBA, FBA, SBA, and DER, the Commission error, and the Omission error. RF resulted in a higher DBA value in comparison to SVM's DBA, approaching more the Copernicus Grading validation product. Regarding the FBA and SBA values, RF performed better than SVM demonstrating a lower false and skipped area value accordingly. According to the detection efficiency rate, RF outperformed SVM resulting in higher effectiveness in burnt area mapping. As for the subsequent errors, the commission error was low in both classification techniques whereas the omission was slightly low. In each case, RF distributed better results in overall burnt area detection concerning the study area.

Differences in burnt area estimates can be attributed in several factors, such as the classifiers operation regarding the spectral information. The high classification accuracy presented in this study is a result of the distinct spectral signatures of each selected class. More specifically, the comparison between burnt and unburnt land outputs provides a solid example. Unhealthy vegetation reflects more radiation in the visible and shortwave infrared while it absorbs in the near infrared, unlike healthy vegetation (Amos et al., 2018; Petropoulos et al., 2014). ML-based algorithms tent to follow some similar patterns in terms of error detection during their implementation (Arjasakusuma et al., 2022).

All in all, RF is suggested to obtain better results in burnt area mapping concerning the specific Greek setting as a case study. This could be attributed to its mechanism, which provides an assembly method constituting of multiple learning trees to acquire better predictions. Altogether, the selection of a suitable ML-based classifier for burnt area mapping assessment consists of a challenging task due to its model's specifications regarding their mechanisms. Nevertheless, both aforementioned ML models used in this study achieved high accuracies and are recommended for further evaluation in future burnt area mapping researches.

7.6 Concluding remarks

In the present study, a GEE-based approach that exploits ML techniques and ESA's Sentinel-2 imagery is developed with the purpose of automating the mapping of burnt areas, exploiting the capabilities offered today from cloud-based EO platforms. The methodological approach developed was tested in a typical Mediterranean setting, located in Greece using as a case study a wildfire occurred in the summer of 2021.

Results of this study demonstrated that imagery from S2, when combined with different pixel-based classifiers such as the RF and SVM in a GEE environment, can be used for burnt area mapping to provide an accurate, rapid, and cost-effective mapping of burnt area with high accuracy levels. Between the two classifiers, RF provided the highest total classification accuracy and burnt area detection capability, at least this was the case in our study. This is likely because this classifier has been developed to acquire better predictions based on numerous learning tree's forming a forest of random decision trees (DTs) to obtain better results for the specialities of the highly fragmented vegetation ecosystem and landscape structure complexities of the Greek test site of our study. Results particularly for RF, when combined with S2 observations, suggest that can be potentially form a very valuable tool in the future, for land cover-related applications, such as in our case burnt scar delineation. It also implies that RF may even be considered for operational use, subject to solving perhaps the automation of some processes.

All in all, this study in the use of S2 multispectral data provides innovative viewpoints in burnt area mapping; as this is to our knowledge, one of the few studies demonstrating essentially the capability of S2 for burnt area detection. S2 qualifies for cost-efficient and highly accurate mapping of burnt areas, especially when exploited with innovative ML techniques. On the other hand, the use of cloud-based platforms such as that of GEE allows rapid processing of large amount of data in an efficient way, saving costs, and time since there is also no need to locally download and process the EO datasets in specialized software packages committing also own computing resources.

Acknowledgments

Authors gratefully acknowledge the European Space Agency (ESA) and also the Copernicus team for providing free access to the Sentinel-2 satellite images and the Copernicus burnt area product used in this study. Authors wish to also thank the anonymous reviewers for their feedback and recommendations that assisted in improving substantially the originally submitted manuscript.

References

Amos, C., Petropoulos, G. P., & Feredinos, K. P. (2018). Determining the use of Sentinel-2A MSI for wildfire burning and severity detection. *International Journal of Remote Sensing*. Available from https://doi.org/10.1080/01431161.2018.1519284.

Arjasakusuma, S., Kusuma, S. S., Vetrita, Y., Prasasti, I., & Arief, R. (2022). Monthly burned-area mapping using multi-sensor integration of Sentinel-1 and Sentinel-2 and machine learning: Case study of 2019's fire events in South Sumatra Province, Indonesia. *Remote Sensing Applications: Society and Environment*, 100790. Available from https://www.sciencedirect.com/science/article/pii/S2352938522000982.

Bar, S., Parida, B. R., & Pandey, A. C. (2020). Landsat-8 and Sentinel-2 based Forest fire burn area mapping using machine learning algorithms on GEE cloud platform over Uttarakhand, Western Himalaya. *Remote Sensing Applications: Society and Environment, 18*, 100324. Available from https://www.sciencedirect.com/science/article/pii/S2352938520300100.

Borgeaud, M., Drinkwater, M., Silvestrin, P., & Rast, M. (2015). Status of the ESA earth explorer missions and the new ESA earth observation science strategy. In: 2015 IEEE International Geoscience and Remote Sensing Symposium (IGARSS) (pp. 4189–4192). IEEE, https://ieeexplore.ieee.org/abstract/document/7326749.

Breiman, L. (2001). Random forests. *Machine Learning, 45*(1), 5–32. Available from https://link.springer.com/article/10.1023/A:1010933404324.

Brown, R. A., Petropoulos, G. P., & Ferentinos, K. (2018). Appraisal of the Sentinel-1 & 2 use in a large-scale wildfire assessment: A case study from Portugal's fires of 2017. *Applied Geography, 100*, 78–89. Available from https://doi.org/10.1016/j.apgeog.2018.10.004.

Bui, D. T., Bui, Q. T., Nguyen, Q. P., Pradhan, B., Nampak, H., & Trinh, P. T. (2017). A hybrid artificial intelligence approach using GIS-based neural-fuzzy inference system and particle swarm optimization for forest fire susceptibility modeling at a tropical area. *Agricultural and Forest Meteorology, 233*, 32–44. Available from https://www.sciencedirect.com/science/article/abs/pii/S0168192316304269.

Colson, D., Petropoulos, G. P., & Ferentinos, K. (2018). Exploring the potential of Sentinels-1 & 2 of the Copernicus mission in support of rapid and cost-effective wildfire assessment. *International Journal of Applied Earth Observation & Geoinformation, 73*, 262–276. Available from https://doi.org/10.1016/j.jag.2018.06.011.

Congalton, R., & Green, K. (1999). *Assessing the accuracy of remotelysensed data: Principles and practices*. BocaRaton: CRC/Lewis Press.

Cortes, C., & Vapnik, V. (1995). Support-vector networks. *Machine Learning, 20*(3), 273–297. Available from https://link.springer.com/article/10.1007/BF00994018.

Evans, A., Lamine, S., Kalivas, D., & Petropoulos, G. P. (2018). Exploring the potential of EO data and GIS for ecosystem health modelling in response to wildfire: A case study in Central Greece. *Environmental Engineering & Management*.

Evelpidou, N., Tzouxanioti, M., Gavalas, T., Spyrou, E., Saitis, G., Petropoulos, A., & Karkani, A. (2021). Assessment of fire effects on surface runoff erosion susceptibility: The case of the summer 2021 forest fires in Greece. *Land, 11*(1), 21. Available from https://doi.org/10.3390/land11010021.

Giannaros, T. M., Papavasileiou, G., Lagouvardos, K., Kotroni, V., Dafis, S., Karagiannidis, A., & Dragozi, E. (2022). Meteorological analysis of the 2021 extreme wildfires in Greece: Lessons learned and implications for early warning of the potential for pyroconvection. *Atmosphere, 13*(3), 475. Available from https://www.mdpi.com/2073-4433/13/3/475.

Gorelick, N., Hancher, M., Dixon, M., Ilyushchenko, S., Thau, D., & Moore, R. (2017). Google Earth Engine: Planetary-scale geospatial analysis for everyone. *Remote Sensing of Environment, 202*, 18–27. Available from https://www.sciencedirect.com/science/article/pii/S0034425717302900.

Iban, M. C., & Sekertekin, A. (2022). Machine learning based wildfire susceptibility mapping using remotely sensed fire data and GIS: A case study of Adana and Mersin provinces, Turkey. *Ecological Informatics, 69*, 101647. Available from https://www.sciencedirect.com/science/article/abs/pii/S1574954122000966.

Ireland, G., & Petropoulos, G. P. (2015). Exploring the relationships between post-fire vegetation regeneration dynamics, topography and burn severity: A case study from the Montane

Cordillera Ecozones of Western Canada. *Applied Geography*, *56*, 232–248. Available from https://doi.org/10.1016/j.apgeog.2014.11.016.

Jayaweera, K. O. L. F., & Ahlnas, K. (1974). Detection of thunderstorms from satellite imagery for forest fire control. *Journal of Forestry*, *72*(12), 768–770. Available from https://academic.oup.com/jof/article-abstract/72/12/768/4660793.

Kalivas, D., Petropoulos, G. P., Athanasiou, I., & Kollias, V. (2013). An intercomparison of burnt area estimates derived from key operational products: Analysis of Greek wildland fires 2005–2007. *Non-linear Processes in Geophysics*, *20*, 1–13. Available from https://doi.org/10.5194/npg-20-1-2013.

Karamesouti, M., Petropoulos, G. P., Papanikolaou, I. D., Kairis, O., & Kosmas, K. (2016). An evaluation of the PESERA and RUSLE in predicting erosion rates at a Mediterranean site before and after a wildfire: Comparison & implications. *Geoderma*, *261*, 44–58. Available from https://doi.org/10.1016/j.geoderma.2015.06.025.

Kontoes, C. C., Poilvé, H., Florsch, G., Keramitsoglou, I., & Paralikidis, S. (2009). A comparative analysis of a fixed thresholding vs. a classification tree approach for operational burn scar detection and mapping. *International Journal of Applied Earth Observation and Geoinformation*, *11*(5), 299–316. Available from https://www.sciencedirect.com/science/article/pii/S0303243409000348.

Mohajane, M., Costache, R., Karimi, F., Pham, Q. B., Essahlaoui, A., Nguyen, H., & Oudija, F. (2021). Application of remote sensing and machine learning algorithms for forest fire mapping in a Mediterranean area. *Ecological Indicators*, *129*, 107869. Available from https://www.sciencedirect.com/science/article/pii/S1470160X21005343.

National's Observatory of Athens BEYOND. (2021). Estimating burned area, Varympompi Attica. Available at: http://beyond-eocenter.eu/index.php/thematic-areas/disasters/fires/446-ektimisi-kamenis-ektasis-varibobi-attikis. Accessed July 12, 2022.

Pandey, P. C., Koutsias, N., Petropoulos, G. P., Srivastava, P. K., & Dor, E. B. (2019). Land use/land cover in view of Earth observation: Data sources, input dimensions and classifiers—A review of the state of the art. *Geocarto International*, [in press].

Petropoulos, G. P., Kalaitzidis, C., & Vadrevu, K. P. (2012). Support vector machines and object-based classification for obtaining land-use/cover cartography from Hyperion hyperspectral imagery. *Computers & Geosciences*, *41*, 99–107. Available from https://www.sciencedirect.com/science/article/pii/S0098300411002871.

Petropoulos, G., Knorr, W., Scholze, M., Boschetti, L., & Karantounias, G. (2010). Combining ASTER multispectral imagery analysis and support vector machines for rapid and cost-effective post-fire assessment: A case study from the Greek fires of 2007. *Natural Hazards and Earth Systems Science*, *10*, 305–317. Available from https://doi.org/10.5194/nhess-10-305-2010.

Petropoulos, G. P., Kontoes, C., & Keramitsoglou, I. (2011). Burnt area delineation from a unitemporal perspective based on landsat TM imagery classification using support vector machines. *International Journal of Applied Earth Observation and Geoinformation*, *13*, 70–80. Available from https://doi.org/10.1016/j.jag.2010.06.008.

Petropoulos, G. P., Griffiths, H. M., & Kalivas, D. (2014). Quantifying spatial and temporal vegetation recovery dynamics following a wildfire event in a Mediterranean landscape using EO data and GIS. *Applied Geography*, *50*, 120–131. Available from https://doi.org/10.1016/j.apgeog.2014.02.006.

Pourtaghi, Z. S., Pourghasemi, H. R., Aretano, R., & Semeraro, T. (2016). Investigation of general indicators influencing on forest fire and its susceptibility modeling using different data mining techniques. *Ecological Indicators*, *64*, 72–84.

Qiu, L., Chen, J., Fan, L., Sun, L., & Zheng, C. (2022). High-resolution mapping of wildfire drivers in California based on machine learning. *Science of The Total Environment, 833*, 155155. Available from https://www.sciencedirect.com/science/article/pii/S0048969722022483.

Quintano, C., Fernández-Manso, A., & Fernández-Manso, O. (2018). Combination of Landsat and Sentinel-2 MSI data for initial assessing of burn severity. *International Journal of Applied Earth Observation and Geoinformation, 64*, 221–225. Available from https://www.sciencedirect.com/science/article/pii/S0303243417302039.

Seydi, S. T., Hasanlou, M., & Chanussot, J. (2022). Burnt-Net: Wildfire burned area mapping with single post-fire Sentinel-2 data and deep learning morphological neural network. *Ecological Indicators, 140*, 108999. Available from https://www.sciencedirect.com/science/article/pii/S1470160X22004708.

Sheykhmousa, M., Mahdianpari, M., Ghanbari, H., Mohammadimanesh, F., Ghamisi, P., & Homayouni, S. (2020). Support vector machine versus random forest for remote sensing image classification: A meta-analysis and systematic review. *IEEE Journal of Selected Topics in Applied Earth Observations and Remote Sensing, 13*, 6308–6325. Available from https://ieeexplore.ieee.org/abstract/document/9206124.

Tsatsaris, A., Kalogeropoulos, K., Stathopoulos, N., Louka, P., Tsanakas, K., Tsesmelis, D. E., Brow, V., Pappas, V., & Chalkias, C. (2021). Geoinformation technologies in support of environmental hazards monitoring under climate change: An extensive review. *ISPRS International Jouranl of Geo-Information, 10*(94), 1–32. Available from https://doi.org/10.3390/ijgi10020094.

Tselka, I., Krassakis, P., Rentzelos, A., Koukouzas, N., & Parcharidis, I. (2021). Assessing post-fire effects on soil loss combining burn severity and advanced erosion modeling in Malesina, Central Greece. *Remote Sensing, 13*(24), 5160. Available from https://www.mdpi.com/2072-4292/13/24/5160.

Yang, L., Driscol, J., Sarigai, S., Wu, Q., Chen, H., & Lippitt, C. D. (2022). Google Earth Engine and artificial intelligence (AI): A comprehensive review. *Remote Sensing, 14*(14), 3253. Available from https://www.mdpi.com/2072-4292/14/14/3253.

Chapter 8

On volunteered geographic information quality: a framework for sharing data quality information

Vyron Antoniou
Hellenic Army Academy, Athens, Greece

8.1 Introduction

Spatial data quality is a subject that has captured the interest of researchers in the geospatial domain for a long time. The research in this field gained momentum with the proliferation of Geographic Information Systems (GIS) and digital datasets. Indeed, extensive literature reviews on the subject can be found in Shi et al. (2002), Van Oort (2005), and Devillers and Jeansoulin (2006), or in handbooks for GIS such as the one provided by Longley et al. (2005), just to name but a few. Since then, few things have changed in terms of concepts and principles that apply in the field of spatial data quality. This is also due to the fact that the research on spatial data quality has been supported and endorsed by efforts and publications of the Open Geospatial Consortium and the International Organization for Standardization (ISO) which provided detailed frameworks for assessing spatial data quality and developed evaluation processes and methods that should be followed for such tasks. With these apparatuses at hand, the producers of authoritative spatial data were well-equipped to monitor and evaluate their in-house created spatial datasets. Equally, consumers of spatial data were able to assess the quality and fitness for purpose of the acquired data.

However, today, the proliferation of volunteered geographic information (VGI) and crowdsourced geospatial content has changed the way that GI is created and published on the Web. In most cases, VGI sources provide open data with flexible licensing agreements that allow both individual and commercial use, in contrast with the practices of existing authoritative sources. Fundamentally different is also the way that, nowadays, GI consumers have

Geoinformatics for Geosciences. DOI: https://doi.org/10.1016/B978-0-323-98983-1.00009-0

access to spatial content. In addition to purchasing GI from authoritative producers, those who seek spatial information can extract data directly from VGI sources in bulk downloads, through intermediary private companies or through Application Programming Interfaces (APIs) both from explicit and implicit VGI sources (Antoniou et al., 2010; Craglia et al., 2012) such as Facebook, Twitter or Foursquare, among others. This new breed of GI has managed to penetrate the Geoinformatics domain and become a key factor in many geospatial processes, products, and services. For example, VGI has been used in the development and maintenance of cadastre (Basiouka & Potsiou, 2012) and 3D cadastre datasets (Gkeli et al., 2020), crisis management (Haworth, 2018) and crisis mapping (Tavra et al., 2021), in disaster monitoring (Panteras & Cervone, 2018), flood forecasting (Annis & Nardi, 2019), the development of Spatial Data Infrastructures (Demetriou et al., 2017), the calibration, validation, and verification of land cover maps (Fonte et al., 2015; Antoniou et al., 2016) or in map production (Olteanu-Raimond et al., 2017), to name a few cases. Given the variety of applications that VGI data can support, it is not surprising that local authorities, government agencies, and non-governmental organizations have started to explore this novel domain (Haklay et al., 2014; Haklay et al., 2017). Similarly, the private sector has embraced VGI data by building products and services directly on top of VGI data or by replacing proprietary data with VGI-generated maps and datasets, or actively contributing to data creation (Anderson et al., 2019; OSM, 2022). Moreover, the intertwining of VGI with Citizen Science creates a strong outcome since VGI is well suited to support GI large projects that need extended participation from volunteers for data gathering and monitoring and are able to help in addressing global-wide problems (Fritz et al., 2019).

8.2 Volunteered geographic information and spatial data quality

The emergence of VGI created a totally new era of GI production and along with the numerous advantages and opportunities from such data a completely new set of challenges regarding data quality evaluation has surfaced, that cannot be addressed by previous methods. There are several reasons why the existing and well-established quality evaluation methods and practices are either difficult or impossible to be implemented in the case of VGI data. The first set of reasons stems from the way that such datasets are generated. The collaborative way of data production by a pool of heterogeneous participants, that is, lay persons with or without formal education in the field of spatial data and with various motives, aims, commitment, and patterns of participation (Haklay, 2013) creates a patchwork that differs fundamentally from, for example, the outcome created by the staff of a National Mapping Agency which is trained and hand-picked to perform the task of spatial data collection following rigid rules that ensure quality. The second is the very nature of VGI data. VGI embeds the local knowledge of contributors which can lead either to totally new or much richer datasets than those found in authoritative databases and portfolios. Moreover, the timely nature of VGI, both in terms of the short time needed from data creation to data publication and the continuous update of information is fundamentally different from the release cycles of authoritative data. All these constitute a major change in how VGI data are created, managed, curated, disseminated, and consumed compared to authoritative data. The depth, spread, and velocity of GI generated through crowdsourcing do not allow for established spatial quality evaluation methods to be implemented in VGI datasets. For example, ISO standards require a reference dataset or concrete specifications against which a new dataset is compared. However, when it comes to VGI, this reference dataset might very well not be available since VGI datasets might be on a larger scale (i.e., more detailed) than any available authoritative dataset or there might not be any strict specifications against which the data can be validated.

In other words, although there is considerable interest in a broad spectrum of VGI applications, the quality of the data remains one of the most challenging issues surrounding its uptake and use. The main issue remains the lack of a standardized framework to allow for meaningful evaluation of the quality. In contrast to professionally produced GI, there are often few rigid protocols in place that govern the data collection process (Mooney et al., 2016). Data specifications are either nonexistent or frequently changing wiki-based instructions, which function more as best-practice guidelines and not as strict rules. In addition, the contribution process introduces new sources of uncertainty that can be related to socio-economic factors. For example, Haklay (2010) showed that there are biases in the completeness of OSM data that correlate with the Index of Multiple Deprivation for the United Kingdom, that is, there is a positive relationship between the lack of OSM data coverage and deprivation in those areas. Similarly, the positional accuracy of individual OSM features is correlated to the number of contributors editing each feature (Haklay et al., 2010). Antoniou and Schlieder (2014) showed that participation patterns in OSM suffer from several biases with respect to the preferred areas of contribution and the type of features that gather contributors' interest. All these factors are new sources of uncertainty that do not affect authoritative spatial data.

To address these challenges, researchers have used novel methods to evaluate VGI quality based both on measures and indicators (Antoniou & Skopeliti, 2015). For example, there are several cases where the quality of VGI was tested against reference data (see e.g., Haklay, 2010; Fan et al., 2014; Forghani & Delavar, 2014; Dorn et al., 2015; Zhang & Malczewski, 2017; Touya et al., 2017). However, as explained, the direct evaluation of VGI data against reference data is not always possible due to data availability or differences in scope and scale between two types of datasets. Hence, many researchers focused on the discovery, documentation, and use of indicators that could credibly predict the intrinsic quality of a VGI dataset. For example, Barron et al. (2014) presented a quality evaluation framework that consists of more than 25 methods and indicators derived from data history. Vahidi et al. (2018) presented a fuzzy trust model for intrinsic quality assessment of species occurrences obtained from biodiversity monitoring programs which evaluate the thematic and positional quality of the observations using as indicators the observation's consistency with habitat, surroundings, and contributor reputation. Jacobs and Mitchell (2020) used the history of the OSM dataset to extract mapping characteristics and metrics which can help to understand the OSM contributor experiences and thus the quality of its contribution.

However, despite the considerable effort and progress in the field of VGI quality evaluation, there is still no framework or solid methodology that can provide concrete results on the quality of VGI data in a holistic and undisputable way as is the case with the ISO specifications for authoritative data. Hence, a question regarding the quality of VGI data and its fitness for purpose will always be present for the end users. One way out of this impasse has been suggested by Devillers et al. (2010) who, although referring to a different context, highlighted that there is plenty of important research in nonspatial domains regarding data quality and despite the specificities of spatial data, still, researchers in the geospatial domain can find a lot of useful insight on the subject of quality. This chapter elaborates further on this idea. It aims to provide a new quality evaluation framework that addresses the main obstacle between producers and consumers of VGI, that is, the existence of imperfect information on VGI data quality, through the lenses of theory and research developed in the domain of economics. It is important to note that this framework could apply to several VGI sources with different scopes, spatial footprints (from local to global), or types of data (e.g., vectors, geotagged imagery, drone imagery, and video).

8.3 Lessons from economic transaction theory

The 2001 Nobel Prize in Economics was awarded to Akerlof, Spence, and Stiglitz for their analysis of markets with asymmetric information (Akerlof, 1970; Spence, 1973; Stiglitz & Rothschild, 1976). Asymmetric information refers to the situation in which one of the two parties of a transaction has considerably more information and insight about the quality of the product or service of the transaction compared to the other part. Importantly, this asymmetry in information exists because the more informed part is not sharing this extra information with its counterpart; hence, there is a substantial imbalance in the overall knowledge that these two parts possess over a specific transaction.

The problem definition, as framed by Akerlof, refers to a market in which a specific product exists in varying qualities, that is, low and high. When the seller of the product is in possession of the information on how good the product is, but the buyer has no such information, the buyer is not able to differentiate between the different products, and so, the information on quality is asymmetrically shared between buyer and seller. In this type of market, Akerlof (1970) proved that the transactions will suffer from bias and favor certain sellers. More specifically, because the buyers are aware that they do not have proper information about the quality of the product, they treat any available product as being of the lowest quality and thus bid down their prices. It was proved that the result of asymmetric information on quality is that either high-quality products will stop being produced since low-quality products will prevail in the market (a situation known as adverse selection), or the transactions for this specific product will stop altogether. Of course, this is against the best interest of sellers who are willing to create a product of known quality and of buyers who are willing to pay a fair price for it.

While Akerlof's contribution framed, documented, and explained the problem of asymmetric information, Spence, Stiglitz, and Rothchild offered possible solutions. For his part, Spence (1973) showed that the sellers can help in balancing the asymmetry regarding quality information. Spence explained that the party that has better information should communicate this information to the less informed one via the use of signals. These signals should serve as quality certifications and thus enable the less informed parties to overcome their previous uncertainty and help them to make an informed decision regarding whether to acquire the product. The key element of Spence's arguments is that the overall cost of each signal should not be the same for all sellers. In fact, there should be a negative correlation between the signaling cost and the quality of the product. In other words, the lower the quality of the product offered, the higher the signaling cost should be (and thus more difficult to achieve). This negative correlation will enable sellers of high-quality products to proceed easily to this step while at the same time deter the sellers of low-quality products. The same principle applies to those sellers that aim to have a permanent presence in the market compared to the ones that plan on infrequent or one-off transactions. As an example of such signals, Spence used higher education degrees in the job market. In brief, he argued that acquiring a higher education degree is easier for capable (and potentially more productive) employees while the difficulty rises for less capable ones. This signal allows employers to have a better understanding of the quality of the candidates and thus facilitates their decision in employing a specific person.

On the other hand, Stiglitz and Rothschild (1976) focused on the actions of buyers to improve the asymmetry in quality information through a method known as "self-screening." More specifically, they explained that by using incentives, the less informed party can force the more informed one to reveal information about the quality of the product. This result will balance out the asymmetry in information and will facilitate the transaction of the product. As an example of this theory, the personal insurance market was used by the authors. They argued that an insurance company can provide a list of insurance contracts that differ in their prices and benefits. By doing so, the individuals that know that they belong to a high-risk category will voluntarily choose the expensive contract that has more benefits (thus, in fact, revealing information that only they have about their personal health) whereas the low-risk individuals will be happy to settle for a cheaper contract.

8.4 A proposed framework for volunteered geographic information quality evaluation

The chapter supports that the use of VGI can be viewed as a transaction between two parties. On the one hand are the data providers, that is, VGI sources who store, manage, and disseminate the data and, on the other, those who use the data for their applications, that is, NMAs, local authorities, private industry, individuals, etc. It is also supported that these two counterparts have asymmetric information regarding the quality of the spatial data. Hence the Nobel-awarded economic theories on asymmetric information can be applied in developing a framework for addressing the challenge of VGI quality evaluation.

More specifically, when it comes to accepting, trusting, and using VGI data, all factors boil down to the quality assessment and evaluation of the data at hand. This is even more important when VGI is to be diffused to and integrated with authoritative databases. Any such kind of transaction, between VGI providers and consumers, needs to be established on a clear understanding of the VGI quality. It is important to note that both parties have a mutual interest to facilitate such transactions. VGI providers can gain from the support (technical or financial) of the authorities but more importantly, the public can gain as open and grassroots data can promote good governance, democratization, local government reform, public accountability, and transparency (Manor, 2004). On the other hand, VGI consumers can exploit the comparative advantages of such datasets.

However, given the nature of VGI data, if no extra effort has been put forward, none of the two parties, VGI providers or consumers, actually knows if the data is of any real value and therefore both parties are in mutual ignorance. On the other hand, assuming there are efforts for VGI quality assessment and evaluation, it is reasonable to expect that due to the dynamic nature of VGI and the direct access that VGI providers (i.e., administrators or governing bodies) have on the entire datasets, they will be the better-informed part compared to the VGI consumers as it is not possible for the latter to have the same kind information at any given time. Indeed, while a potential VGI consumer has no clear view of VGI quality, the administrators of VGI projects are in a position to recognize and cure erroneous processes, analyze GI production, and study participation patterns so as to gain a better insight into their spatial data quality. Thus there is either a mutual lack of information or imbalanced information regarding the true quality of the data in a VGI repository; information needed to make the final decision on the realization of the transaction. The absence of an environment with symmetric information, where the interested parties will be mutually clearly informed of the spatial data quality, is the fundamental problem when considering VGI data quality.

It is not difficult to spot common elements between the economic theories on transactions and the current situation in VGI. As noted, the specific product in the market is the VGI datasets, the "sellers" are the VGI providers and the "buyers" are the potential VGI consumers while at the same time there is asymmetric information regarding the quality of the product, exactly as described by Akerlof's model. Thus the question now is how the methodologies suggested by Spence, Stiglitz, and Rothschild can be implemented to distinguish prominent from obscure VGI sources and high-quality from low-quality VGI data.

Starting by using the concepts provided by Spence, the challenge is to spot proper signals that could be provided by the party that has better information (i.e., the VGI providers). One example can be an effort by the VGI provider to evaluate and document the quality of their data. Sporadic academic findings can give an indication of the spatial data quality of selected sources, but they cannot function as a proxy for the entire geographic coverage of these sources or for every VGI source. However, internally assessing and documenting the quality of VGI datasets with well-established and recognized methods (e.g., using ISO standards) is a cumbersome effort. Prominent VGI sources with a large pool of contributors that use standardization processes and closely monitor their data repositories can more easily deliver on this task compared with obscure or not-so-methodical VGI sources, which are possible of low quality. This negative correlation between performing internal quality evaluation and the credibility of VGI sources (and consequently of their VGI datasets) can provide the necessary input to VGI consumers to make informed decisions regarding the acceptance of VGI datasets. Another signal toward this direction could be the design and implementation of long-term investments (in terms of time and effort) in creating web services, improving usability, enhancing interactivity, and optimizing tools so that contributors can provide content in more consistent and error-free ways. Again, solid and longstanding VGI sources have the means to generate such signals easier, just by mobilizing their pool of contributors compared to obscure VGI sources.

On the other hand, Stiglitz and Rothschild (1976) transpose the ability to improve an information asymmetric environment to the less informed party through incentives that lead to the "self-screening" of the more informed party. In the Geomatics domain, the key is to recognize incentives that VGI consumers can create so as to make VGI providers enter a "self-screening process" in order to release information about the quality of their VGI data. One such case could be for the VGI consumers to offer support or a premium for datasets for which the fulfillment of certain quality standards can be proven. This can motivate VGI providers, in an effort to get the higher possible premium, to voluntarily release needed information about a dataset's quality. For example, in order for possible cooperation to be established between an authority (e.g., a NMA) and a VGI source, the authorities could provide financial or technical support if datasets are accompanied with proper metadata or if the VGI source can share participation and contribution statistics. VGI providers that are managing well their data repositories will find it easier to generate and provide such information so to get the highest possible support or premium from possible cooperation. In any case, VGI consumers can develop a balanced strategy against VGI datasets of different quality levels that can even lead to direct involvement in the data collection process as is the case with OSM and several private companies (Anderson et al., 2019).

8.5 Application of the framework to current volunteered geographic information initiatives

Although the above-suggested framework, which describes a combination of signals and incentives has not been applied by a VGI providers or consumers, there are scattered cases where efforts are made toward this direction. Using as a case study the OSM, which is perhaps the most successful case of VGI source, it can be seen that there is a substantial effort from the OSM community to bring to the surface any quality issues and thus safeguard the overall value of data, products, and services. For example, Minghini and Frassinelli (2019) presents a tool that enables the quick assessment of OSM intrinsic quality based on the object history. Sehra et al. (2017) present an extension of the Quantum GIS (QGIS) processing toolbox to enable the assessment of the completeness of spatial data using intrinsic indicators. Similarly, Raifer et al. (2019) presented the OpenStreetMap History Database, a framework for spatiotemporal analysis of OpenStreetMap history data which can facilitate the quality assessment of OSM data by using intrinsic measures. Moreover, an updated list of OSM quality assurance tools can be found on the dedicated wiki-page of the project (https://wiki.openstreetmap.org/wiki/Quality_assurance). A similar process can be traced at the iNaturalist project which classifies observations into "Casual," "Needs ID," and "Research Grade" as a measure to indicate the quality of the observations submitted or at research efforts that design quality frameworks for the use of smartphone application for improving data collection and verification (Wittmann et al., 2019). Observations classified as Research Grade are of the highest quality level as they must meet several predefined criteria, including georeferenced data (iNaturalist, 2022).

This systematic effort of successful platforms such as OSM and iNaturalist can be seen as indirect proof that the Nobel-awarded economic theories can be implemented in the Geoinformatics domain. The high-quality data sources with large pools of contributors can develop the processes and infrastructure required to generate quality contributions and signal this to data consumers. On the flip side, the preference and support that data consumers (from individuals to NMAs, to private companies or research institutions) show for these sources is generated by a self-screening process that motivates data providers to show their effort and achievements in safeguarding their data quality.

8.6 Conclusion

The chapter supported that there is a need to establish a framework that, while taking the specific characteristics of VGI into account, it will enable end users to make informed decisions on the fitness for purpose of the crowdsourced data. The chapter also supported that one of the fundamental pillars of such a transaction is the understanding of the VGI data quality. VGI data is unique in many senses. VGI is both spatial and social and its quality assessment riddle extends beyond the borders of geospatial theories. Crowdsourced datasets gain valuable characteristics from their grassroots mechanisms, but they also suffer from socio-economic biases. Hence, existing methods and practices for quality assessment and evaluation cannot be applied here, at least not in a straightforward manner. Stemming from this need, the chapter presented a novel framework for data quality information sharing using prominent theories developed in the domain of Economics. This framework takes into account both the need of data consumers for data of known quality and the fuzziness that surrounds the quality of crowdsourced datasets and suggests a way to implement theories from another domain. Common elements have been recognized, parallels have been highlighted, and opportunities for implementation have been described. By doing so, prominent economic theories are suggested as the basic components of a new framework to remedy the asymmetric availability of information on the quality of VGI.

References

Akerlof, A. (1970). The market for 'Lemons': Quality uncertainty and the market mechanism. *The Quarterly Journal of Economics*, *84*(3), 488–500.

Anderson, J., Sarkar, D., & Palen, L. (2019). Corporate editors in the evolving landscape of OpenStreetMap. *ISPRS International Journal of Geo-Information*, *8*(5), 232.

Annis, A., & Nardi, F. (2019). Integrating VGI and 2D hydraulic models into a data assimilation framework for real time flood forecasting and mapping. *Geo-spatial Information Science*, *22* (4), 223–236.

Antoniou, V., et al. (2016). Investigating the feasibility of geo-tagged photographs as sources of land cover input data. *ISPRS International Journal of Geo-Information*, *5*(5), 64.

Antoniou, V., Haklay, M., & Morley, J. (2010). Web 2.0 geotagged photos: Assessing the spatial dimension of the phenomenon. *Geomatica*, *64*(1), 99–110.

Antoniou, V., & Schlieder, C. (2014). *Participation patterns, VGI and gamification. Castellon, Agile.*

Antoniou, V., & Skopeliti, A. (2015). Measures and indicators of VGI quality: An overview. *ISPRS Annals of Photogrammetry, Remote Sensing & Spatial Information Sciences*, *2*.

Barron, C., Neis, P., & Zipf, A. (2014). A comprehensive framework for intrinsic OpenStreetMap quality analysis. *Transactions in GIS*, *18*(6), 877–895.

Basiouka, S., & Potsiou, C. (2012). VGI in Cadastre: A Greek experiment to investigate the potential of crowd sourcing techniques in Cadastral Mapping. *Survey Review*, *44*(325), 153–161.

Craglia, M., Ostermann, F., & Spinsanti, L. (2012). Digital Earth from vision to practice: Making sense of citizen-generated content. *International Journal of Digital Earth*, *5*(5), 398–416.

Demetriou, D, Campagna, M, Racetin, I, & Konecny, M (2017). *Integrating Spatial Data Infrastructures (SDIs) with Volunteered Geographic Information (VGI) creating a Global GIS platform. Mapping and the citizen sensor* (pp. 273–297). London: Ubiquity Press.

Devillers, R., & Jeansoulin, R. (2006). *Fundamentals of spatial data quality elements.* London: ISTE.

Devillers, R., et al. (2010). Thirty years of research on spatial data quality: achievements, failures, and opportunities. *Transactions in GIS*, *14*(4), 387–400.

Dorn, H., Törnros, T., & Zipf, A. (2015). Quality evaluation of VGI using authoritative data: A comparison with land use data in Southern Germany. *SPRS International Journal of Geo-Information*, *4*(3), 1657–1671.

Fan, H., Zipf, A., Fu, Q., & Neis, P. (2014). Quality assessment for building footprints data on OpenStreetMap. *International Journal of Geographical Information Science*, *28*(4), 700–719.

Fonte, C., et al. (2015). Usability of VGI for validation of land cover maps. *International Journal of Geographical Information Science*, *29*(7), 1269–1291.

Forghani, M., & Delavar, M. (2014). A quality study of the OpenStreetMap dataset for Tehran. *ISPRS International Journal of Geo-Information*, *3*(2), 750–763.

Fritz, S., et al. (2019). Citizen science and the United Nations sustainable development goals. *Nature Sustainability*, *2*(10), 922–930.

Gkeli, M., Potsiou, C., & Ioannidis, C. (2020). Design of a crowdsourced 3D cadastral technical solution. *International Archives of the Photogrammetry, Remote Sensing and Spatial Information Sciences*, *XLIII*(B4), 269–276.

Haklay, M. (2010). How good is volunteered geographical information? A comparative study of OpenStreetMap and Ordnance Survey datasets. *Environment and Planning B: Planning and Design*, *37*(4), 682–703.

Haklay, M. (2013). *Citizen science and volunteered geographic information: Overview and typology of participation. Crowdsourcing geographic knowledge* (pp. 105–122). Dordrecht: Springer.

Haklay, M., et al. (2014). *Crowdsourced geographic information use in government.* London: World Bank.

Haklay, M., et al. (2017). *Identifying success factors in crowdsourced geographic information use in government.* London: World Bank.

Haklay, M., Basiouka, S., Antoniou, V., & Ather, A. (2010). How many volunteers does it take to map an area well? The validity of Linus' law to volunteered geographic information. *The Cartographic Journal, 47*(4), 315–322.

Haworth, B. (2018). Implications of volunteered geographic information for disaster management and GIScience: A more complex world of volunteered geography. *Annals of the American Association of Geographers, 108*(1), 226–240.

iNaturalist. (2022). Help [Online]. Available at: http://www.inaturalist.org/pages/help#quality (Accessed 23 October 2022).

Jacobs, K., & Mitchell, S. (2020). OpenStreetMap quality assessment using unsupervised machine learning methods. *Transactions in GIS, 24*(5), 1280–1298.

Longley, P., Goodchild, M., Maguire, D., & Rhind, D. (2005). *Geographic information system and Science*. Oxford: John Wiley & Sons.

Manor, J. (2004). Democratisation with inclusion: Political reforms and people's empowerment at the grassroots. *Journal of Human Development, 5*(1), 5–29.

Minghini, M., & Frassinelli, F. (2019). OpenStreetMap history for intrinsic quality assessment: Is OSM up-to-date? *Open Geospatial Data, Software and Standard, 4*(1), 1–17.

Mooney, P., et al. (2016). Towards a protocol for the collection of VGI vector data. *ISPRS International Journal of Geo-Information, 5*(11), 217.

Olteanu-Raimond, A.-M., et al. (2017). The scale of VGI in map production: a perspective on European National Mapping Agencies. *Transactions in GIS, 21*(1), 74–90.

OSM. (2022). Major OpenStreetMap consumers [Online]. Available at: https://wiki.openstreetmap.org/wiki/Major_OpenStreetMap_consumers (Accessed 23 October 2022)

Panteras, G., & Cervone, G. (2018). Enhancing the temporal resolution of satellite-based flood extent generation using crowdsourced data for disaster monitoring. *International Journal of Remote Sensing, 39*(5), 1459–1474.

Raifer, M., et al. (2019). OSHDB: A framework for spatio-temporal analysis of OpenStreetMap history data. *Open Geospatial Data, Software and Standards, 4*(1), 1–12.

Sehra, S., Singh, J., & Rai, H. (2017). Assessing OpenStreetMap data using intrinsic quality indicators: An extension to the QGIS processing toolbox. *Future Internet, 9*(2), 15.

Shi, W., Fisher, F., & Goodchild, M. (2002). *Spatial data quality*. New York: Taylor and Francis.

Spence, M. (1973). Job market signaling. *The Quarterly Journal of Economics, 87*(3), 355–374.

Stiglitz, J., & Rothschild, M. (1976). Equilibrium in competitive insurance markets. *Quarterly Journal of Economics, 90*(4), 629–649.

Tavra, M., Racetin, I., & Peroš, J. (2021). The role of crowdsourcing and social media in crisis mapping: A case study of a wildfire reaching Croatian City of Split. *Geoenvironmental Disasters, 8*(1), 1–16.

Touya, G., Antoniou, V., Olteanu-Raimond, A.-M., & Van Damme, M. (2017). Assessing crowdsourced POI quality: Combining methods based on reference data, history, and spatial relations. *ISPRS International Journal of Geo-Information, 6*(3), 80.

Vahidi, H., Klinkenberg, B., & Yan, W. (2018). Trust as a proxy indicator for intrinsic quality of volunteered geographic information in biodiversity monitoring programs. *GIScience & Remote Sensing, 55*(4), 502–538.

Van Oort, P. (2005). *Spatial data quality: From description to application. PhD thesis.* Wageningen University.

Wittmann, J., Girman, D., & Crocker, D. (2019). Using iNaturalist in a coverboard protocol to measure data quality: Suggestions for project design. *Citizen Science: Theory and Practice*, *4*(1).

Zhang, H., & Malczewski, J. (2017). Accuracy evaluation of the Canadian OpenStreetMap road networks. *International Journal of Geospatial and Environmental Research*, *5*(2).

Section 3

GIS & Remote sensing applications

Chapter 9

Natural disaster monitoring using ICEYE SAR data

Penelope Kourkouli
ICEYE Oy, Espoo, Finland

9.1 Introduction

Natural disasters are severe and extreme natural events that can have a detrimental impact on human lives and properties. In recent decades, natural catastrophes are happening more frequently, as shown in Fig. 9.1. The most frequent natural disasters are reported in Table 9.1 and include floods, droughts, extreme rainfall, wildfires, extreme temperature, volcanic activity, earthquakes, landslides, tsunamis, avalanches, and droughts (Barrett & Curtis, 1999). Climate change increases the probability of climate-induced disasters such as heatwaves and wildfires as well as the probability of flooding.

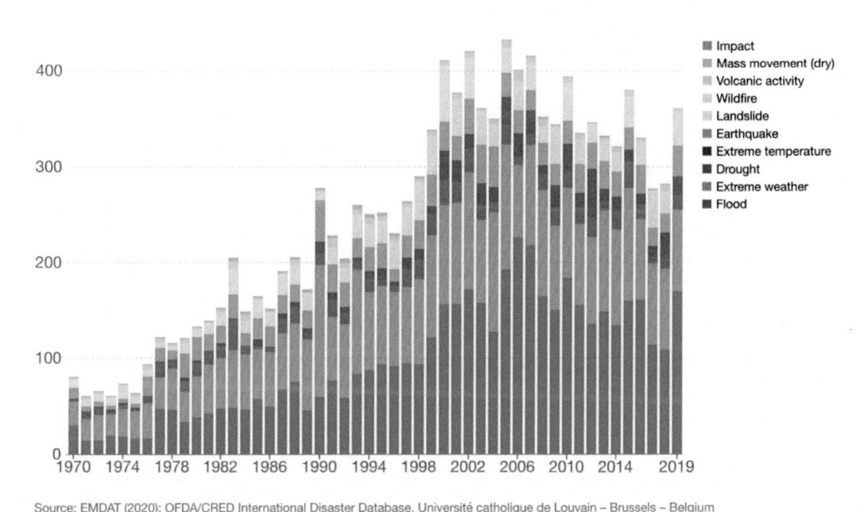

FIGURE 9.1 Global reported natural disasters by type (1970–2019). *Data from EM-DAT (2020), figure taken from Our World in Data.*

Geoinformatics for Geosciences. DOI: https://doi.org/10.1016/B978-0-323-98983-1.00010-7

TABLE 9.1 The most frequent natural disaster.

Geology	Meteorology	Hydrology	Vegetation	Oceanography
Earthquakes	Extreme rainfall	Floods	Wildfires	Coastal floods
Volcanic eruptions	Tropical rainstorms	Droughts	Droughts	Sea level rise
Landslides	Ice storm	Avalanches		
Tsunami		Snow		
Erosion				

Source: Modified from Cartalis and Feidas (2006).

In 2021 the Emergency Event Database (EM-DAT) recorded 432 disastrous events related to natural catastrophes worldwide. Overall, 9,492 deaths have been recorded, impacting 91.8 million people and causing approximately 252.1 billion US$ of economic losses (EM-DAT, 2021). During the most recent decades, satellites have played a vital role in monitoring natural hazards. Due to the fact that the majority of such events are large-scale, the only way to monitor and map their impact rapidly and cost-effectively is via satellite remote sensing techniques. Especially, the occurrence of a hazard in remote and inaccessible areas—and in many cases in different areas—makes natural disaster observation from space the most suitable means for hazard monitoring and risk assessment.

Synthetic Aperture Radar (SAR) satellites have been used more frequently, as they are independent of weather and light conditions and can penetrate clouds, rain, fog, and smoke. All of this makes SAR one of the most reliable Earth observation technology, and SAR imagery is valuable in sudden crisis situations requiring a fast response. One of the main challenges is the large temporal gaps between acquisitions. In crisis situations, fast response and rapid monitoring is very crucial. ICEYE has built the world's first small and largest SAR satellite constellation with a network of sensors designed to capture daily or even subdaily imagery, enabling a new level of change detection.

In the following sections, we are going to present how and why SAR daily images can help us to monitor natural catastrophes and will give a few examples per different peril.

9.2 ICEYE SAR data

SAR community utilizes a big variety of data that has been proven very useful in understanding natural disasters. However, the lack of timely data is still a challenge for disaster monitoring. ICEYE is the first to achieve daily coherent ground track repeat capability which enables change detection on a daily basis and with very high precision.

ICEYE SAR satellites carry X-band and provide three different SAR imaging modes with the image products Spot, Spot Extended Area, Strip, and Scan. Depending on the type of the phenomenon and the scale of the disaster, each mode gives the flexibility to monitor very large-scale areas and focus on areas of interest at very high resolutions (Fig. 9.2).

ICEYE Strip images have a ground resolution of 3 m in range (azimuth) and cover an area of 30 km (range) by 50 km (azimuth). The strip length can be tailored up to a length of 500 km, in increments of 50 km. This mode provides the ability to monitor wide areas and a moderate resolution that are useful for situational awareness, and it can be used to assess the impact of floods, wildfires, earthquakes, or volcanic eruptions.

ICEYE's Spot collection covers an area of 5 km x 5 km with a 1 m ground resolution for multilooked amplitude images. These are formed from

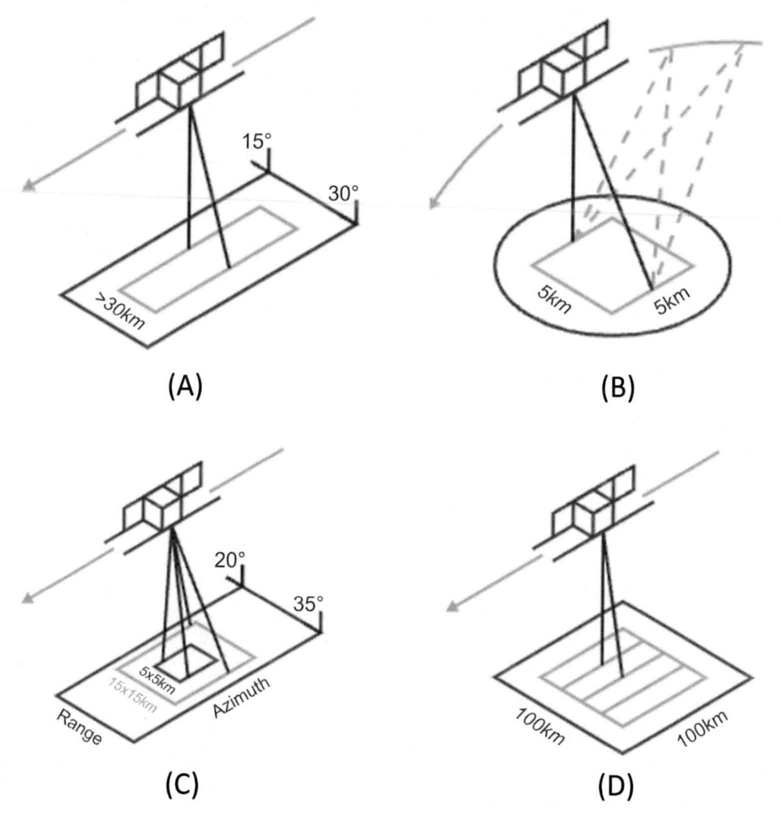

FIGURE 9.2 ICEYE SAR imagery products, (A) Strip, (B) Spot, (C) Spot Extended Area, and (D) Scan.

four independent looks to suppress speckle and increase image quality. The Spot Extended Area collection has a scene size of 15 km x 15 km at 1 m ground resolution with no multilooking applied. These modes provide the finest resolution, and hence, Spot images are useful to capture more detailed information. These can be mainly used to discriminate between different types of objects such as buildings or infrastructures.

Last, the Scan mode provides images that cover large-scale areas of 90 km x 90 km with a coarser resolution than the above ones, which is better than 15 m. The length of a Scan image can be increased to 300 km. Scan acquisitions are suited to wide area monitoring such as oceans or even monitoring large-scale disasters on a modest resolution, such as the extent of floods or wildfires.

In the next section, we are going to demonstrate mostly how the ICEYE SAR Strip daily ground track repeat data provides useful information regarding natural catastrophe monitoring.

9.3 Flood and wildfire monitoring using ICEYE constellation

For estimating the aftermath of a large-scale natural disaster and especially its impact on properties, there is a demand for immediate response to severely damaged areas. Besides the precision level of detecting damages, the timely response is equally significant. Rapid damage assessment after natural hazards leads to better preparedness, allocation of the search and rescue teams, and fast identification of property loss.

Disasters, despite their different source of mechanisms, have the same result; damages, such as partially or fully collapsed buildings. The most common remote sensing technique for damage assessment is change detection. To monitor the impact of a disaster, at least a preevent and postevent image is needed. Two main change detection methods are usually applied, (1) based on the amplitude, (2) based on the interferometric coherence, and (3) the combination of both. The latter one is the sweet spot as it combines both amplitude and phase information for detecting slight to large-scale damages with a building-level precision.

9.3.1 Flood monitoring

Climate change is increasing the occurrence of flood events around the world, affecting lives and properties. Flooding can be caused by phenomena like typhoons, hurricanes, cyclones, and extreme rainfall. Near-time monitoring of floods is crucial for situational awareness, allocating resources, and ensuring a fast response. Moreover, insurance companies need timely information on property losses for financial allocation after a disaster.

Several studies have demonstrated the effectiveness of SAR data for flood mapping. Sentinel-1 has paved the ground for the use of SAR for flood mapping applications (ICEYE Oy, 2021). ICEYE SAR constellation provides timely data with enhanced spatial resolution, enabling the capability of multiple observations only a few hours after a flood event.

ICEYE experts mapped over 60 flood events during 2019 and 2020, with a focus on events impacting the urban environment (Ardila et al., 2022). A piece of detailed information about each flood event is provided in the flood briefings that are available at https://www.iceye.com/downloads/flood-briefings. The spatial distribution of the flood events mapped using ICEYE data—mainly using Strip mode—is shown in Fig. 9.3. Three outputs are generated per flood event, (1) the flood footprint of the affected area, (2) the flood depth raster surface, representing the depth of water, and (3) the affected buildings assigned with the flood depth estimation.

Core methodologies used for generating the above flood products are machine learning, supervised image interpretation of amplitude images, and geospatial analysis. In addition, a variety of third party data are used to calibrate the final outputs, such as tidal gauges, ground sources, river gauges, and rainfall data (Ardila et al., 2022).

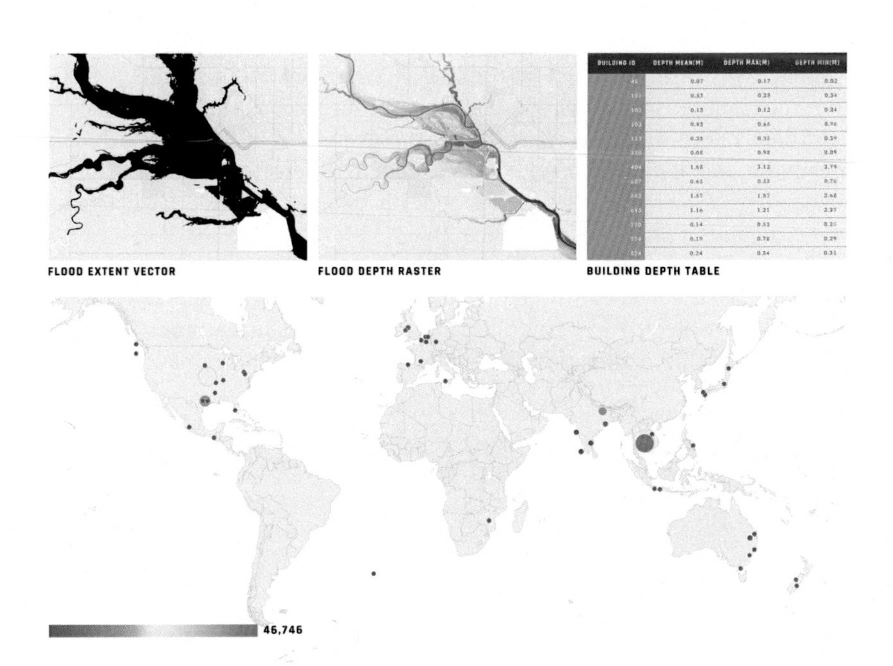

FIGURE 9.3 Three outputs are generated per flood event (flood extent, flood depth, and building depth table) (top panel) and the flood event distribution globally, with the green color representing the largest flooded area mapped (km^2) (bottom panel). *From Ardila, J., Laurila, P., Kourkouli, P., Strong, S. (2022). Persistent monitoring and mapping of floods globally based on the ICEYE SAR Imaging constellation.*

ICEYE SAR satellites and data demonstrated efficient targeting as well as mapping of a big variety of flood types, covering different topographies and land cover types (Ardila et al., 2022). The first results are very promising; however, new methodologies are needed to enhance the outputs, considering different terrain conditions, flood types, and hydrological models.

9.3.2 Wildfire monitoring

Climate change and especially extreme weather are increasing the occurrence of wildfires globally. Wildfires are often caused by human activity or a natural phenomenon such as lightning, and they can happen at any time or anywhere. In 50% of wildfires recorded, it is not known how they started (World Health Organization, n.d.).

Wildfires can cause loss of properties, lives, crops, forests, and animals. Similar to floods, near-time monitoring of wildfires is crucial for situational awareness, allocating resources, and ensuring a fast response. In addition, insurance companies need timely information on property losses for financial allocation after a wildfire event.

In the present section, we are going to demonstrate a wildfire case study using ICEYE Strip data. In July 2021, an enormous wildfire called Dixie Fire took place in California. Dixie Fire recorded the largest one in California's history. The fire damaged several small towns including Greenville and Plumas County. ICEYE acquired daily ground track repeat images over Greenville, on the 4th and 5th of August 2021. Both images were acquired in Strip mode (Table 9.2).

To ensure capturing damages caused by the crisis event, at least one pair of images is needed. However, the more the preimagery, the fewer false positives are expected to be introduced to the final damage map. The logic behind this is the assumption that nothing has changed dramatically on a

TABLE 9.2 ICEYE Strip mode characteristics.

Frequency	X-band
Polarization	VV
Imaging mode	Strip
Incidence angle range	15–35 degrees
Ground spatial resolution	2.5 m
Nominal swath width	30 km

property level before a catastrophic event. Thus, when comparing a preimage with a postimage, the change in the urban cell, after a natural catastrophe, should be attributed mainly to the damages caused by the event. To detect such changes, interferometric coherence can be used. Coherence refers to the amplitude of the complex correlation coefficient between two SAR images (see equation), indicating values between 0 and 1. High coherence values—close to 1—are expected to show no change, and low values—close to 0—represent dramatic change between two dates.

$|\gamma| = \frac{|E\{S_1 S_2^*\}|}{\sqrt{E\{|S_1|^2\}E\{|S_2|^2\}}}$. where S_1 and S_2 represent SAR images, * denotes complex conjugate, and $E(x)$ represents the ensemble average.

For the Dixie wildfire case, a preimage was acquired on August 4, 2021 and a postimage on Aug 5, 2021. For estimating coherence, we ensured a coregistration with subpixel level accuracy. Then, the interferometric coherence output is generated (Fig. 9.4). To indicate the changes due to the wildfire, we set a threshold of 0.3 to mask out the areas with the most dramatic changes. Last, the wildfire footprint combined with the building footprints were used to extract the damaged and undamaged properties (Fig. 9.5). For validation, FEMA conducted a field damage inspection and provided a map with the severity of damages per building. This preliminary result has 83% accuracy, 96% recall, and 76% precision. The high number of false positives is reasonable as coherence is very sensitive to changes.

Preliminary results with ICEYE data on wildfire monitoring with just a pair of coherent images led to an accurate extent of the burnt area. In addition, the current approach resulted in high accuracy in estimating the number of damaged and undamaged properties. However, more investigation needs to be done for understanding false positives and false negatives by introducing auxiliary data, for example, land cover maps.

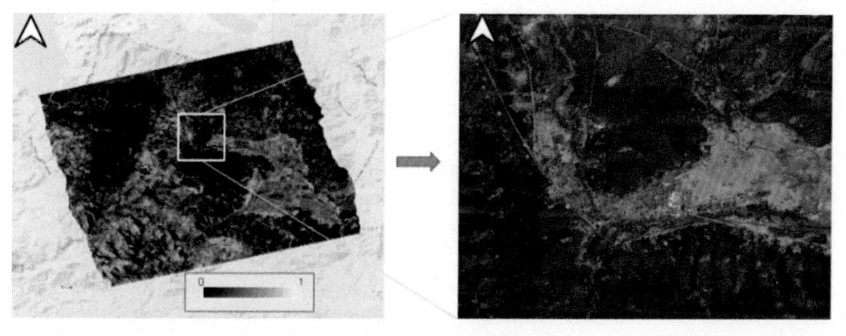

FIGURE 9.4 Coherence output of Dixie wildfire event with a yellow box indicating Greenville town (left). The red color indicates the wildfire footprint over Greenville town (right)

FIGURE 9.5 Accuracy performance of building damage assessment over Greenville indicating true positives (TP), true negatives (TN), false positives (FP), and false negatives (FN).

9.4 Conclusions

Using the ICEYE constellation of SAR small satellites, we demonstrated the capability to monitor natural disasters, at a time with very high resolution, and provide actionable information on disasters such as floods and wildfires, in urban, agricultural, forests, and vegetated areas. The agility and the tasking efficiency enable ICEYE satellites to acquire images of impacted areas before and after an event. The existence of other modes, Spot and Scan, is likely to improve the accuracy and coverage of large-scale events.

References

Ardila, J., Laurila, P., Kourkouli, P., & Strong, S. (2022). *Persistent monitoring and mapping of floods globally based on the ICEYE SAR Imaging constellation.*

Barrett, E. C., & Curtis, L. F. (1999). *Introduction to environmental remote sensing.*

EM-DAT. (2021). Unpublished content EM-DAT Report. 2021 Disasters in numbers: extreme events defining our lives. Available from: https://cred.be/sites/default/files/2021_EMDAT_report.pdf.

Cartalis, C., & Feidas, C. (2006). *Principles and applications of remote sensing*. Giourdas.

ICEYE Oy. (2021). *The Claims Landscape*. Available from: https://www.iceye.com/solutions/insurance-industry.

World Health Organization. Wildfires. Available from: https://www.who.int/health-topics/wildfires#tab = tab_1

Chapter 10

Oil spill detection using optical remote sensing images and machine learning approaches (case study: Persian Gulf)

Nadia Abbaszadeh Tehrani[1], Milad Janalipour[1] and Farzaneh Shami[2]
[1]*Aerospace Research Institute, Ministry of Science, Research and Technology, Tehran, Iran,*
[2]*Department of Surveying Engineering, Islamic Azad University, South Tehran Branch, Tehran, Iran*

10.1 Introduction

According to reports, a large part of the world's population lives in coastal areas (Creel, 2003). There are some important factors that can damage coastal ecosystems, including plastic debris, sewage, effluents, oil spills, and nonpoint sources (Vikas & Dwarakish, 2015). The oil spill is one of the major outcomes of human activities that can damage coastal ecosystems, in particular, its biological zone includes Zooplankton, fish, and benthic (ITOPF, 2011; Teal & Howarth, 1984). Some factors such as shipping routes, offshore installations, and tanker accidents can cause oil pollution in the coastal areas (Brekke & Solberg, 2005). Remote sensing (RS) is the science and technology of extracting information from earth's objects without direct contact with them (Schowengerdt, 2006). It is applicable for different studies such as forestry, agriculture, urban, water management, and so on (Pohl & Van Genderen, 1998). Due to good revisit time, digital format, different sensors, and synoptic view, oil spill detection using space-borne RS data is very cost-effective in terms of time and cost. Different sensors can be used to detect oil spills and their characteristics (more details can be found in Brekke & Solberg, 2005; Fingas & Brown, 2011, 2014; Fingas & Brown, 1997). In developing countries, it is possible to use free space-borne RS data including optical and radar images for detecting oil spills in coastal areas. Radar data can be useful for all weather conditions, but its speckle noise may affect on the accuracy of oil spill detection (Liu et al., 2019; Liu et al., 2021; Richards, 2009). In contrast, optical images can be suitable for monitoring oil spills in good weather conditions.

Geoinformatics for Geosciences. DOI: https://doi.org/10.1016/B978-0-323-98983-1.00011-9

Therefore, in this study, we focus on optical RS images and their application in monitoring oil spills in the Persian Gulf. Hu et al. (2021) discussed the spectral behavior of oil slick and the limitations of detecting it from optical images. Based on outputs, oil spills can be detected from optical RS images using their spectral behavior. Mohammadi et al. (2021) examined an object-based image analysis method to detect oil spills from Sentinel-1 and Sentinel-2 images. They concluded that optical image is more robust than radar one to detect oil spills (Mohammadi et al., 2021). Li et al. (2021) employed four machine learning methods including random forest (RF), support vector machine (SVM), and two deep learning methods to identify oil spill types from hyperspectral images. They concluded that the deep learning method is more accurate than the other ones (Li et al., 2021). Seydi et al. (2021) proposed an oil spill detection method based on a deep learning approach and the Landsat-5 image. The proposed method can detect oil spills with an overall accuracy of 95% (Seydi et al., 2021). Rajendran et al. (2021) proposed some indicators to identify the oil spill from Sentinel-2 images. They concluded that bands 2, 3, and 4 are more appropriate for this purpose (Rajendran et al., 2021). Kolokoussis and Karathanassi (2018) proposed an object-based image analysis method to extract the oil spill from the Sentinel-2 image. Talebpour et al. (2018) detected oil spills from normalized difference water index obtained from Sentinel-2 image and thresholding method. In this study, a framework was introduced to extract oil spills using space-borne optical images. Two machine learning approaches including SVM and RF are employed. Moreover, sensitivity analysis of parameters of the machine learning approaches was performed to find optimized parameters. Furthermore, the capabilities of Landsat-8 and Sentinel-2 images to detect oil spills were compared.

10.2 Study area and data used

According to Fig. 10.1, the South Pars region in the Persian Gulf was selected as a study area, since an oil slick was observed in the region due to the activity of the drilling rig. The study area location is 26°45′40″ N and 52°16′43″ E. After the event, Landsat-8 and Sentinel-2 images were acquired for two dates: 2016-02-15 and 2016-02-15, respectively.

10.3 Methodology

The proposed method was performed in four stages on Sentinel-2 and Landsat-8 bands separately. In the first stage, radiometric corrections were applied to Landsat-8 or Sentinel-2 bands. In the second stage, some features were extracted from Gray Level Co-occurrence Matrix (GLCM), minimum noise fraction (MNF), and independent component analysis (IC) approaches, and appropriate features were selected based on visual analysis. In the third

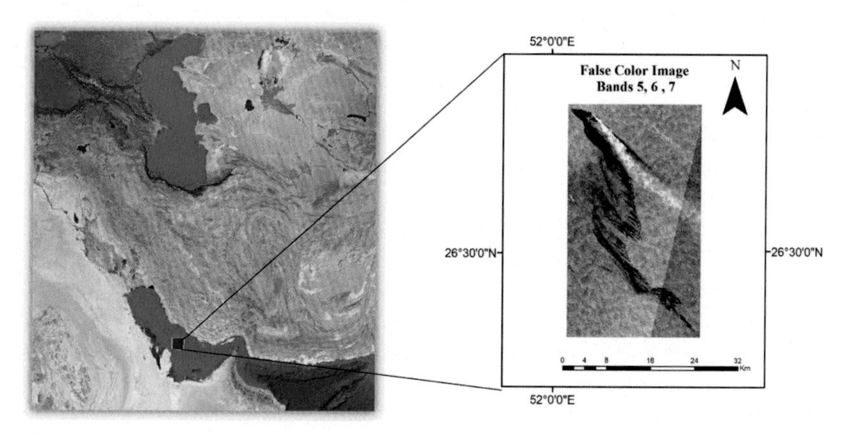

FIGURE 10.1 Case study. The study area and false color image used in the study permitting the Sentinel-2 image of the study area used in this research.

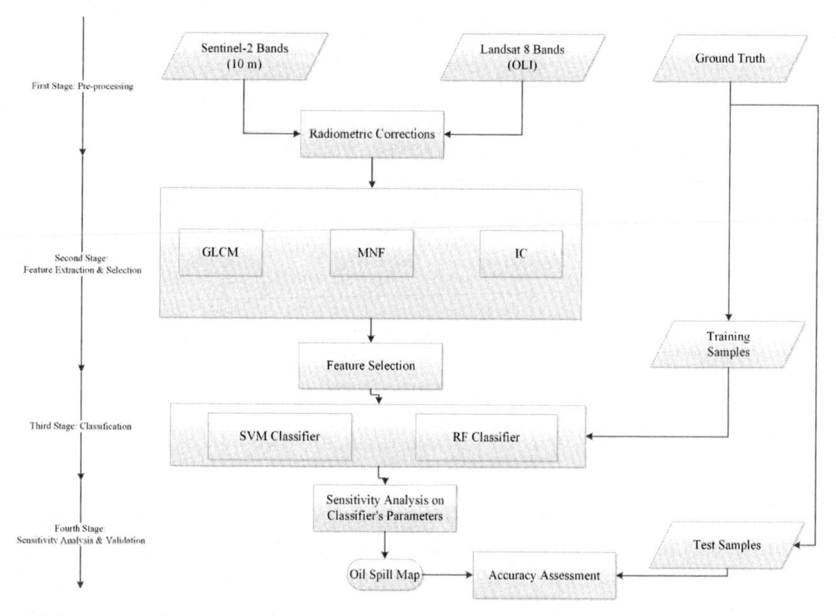

FIGURE 10.2 The proposed method. The workflow of the proposed method.

stage, the oil spill and water body were classified from the specified features using SVM and RF classifiers. For analyzing the robustness of the results, a sensitivity analysis regarding the classifier's parameters was performed in the fourth stage. Finally, the oil spill map was assessed using test samples. The flowchart of the study is presented in Fig. 10.2.

10.3.1 First stage: preprocessing

Preprocessing steps for images of Sentinel-2 and Landsat-8 were performed separately. For preprocessing the Landsat-8 image, the digital number (*DN*) was converted into radiance (*R*) value, using gain and offset coefficients according to Eq. (10.1) (Chander et al., 2007).

$$R = \text{gain} \times DN + \text{offset} \tag{10.1}$$

For preprocessing the Sentinel-2 image, at first, geo-referencing was performed on all bands using IMPACT software. Then, DN was converted into surface reflectance using Eq. (10.2). Afterward, by using Quick Atmospheric Correction (QUAC), the atmospheric correction was applied to visible and infrared bands (Bernstein et al., 2012).

$$X = \begin{cases} 0 \text{if} DN < 0 \\ 1 \text{if} DN > 10,000 \\ \dfrac{DN}{10,000} \text{if} 0 < DN < 10000 \end{cases} \tag{10.2}$$

10.3.2 Second stage: feature extraction

In addition to spectral bands, to separate oil spills from water bodies, three feature extraction methods including GLCM, MNF, and IC based on their capability were selected to extract proper features. GLCM is a powerful texture extraction method that was widely used in different applications (Huang et al., 2014; Marceau et al., 1990; Zhang et al., 2017). In this method, at first, a probability matrix was computed and then texture features were extracted from the related equations. For more details, study Haralick et al. (1973) and Theodoridis and Koutroumbas (2009).

MNF transformation is a feature extraction method that converts the feature space into a new one by maximizing the signal-to-noise ratio (Xia et al., 2013). It can be used to create independent features with minimum noise. For this reason, we employed it for feature extraction in this study (Green et al., 1988; Joseph, 1994).

IC transformation on multispectral or hyperspectral RS images can be used to transform a set of mixed, random signals into bands that are mutually independent. IC is based on the non-Gaussian assumption of the independent sources and uses higher-order statistics to reveal interesting features in RS data sets (Hyvarinen, 1999; Hyvärinen & Oja, 2000). For feature selection, a visual analysis of the extracted features was done by an expert to select the most appropriate features.

10.3.3 Third stage: classification

Classification is a common method of identifying targets in RS. SVM and RF classifiers were used to separate oil spills from water in RS data sets in this study since they are the most applicable and highly accurate approaches in RS data analysis (Belgiu & Drăguţ, 2016; Janalipour & Mohammadzadeh, 2018; Mountrakis et al., 2011).

A surface (optimal hyperplane) was employed in the SVM classifier to separate classes from each other in the feature space. SVM maximizes the margin between the classes in an optimization procedure. Training samples are essential for defining the optimal hyperplane. It is possible to use different surfaces such as linear, polynomial, and so on in SVM (Chih-Wei et al., 2003; Richards & Richards, 1999). Nonlinear decision surfaces have some unknown parameters, which must be set by users. In this study, an SVM classifier with a radial basis function was used and its parameters were set in the sensitivity analysis stage.

RF is another classifier that was used in this study. RF is a decision tree-type classifier. It randomly creates some trees in a forest and classifies the input image with each of them. Then, the results of trees are integrated using majority analysis. In fact, RF benefits some classifiers instead of one. The most important parameter of RF is the number of trees which is set in the sensitivity analysis stage (Belgiu & Drăguţ, 2016).

10.3.4 Fourth stage: sensitivity analysis and validation

For sensitivity analysis, parameters of RF and SVM classifiers were changed in a grid partitioning form to classify features obtained from the study area (Janalipour & Taleai, 2017). The best parameters were selected according to the accuracy of the classification result. The overall accuracy extracted from the confusion matrix was utilized to do the sensitivity analysis. For validation, a number of test samples from oil and water classes were obtained by an expert.

10.4 Results and discussion

A false-color Sentinel-2 image was presented in Fig. 10.1. According to the visual interpretation, it seems that an oil slick can be detected in these images. After the first stage, appropriate features were selected visually according to Tables 10.1 and 10.2. According to the results, the mean of bands was the best feature for the oil slick detection. Moreover, some features of IC and MNF were appropriate for the purpose of this study.

For applying classification methods and validation, training and test samples were employed. According to Fig. 10.3, the training and test samples were collected by an expert from the images. The number of test samples was higher than the training ones to have a comprehensive analysis. To this

TABLE 10.1 Appropriate features selected from Sentinel-2 image.

Bands	B1	B2	B3	B4	B5	B6	B7	B8	B9	B10	B11	B12
Data range	-	-	-	-	-	-	-	-	-	-	-	-
Mean	√	√	√	√	√	√	√	√	√	-	√	√
Variance	-	-	-	-	-	-	-	-	-	-	-	-
Entropy	-	-	-	-	-	-	-	-	-	-	-	-
Skewness	-	-	-	-	-	-	-	-	-	-	-	-
IC	√	-	-	-	-	-	-	-	-	-	-	-
MNF		√	-	-	-	-	-	-	-	-	-	-

TABLE 10.2 Appropriate features selected from Landsat-8 image.

Bands	B1	B2	B3	B4	B5	B6	B7
Data range	-	-	-	-	-	-	-
Mean	-	-	-	√	√	√	√
Variance	-	-	-	-	-	-	-
Entropy	-	-	-	-	-	-	-
Skewness	-	-	-	-	-	-	-
IC	-	-	-	-	-	-	-
MNF	-	-	-	√	-	-	-

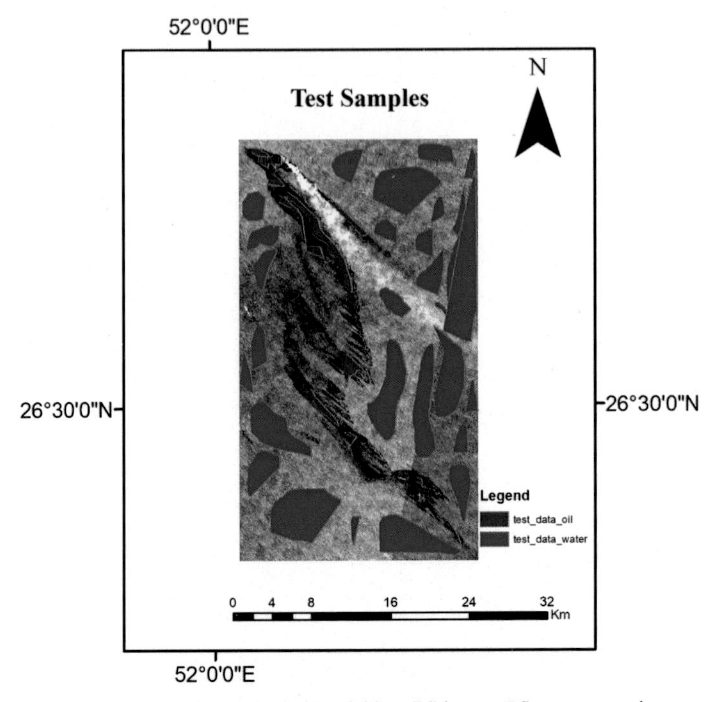

FIGURE 10.3 Training and test samples overlaid on RS image. *RS*, remote sensing.

end, 667 oil spills and 8040 water samples for training, and 6668 oil spills and 80397 water samples for testing were selected.

Sensitivity analysis on SVM's parameters is presented here. Table 10.3 presents the effect of gamma and *C* parameters on the overall accuracy of the SVM classifier applied to Sentinel-2 data. According to this table, *C* was an effective parameter for oil slick detection. The overall accuracy of the SVM classifier for the *C* parameters of 0.031 and 128 was, respectively, 98.53% and 99.38%. Therefore it is necessary to set it to achieve more robust results. According to Table 10.3, increasing the *C* parameter can improve the overall accuracy of Sentinel-2 classification in detecting oil slicks. Based on Table 10.4, the *C* parameter was also effective in the results of Landsat-8 image classification. In comparison with Landsat-8, it seems the accuracy of the SVM classifier in detecting oil slicks from Sentinel-2 was more accurate. For example, the best overall accuracies for Landsat-8 and Sentinel-2 were 98.98% and 99.38%, respectively.

Similar to the SVM classifier, a sensitivity analysis was performed on a number of trees in the RF classifier. Table 10.4 shows the effect of a number of trees on the overall accuracy of Sentinel-2 and Landsat-8 images. According to this table, it seems that the output of RF was not more sensitive to the number of trees. Unlike the output of the SVM classifier, the accuracy

TABLE 10.3 Sensitivity analysis on parameters of SVM classifier applied on Sentinel-2 (S2) and Landsat-8 (L8).

Images	S2	L8	S2	L8	S2	L8	S2	L8	S2	L8	S2	L8
16	98.53	98.47	98.89	98.55	98.97	98.56	99.23	98.59	99.28	98.53	99.38	98.73
8	98.53	98.47	98.89	98.55	98.97	98.56	99.23	98.59	99.28	98.53	99.38	98.73
4	98.53	98.47	98.89	98.55	98.97	98.56	99.23	98.59	99.28	98.53	99.38	98.73
2	98.53	98.47	98.89	98.55	98.97	98.56	99.23	98.59	99.28	98.53	99.38	98.73
1	98.53	98.47	98.89	98.55	98.97	98.56	99.23	98.59	99.28	98.53	99.38	98.73
0.5	98.53	98.47	98.89	98.55	98.97	98.56	99.23	98.59	99.28	98.53	99.38	98.73
0.25	98.53	98.47	98.89	98.55	98.97	98.56	99.23	98.59	99.28	98.53	99.38	98.73
0.125	98.53	98.47	98.89	98.55	98.97	98.56	99.23	98.59	99.28	98.53	99.38	98.73
0.062	98.53	98.47	98.89	98.55	98.97	98.56	99.23	98.59	99.28	98.53	99.38	98.73
0.031	98.53	98.47	98.89	98.55	98.97	98.56	99.23	98.59	99.28	98.53	99.38	98.73
Gamma/*C*	0.031		0.5		1		16		64		128	

TABLE 10.4 Sensitivity analysis on parameters of RF classifier applied on Landsat-8 and Sentinel-2.

Sentinel-2		Landsat-8	
Number of trees	Overall accuracy (%)	Number of trees	Overall accuracy (%)
1000	99.52	1000	99.65
900	99.51	900	99.65
800	99.52	800	99.65
700	99.51	700	99.65
600	99.51	600	99.65
500	99.52	500	99.65
400	99.52	400	99.65
300	99.52	300	99.65
200	99.51	200	99.65
100	99.51	100	99.65

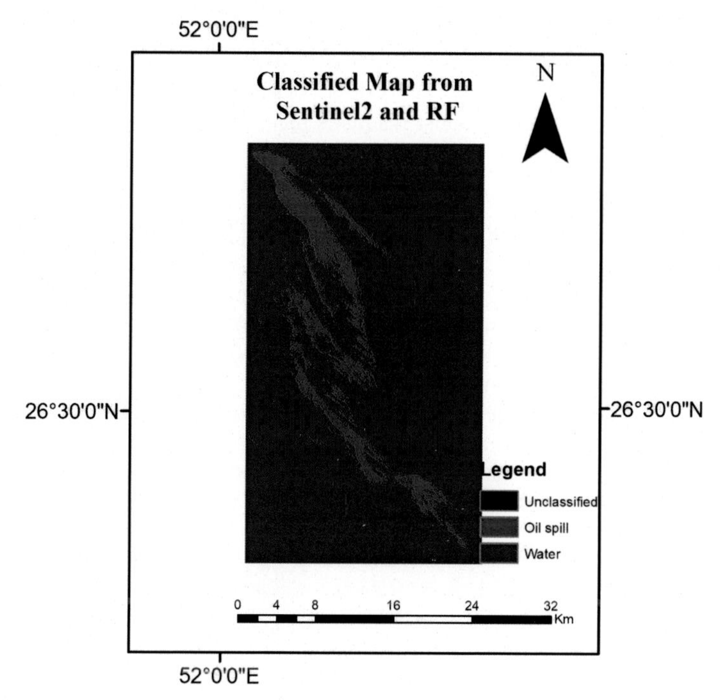

FIGURE 10.4 Classified maps obtained from SVM and RF classifiers on Sentinel-2 and Landsat-8 images, (A) Sentinel-2 and SVM, (B) Landsat-8 and SVM, (C) Sentinel-2 and RF, (D) Landsat-8 and RF. *SVM*, support vector machine; *RF*, random forest.

of the Landsat-8 image for identifying oil slicks was higher than Sentinel-2. For example, the overall accuracy of Sentinel-2 and Landsat-8 was 99.52% and 99.65%, respectively.

According to the outputs of the classifiers, it can be deduced that Sentinel-2 and Landsat-8 images and also SVM and RF classifiers can be used to extract oil slick in a robust manner. From the perspective of overall accuracy, it can be concluded that RF was more accurate than SVM classifier in identifying oil slicks. Output maps of SVM and RF classifiers are presented in Fig. 10.4.

10.5 Conclusion

In this chapter, an oil spill detection method based on two machine learning approaches including RF and SVM was presented. To detect the oil spill, Landsat-8 and Sentinel-2 optical images were employed. Based on the results, the mean of spectral bands is appropriate to feature for oil spill detection. Moreover, it seems that Sentinel-2 is more accurate than Landsat-8 for identifying oil spills. Furthermore, the overall accuracy of RF and SVM in detecting the oil spill is higher than 98% which shows their high reliability. Based on the outcomes of the sensitivity analysis stage, it seems the SVM classifier depends on its input parameters to achieve more accurate results. For future studies, it is recommended to use deep learning approaches to obtain more accurate results. Moreover, integrating optical and radar images to make more robust results is recommended.

References

Belgiu, M, & Drăguţ, L (2016). Random forest in remote sensing: A review of applications and future directions. *ISPRS Journal of Photogrammetry and Remote Sensing*, *114*, 24–31.

Bernstein, LS, Jin, X, Gregor, B, & Adler-Golden, SM (2012). Quick atmospheric correction code: Algorithm description and recent upgrades. *Optical Engineering*, *51*(11), 111719.

Brekke, C, & Solberg, AHS (2005). Oil spill detection by satellite remote sensing. *Remote Sensing of Environment*, *95*(1), 1–13.

Chander, G, Markham, BL, & Barsi, JA (2007). Revised Landsat-5 thematic mapper radiometric calibration. *IEEE Geoscience and Remote Sensing Letters*, *4*(3), 490–494.

Chih-Wei, H., Chih-Chung, C., Chih-Jen, L. (2003). A practical guide to support vector classification, Taipei. Unpublished content A practical guide to support vector classification 1651914416693

Creel, L (2003). *Ripple effects: Population and coastal regions.* Population Reference Bureau Washington, DC.

Fingas, M, & Brown, CE (2011). Oil spill remote sensing: A review. *Oil Spill Science and Technology*, 111–169.

Fingas, M, & Brown, C (2014). Review of oil spill remote sensing. *Marine Pollution Bulletin*, *83*(1), 9–23.

Fingas, MF, & Brown, CE (1997). Review of oil spill remote sensing. *Spill Science & Technology Bulletin*, *4*(4), 199–208.

Green, AA, Berman, M, Switzer, P, & Craig, MD (1988). A transformation for ordering multispectral data in terms of image quality with implications for noise removal. *IEEE Transactions on Geoscience and Remote Sensing*, *26*(1), 65–74.

Haralick, RM, Shanmugam, K, & Dinstein, Its' Hak (1973). Textural features for image classification. *IEEE Transactions on Systems, Man, and Cybernetics* (6), 610–621.

Hu, C, Lu, Y, Sun, S, & Liu, Y (2021). Optical remote sensing of oil spills in the ocean: What is really possible? *Journal of Remote Sensing, 2021.*

Huang, X, Liu, X, & Zhang, L (2014). A multichannel gray level co-occurrence matrix for multi/hyperspectral image texture representation. *Remote Sensing*, *6*(9), 8424–8445.

Hyvarinen, A (1999). Fast and robust fixed-point algorithms for independent component analysis. *IEEE Transactions on Neural Networks*, *10*(3), 626–634.

Hyvärinen, A, & Oja, E (2000). Independent component analysis: Algorithms and applications. *Neural Networks*, *13*(4-5), 411–430.

ITOPF. (2011). Effects of oil pollution on the marine environment. Unpublished content Effects of Oil Pollution on the Marine Environment, 1651907277619

Janalipour, M, & Taleai, M (2017). Building change detection after earthquake using multicriteria decision analysis based on extracted information from high spatial resolution satellite images. *International Journal of Remote Sensing*, *38*(1), 82–99.

Janalipour, M, & Mohammadzadeh, A (2018). Evaluation of effectiveness of three fuzzy systems and three texture extraction methods for building damage detection from post-event LiDAR data. *International Journal of Digital Earth*, *11*(12), 1241–1268.

Joseph, W. (1994). Proc. Tenth Thematic Conference on Geologic Remote Sensing, Environmental Research Institute of Michigan, 1407-1418, Automated spectral analysis: A geologic example using AVIRIS data, north Grapevine Mountains, Nevada. Unpublished content Automated spectral analysis: A geologic example using AVIRIS data, north Grapevine Mountains, Nevada, 1651913412868

Kolokoussis, P, & Karathanassi, V (2018). Oil spill detection and mapping using sentinel 2 imagery. *Journal of Marine Science and Engineering*, *6*(1), 4.

Li, Y, Yu, Q, Xie, M, Zhang, Z, Ma, Z, & Cao, K (2021). Identifying oil spill types based on remotely sensed reflectance spectra and multiple machine learning algorithms. *IEEE Journal of Selected Topics in Applied Earth Observations and Remote Sensing*, *14*, 9071–9078.

Liu, P, Li, Y, Xu, J, & Wang, T (2019). Oil spill extraction by X-band marine radar using texture analysis and adaptive thresholding. *Remote Sensing Letters*, *10*(6), 583–589.

Liu, P, Zhao, Y, Liu, B, Li, Y, & Chen, P (2021). Oil spill extraction from X-band marine radar images by power fitting of radar echoes. *Remote Sensing Letters*, *12*(4), 345–352.

Marceau, DJ, Howarth, PJ, Dubois, J-MM, & Gratton, DJ (1990). Evaluation of the grey-level co-occurrence matrix method for land-cover classification using SPOT imagery. *IEEE Transactions on Geoscience and Remote Sensing*, *28*(4), 513–519.

Mohammadi, M, Sharifi, A, Hosseingholizadeh, M, & Tariq, A (2021). Detection of oil pollution using SAR and optical remote sensing imagery: A case study of the Persian Gulf. *Journal of the Indian Society of Remote Sensing*, *49*(10), 2377–2385.

Mountrakis, G, Im, J, & Ogole, C (2011). Support vector machines in remote sensing: A review. *ISPRS Journal of Photogrammetry and Remote Sensing*, *66*(3), 247–259.

Pohl, C, & Van Genderen, JL (1998). Review article multisensor image fusion in remote sensing: Concepts, methods and applications. *International Journal of Remote Sensing*, *19*(5), 823–854.

Rajendran, S, Vethamony, P, Sadooni, FN, Al-Kuwari, H. Al-S, Al-Khayat, JA, Govil, H, & Nasir, S (2021). Sentinel-2 image transformation methods for mapping oil spill—A case study with Wakashio oil spill in the Indian Ocean, off Mauritius. *MethodsX*, *8*, 101327.

Richards, JA (2009). *Remote sensing with imaging radar*. Springer.

Richards, JA, & Richards, JA (1999). *Remote sensing digital image analysis*. Springer.

Schowengerdt, RA (2006). *Remote sensing: models and methods for image processing*. Elsevier.

Seydi, ST, Hasanlou, M, Amani, M, & Huang, W (2021). Oil spill detection based on multiscale multidimensional residual CNN for optical remote sensing imagery. *IEEE Journal of Selected Topics in Applied Earth Observations and Remote Sensing*, *14*, 10941–10952.

Talebpour, N, Safarrad, T, Akbarinasab, M, & Rasolian, M (2018). Investigation of proper index of oil spill detection using space-borne Sentinel-2 (case study: The Persian Gulf, 15 Feb 2016). *Journal of Oceanography*, *9*(33), 31–40.

Teal, JM, & Howarth, RW (1984). Oil spill studies: A review of ecological effects. *Environmental Management*, *8*(1), 27–43.

Theodoridis, S, & Koutroumbas, K (2009). Chapter 7—Feature generation II. *Pattern Recognition*, 411–479.

Vikas, M, & Dwarakish, GS (2015). Coastal pollution: A review. *Aquatic Procedia*, *4*, 381–388.

Xia, J, Du, P, He, X, & Chanussot, J (2013). Hyperspectral remote sensing image classification based on rotation forest. *IEEE Geoscience and Remote Sensing Letters*, *11*(1), 239–243.

Zhang, X, Cui, J, Wang, W, & Lin, C (2017). A study for texture feature extraction of high-resolution satellite images based on a direction measure and gray level co-occurrence matrix fusion algorithm. *Sensors*, *17*(7), 1474.

Chapter 11

Remote sensing and geospatial analysis

Emmanouil Oikonomou
Department of Surveying & Geoinformatics Engineering, University of West Attica, Athens, Greece

11.1 Introduction

Satellite remote sensing (RS) for the last 60 years has increasingly provided large amounts of Earth Observation (EO) data, thus contributing to major development for monitoring and understanding the climate, its changes, and several complex processes at various spatio-temporal scales occurring in the atmosphere, land, and oceans. According to the Union of Concerned Scientists which keeps a record of the operational satellites, as of May 2022, there were a total of 5465 satellites orbiting the planet, of which 1113 were dedicated to EO, although it is considered that more than half may not be even active. CEOS (Committee on EO Satellites) claims that about 201 EO satellites are currently operational belonging to about 63 different public agencies, with an additional 19 EO commercial ones.

This continuously expanding amount of satellite data demands the implementation of Big Earth Data cloud processing platforms, that have undoubtfully transformed the user community and applications of geospatial data. As a result of these advancements in space, due to computing and image-processing technology, an equivalent 10-fold increase is noted in the EO publications during the last 20 years, reaching about 6500 with over 20,000 citations in 2020 (Zhao et al., 2022). The application of EO data, however, seems to be focused on only a few sensors, with Landsat and MODIS providing about 70% of results, mainly because, on the one hand, Landsat provides the longest global EO record at medium spatial resolution and with open access since 2008 (Zhu, 2017), whereas MODIS is explicitly improved and designed for land applications. In addition, the ESA (European Space Agency) Sentinel missions, with Sentinel-1 initiated in 2014 and a further seven satellites to be launched in the next years, show the most promising trend in advanced RS applications providing short revisit cycles and

Geoinformatics for Geosciences. DOI: https://doi.org/10.1016/B978-0-323-98983-1.00012-0

relatively high spatial resolution. In particular, Sentinel-2 is considered the most appropriate mission for vegetation and forest monitoring and management, with further potential in the synergetic fusion between Sentinel-1 and 2 for natural disasters (Radočaj et al., 2020; Sentinel-2 User Handbook, 2020). These EO data provide significant new aspects mainly in the field of global Land Use/Land Cover (LU/LC) mapping, replacing the applications of previous missions, such as SPOT, TerraSAR, and AVHRR, that seem to decrease especially after around 2005.

An expanding variety of EO satellite data is now available to researchers, such as from multi and hyperspectral, microwave radiometers, spaceborne radar, and synthetic aperture radar (SAR), as well as lasers. In addition to the large multisensor platforms, the number of small Low Earth Orbiting and nanosatellites (Esper et al., 2000), as well as Unmanned Aerial Vehicles (UAVs) and the so-called RS UAV synergy have witnessed literally an explosion of uses, mainly due to the development of robotics, computer vision, and sensor miniaturization (Alvarez-Vanhard et al., 2021).

11.2 Climate change

Global climate change has become one of the most important—and at times controversial—international issues, highly related to national and financial interests and international politics and diplomacy. Large and complex amounts of EO data and interpretation analysis are demanded if a coordinated effort toward environmental protection and sustainable development is ever to be achieved. For this reason, the United Nations Framework Convention on Climate Change (UNFCCC) provides a total of 34 Essential climate variables (ECVs) that require contributions from satellite EO (Table 11.1; Hua-Dong et al., 2015; CEOS, 2012).

In addition, the Global Climate Observing System has introduced a set of Global Climate Indicators describing climate change, without confining only to sea surface temperature (SST) that for many years was the predominant observation: (1) temperature and energy (surface temperature, ocean heat), (2) atmospheric composition of CO_2, (3) ocean acidification and sea level rise (SLR), and (4) cryosphere (glaciers, Artic, and Antarctic Sea Ice extent). The above monitoring variables have shaped the main domains and advancement in EO research, as well as in achieving the sustainable development goals (Anderson et al., 2017). As a result of these in-increasing-demand satellite observations for climate change studies, Big EO data (BEOD) has become a focal issue, mainly due to the rapid development of the Internet, cloud computing, mobile, and Internet of Things, leading to the development of two major systems (Yao et al., 2020):

Cloud computing: A variety of commercial cloud development companies (e.g., Google, Amazon, and Ali Cloud) have provided the appropriate

TABLE 11.1 Essential climate variables (ECVs) that are feasible for global implementation (CEOS, 2012).

Domain		Essential climate variables
Atmospheric	Surface	Air temperature, wind speed and direction, water vapor, pressure, precipitation, and surface radiation budget
	Upper-air	Temperature, wind speed and direction, water vapor, cloud properties, and Earth radiation budget (including solar irradiance)
	Composition	Carbon dioxide, methane, and other long-lived greenhouse gases; ozone and aerosols, supported by their precursors
Oceanic	Surface	Surface sea surface temperature, sea surface salinity, sea level, sea state, sea ice, surface current, ocean color (for biological activity), carbon dioxide partial pressure, and ocean acidity
	Oceanic subsurface	Temperature, salinity, current, nutrients, carbon dioxide partial pressure, ocean acidity, oxygen, tracers, phytoplankton, marine biodiversity, and habitat properties
Terrestrial	River discharge, water use, ground water, lakes, snow cover, glaciers and ice caps, ice sheets, permafrost, albedo, land cover (including vegetation type), fraction of absorbed photosynthetically active radiation (fPAR), leaf area index (LAI), above-ground biomass, soil carbon, fire disturbance, soil moisture, terrestrial biodiversity, and habitat properties	

infrastructure and scientific computing services for handling BEOD, such as parallel mosaic and interpretation algorithms.

Discrete global grid systems: applied to management, analysis, and spatial visualization, with the indexing method based on grid encoding being the most common application mode for BEOD. In addition, Open Data Cube is also increasingly used to systematically manage and process EO data, such as monitoring environmental change and surface waters (Lewis et al., 2017).

The accumulation of hundreds of petabytes in EO climate data can only be managed in a distributed and scalable environment, by using for example on-demand flexible virtual machines. Also, clusters such as Google File System and Hadoop Distributed File System are built from standard servers and each node takes on both compute and storage functions, resulting in leveraging data location and reducing network traffic—one of the major obstacles in advanced RS EO computing applications. Today, several EO data platforms exist (Google Earth Engine, Sentinel Hub, Open Data Cube, SEPAL, JEODPP, OpenEO, and pipsCloud) (Yang et al., 2017).

In addition to climate change studies, there is the need for using RS EO data for continuous monitoring and short-term weather forecasting, which typically require ultra-fast radiative transfer models, high-resolution EO observations, and sophisticated prediction algorithms. Machine learning (ML) and deep learning (DL), combined with the above-distributed computing architectures, seem to gradually adapt to handling more complex instantaneous simulations (i.e., 25 million/min), motion flow techniques for cloud microphysics forecasting, etc. (Gomes et al., 2020). As a result, ML and DL combined with climate and weather prediction models may provide an outstandingly essential approach to forecasting extreme climatic conditions and assisting in decision-making related to many fields; agriculture, economy, tourism, energy management, and above all, human and animal healthcare (Bochenek & Ustrnul, 2022). Those new methods offer several future opportunities in RS climate change research at various spatial and temporal scales, depending strongly on the EO data availability, which over recent years has been constantly improving and increasing in volume.

11.3 Mean sea level rise, coastline changes, and remote sensing bathymetry

Since the start of the 20th century, the global mean sea level is estimated to have risen by about 16–21 cm with more than 7 cm of this occurring since 1993 (Oppenheimer et al., 2019). Coastal regions, usually characterized by high population and infrastructure densification, accommodate about 600 million people living along or in close vicinity to the coast, with this figure expected to double by 2060. Consequently, SLR is projected as a severe threat to coastal ecosystems, which are of great economic, social, and environmental importance.

Hundreds of published articles reported methods of shoreline monitoring and proposed indices and methodologies through an intercombination of historic aerial and satellite imagery of different sensor types, Global Navigation Satellite System surveys, and lately with UAVs (Apostolopoulos & Nikolakopoulos, 2021). These methods indicate a gradually decreasing rate of high-resolution satellite usage for coastline research due to the appearance of new UAV generation, especially since 2015, as a reliable alternative low-cost source and very-high accurate imagery; furthermore, Landsat images remain globally the only RS source related to the distant past and freely available. Recent global assessments of ongoing shoreline changes have confirmed that human interventions are the most obvious cause (Mentaschi et al., 2018), with SLR remaining undetectable in most temperate and tropical areas whilst in other areas coastal changes are suspected due to the combined effect of waves and subsidence, a process also detected by satellite RS (Melet et al., 2020).

Shoreline evolution may be severely influenced by or interact with SST, although the accumulated experience of retrieving SST from satellites in the coastal zone indicates that it tends to be less accurate than in the open ocean, due to a larger variability in atmospheric temperature, water vapor, and aerosol concentrations, or due to potential alteration of the ocean surface emissivity related to contaminants (O'Carroll et al., 2019). As a result, there is the need for an integrated coastal and marine management intercombining several RS techniques with field data (Mahrad El et al., 2020), and with the promising but still restricted in small areas of SST retrieval from thermal UAVs, improving EO SST quality in the coastal zone is a priority in the next decade.

Several other water quality information may be obtained nowadays from satellite sensors, including oil spill detection/pollution from radar sensors (Al-Ruzouq et al., 2020), and from optical sensors the bathymetry and surface particulate matter, as well as chemical compounds. Since the pioneering work of Lyzenga (1978) and Lee et al. (1998), these methods nowadays implement semianalytical and empirical steps to model sea-water inherent (IOPs) and apparent optical properties (AOPs) observed by the free-of-charge Landsat and Sentinel-2 multispectral satellites. Nowadays hybrid approaches implement the relationships of the calculated IOPs and AOPs are then investigated and utilized to classify, even using ML, the study area into subregions with similar water optical characteristics (Mavraeidopoulos et al., 2019). These techniques are particularly useful for regions where no environmental observations have been previously collected, or in locations where bathymetry and sediment content may change considerably according to seasons. Finally, the implementation of different supervised classifiers on satellite optical imagery can assist in accurately mapping and monitoring worldwide the distribution of seagrass species, which are precious ecosystems (Traganos & Reinartz, 2018).

11.4 Remote sensing data fusion

RS data fusion is one of the most commonly used techniques to integrate the information acquired with different spatial and spectral resolutions from sensors mounted on satellites, aircraft, and ground platforms, with the aim to produce fused data that contains more detailed information than each of the sources (Zhang, 2010). The algorithms available currently are divided into three main categories: (1) pixel-level fusion with prevailing techniques being the component substitution, the modulation-based, and the multiresolution analysis; (2) high-level fusion of optical and SAR RS; Lidar and aerial images with particular applications into Digital Surface Model/digital elevation model generation, 3D object extraction and modeling to land cover mapping; optical imagery and GIS; satellite, aerial, and close-range images.

Often the solution to data fusion is provided by the Bayes formula, which is implemented by using Kalman Smoother, (stochastic) Ensemble Kalman Filter, and error Subspace Transform Kalman Smoother (Castanedo, 2013). This approach, however, eventually leads to large dimensions and low computational speed. To increase the speed, we need evolutionary computation solutions, low-memory versions of the filter, and smoother AI algorithms or trend models, along with seasonal effects and covariates (time of measurement, depth). In addition, high-performance embedded architecture (FPGA and/or NC) has been showing encouraging results when used as a prototyping method for the design and implementation of multiclassification algorithms (e.g., MSVM) in real-time imagery, offering a potentially low-cost and reproducible solution on UAV and computer facilities (Ghamisi et al., 2019).

The PAN-sharpened satellite images (PAN) are the most detailed layer of information with use in visualization, classification, and feature extraction. There are currently many methods of fusing PAN with multispectral images, however, no best-performing method can be fixed in an absolute way as the data fusion depends on the characteristics of the scene, and several algorithms must be compared to select one providing suitable results for a defined purpose each time (Alcaras et al., 2021). The state-of-the-art in advanced data fusion indicates that the discontinuous and heterogeneous spectral and spatial characteristics of the multisource very-high-resolution PAN and multispectral images require precise registration and new pan-sharpening techniques. Although several methods have been proposed, problems still exist in computation efficiency and effectiveness. Therefore it remains necessary to create sophisticated fusion approaches that go beyond classical Bayesian estimation and can incorporate fuzzy, imprecise, or incomplete data without losing the ability to deliver statistically optimal estimates for stochastically sound quality and reliability measures (Schmitt & Zhu, 2016). The future trend is thus going far beyond the early horizon described by the combination of images for pan-sharpening or classification purposes.

11.5 Land-use–cover and agriculture

RS sensors and multiple analysis methods are used for assessing LU/LC, in environmental and hazards monitoring, such as deforestation and flooding, as well as agriculture, such as soil properties, crop classification, monitoring, and crop stress, supporting decisions on irrigation, fertilization, and pest management for production. Analytical reviews on the existing issues, latest advances, and future directions in these RS applications are presented by Radočaj et al. (2020) and Huang et al. (2018).

For RS LU/LC typically used spectral indices, namely calculated using arithmetic functions of pixel values from two or more RS spectral bands. Each spectral index aims to extract a certain type of LC or its specific property, which is usually characterized by distinct values relative to the rest of the observed area. Spectral indices are often calculated using normalized differences of reflectance in two spectral bands, with the highest and lowest sensed relative reflectance of a particular land cover property. The advantage of this approach is the repeatability of the LC/LU values and the ability to then construct time-series monitoring. As a result, a series of popular RS spectral indices have been developed: (1) direct vegetation indices: the normalized difference vegetation index, soil-adjusted vegetation index, enhanced vegetation index, normalized difference red edge, and normalized green-red vegetation index; (2) indirect vegetation indices: modified normalized difference water index, which is sensitive to water content in vegetation and soil; normalized difference water index, which was upgraded by reducing the noise of the detection of soil and vegetation water content by using green, instead of the near-infrared spectral band; normalized difference built-up index; and normalized difference soil index (Gómez et al., 2016). Extensive research experience in the last two decades by implementing various unsupervised and supervised image LC/LU classification methods has shown that, for large datasets, k-means and ISODATA algorithms are preferred, as they are less time-consuming, whereas parametric supervised classifiers appear difficult to use with multitemporal data of many and multimodal distributions.

The appearance of Sentinel-2, the increasing use and availability of new hyperspectral sensors the developments in UAV technology, and advances in computing architecture and power led ML algorithms to have completely replaced conventional algorithms of supervised classification from application in land monitoring and conservation. The superior accuracy of classification results, reduced overfitting, and computing efficiency are the main advantages of ML algorithms over conventional supervised classification algorithms (Maxwell et al., 2018). The most commonly used ML algorithms for supervised classification in RS currently are random forest, artificial neural networks, and support vector machines (Yuan et al., 2017).

The current trend of ML algorithms development in RS is deep ML algorithms (Zhu et al., 2017). Their advantage over the ML algorithms described above is the higher classification accuracy and processing capabilities of a larger quantity of training samples, but their characteristics and performance still require additional research. The implementation of these algorithms requires expensive computer hardware, primarily in terms of the graphics processing unit, which currently makes them less available for use in a broad scientific community. Regarding Precision Agriculture, despite the impressively large number of studies on RS applications, however, there is still a general lack of established techniques and/or frameworks that are accurate, reproducible, cost-effective, and applicable under a wide variety of climatic, soil, crop, and management conditions. The known multisource-data methods (satellite, UAV, ground) have accuracies that depend on image resolution, atmospheric, climatic, and weather conditions, crop, and field conditions (e.g., growth stage, land cover), and the analyses technique (e.g., regression-based, ML, physically based modeling) (Sishodia et al., 2020).

11.6 Land subsidence

Land subsidence can occur due to several physical and anthropogenic factors, such as earthquakes, floods, coastal erosion, mining, intensive cropping, and aquifer underground depletion. These phenomena can seriously damage infrastructures (bridges, roads, railways, and dams) and building stability, whereas they may result in severe economic loss of even several $ million/year. In relation to previously discussed climate change and coastline changes, land subsidence has increasingly attracted research interest with the implementation of sophisticated techniques from satellite radar sensors (Solari et al., 2018).

The DInSAR (Differential SAR Interferometry; Strozzi et al., 2001) technique and its development PSinSAR (Persistent Scatterer Interferometric Synthetic Aperture Radar) have been used for mapping the ground deformation from space, providing high-resolution and all-weather monitoring capacity. In recent years, SAR technology enabled to enhance the ability to resolve spatial distribution and temporal evolution of displacements.

So far, many SAR Interferometry methods have been proposed to detect the so-called land-subsidence troughs: Gabor transformation, Hough transformation, template recognition, convolutional neural networks, Circlet Transform, and slope analysis (Franczyk et al., 2022). These methods still exhibit limitations and different efficiency in noisy satellite images, supporting subsidence detection mainly in larger SAR images, but not yet suitable for fully automated operations. Currently, ML and DL techniques have been successfully applied to wrapped interferograms for detecting volcanic deformation, coal mining, and tunneling subsidence (Anantrasirichai et al., 2021), although further research is required to detect very localized deformation and make accurate predictions in the evolution of such phenomena.

References

Alcaras, E, Parente, C, & Vallario, A. (2021). Automation of Pan-sharpening methods for Pléiades images using GIS basic functions. *Remote Sensing*, *13*(8), 1550. Available from https://doi.org/10.3390/rs13081550.

Al-Ruzouq, R, Gibril, MBA, Shanableh, A, Kais, A, Hamed, O, Al-Mansoori, S, & Khalil, MA. (2020). Sensors, features, and machine learning for oil spill detection and monitoring: A review. *Remote Sensing*, *12*(20), 3338. Available from https://doi.org/10.3390/rs12203338.

Alvarez-Vanhard, E, Corpetti, T, & Houet, T (2021). UAV & satellite synergies for optical remote sensing applications: A literature review. *Science of Remote Sensing*, *3*.

Anantrasirichai, N., Biggs, J., Kelevitz, K., Sadeghi, Z., Wright, T., Thompson, J., Achim, A. M., & Bull, D. (2021). Detecting ground deformation in the built environment using sparse satellite InSAR data with a convolutional neural network. *IEEE Transactions on Geoscience and Remote Sensing*, *59*(4), 2940–2950. Available from https://doi.org/10.1109/TGRS.2020.3018315.

Anderson, K., Ryan, B., Sonntag, W., Kavvada, A., & Friedl, L. (2017). Earth observation in service of the 2030 Agenda for Sustainable Development. *Geo-spatial Information Science*, *20*(2), 77–96. Available from https://doi.org/10.1080/10095020.2017.1333230.

Apostolopoulos, D., & Nikolakopoulos, K. (2021). A review and meta-analysis of remote sensing data, GIS methods, materials and indices used for monitoring the coastline evolution over the last twenty years. *European Journal of Remote Sensing*, *54*(1), 240–265. Available from https://doi.org/10.1080/22797254.2021.1904293.

Bochenek, B, & Ustrnul, Z. (2022). Machine learning in weather prediction and climate analyses—Applications and perspectives. *Atmosphere*, *13*(2), 180. Available from https://doi.org/10.3390/atmos13020180.

Castanedo, F (2013). A review of data fusion techniques. *The Scientific World Journal*, 19. Article ID 704504. Available from https://doi.org/10.1155/2013/704504.

Esper, J., Panetta, P. V., Ryschkewitsch, M., Wiscombe, W., & Neeck, S. (2000). NASA-GSFC nano-satellite technology for earth science missions. *Acta Astronautica*, *46*, 287–296.

Franczyk, A, Bała, J, & Dwornik, M. (2022). Monitoring subsidence area with the use of satellite radar images and deep transfer learning. *Sensors*, *22*(20), 7931. Available from https://doi.org/10.3390/s22207931.

Ghamisi, P., et al. (2019). Multisource and multitemporal data fusion in remote sensing: A comprehensive review of the state of the art. *IEEE Geoscience and Remote Sensing Magazine*, *7*(1), 6–39. Available from https://doi.org/10.1109/MGRS.2018.2890023.

Gomes, V. C. F., Queiroz, G. R., & Ferreira, K. R. (2020). An overview of platforms for big earth observation data management and analysis. *Remote Sensing*, *12*, 1253. Available from https://doi.org/10.3390/rs12081253.

Gómez, C., White, J. C., & Wulder, M. A. (2016). Optical remotely sensed time series data for land cover classification: A review. *ISPRS Journal of Photogrammetry and Remote Sensing*, 0924-2716*Volume 116*, 55–72. Available from https://doi.org/10.1016/j.isprsjprs.2016.03.008.

Hu, B., Chen, J., & Zhang, X. (2019). Monitoring the land subsidence area in a Coastal urban area with InSAR and GNSS. *Sensors*, *19*, 3181. Available from https://doi.org/10.3390/s19143181.

Hua-Dong, G, Li, Z, & Zhu, L-W (2015). Earth observation big data for climate change research. *Advances in Climate Change Research*, *6*(2).

Huang, Y., Chen, Z., Yu, T., Huang, X., & Gu, X. (2018). Agricultural remote sensing big data: Management and applications. *Journal of Integrative Agriculture*, *17*, 1915–1931.

Lee, ZP, Carder, KL, Mobley, CD, Steward, RG, & Patch, JS. (1998). Hyperspectral remote sensing for shallow waters: Deriving bottom depths and water properties by optimization. *Applied Optics*, *38*(18), 3831–3843. Available from https://doi.org/10.1364/AO.38.003831.

Lewis, A., Lymburner, L., Purss, M. B. J., Brooke, B., Evans, B., Ip, A., Dekker, A. G., Irons, J. R., Minchin, S., & Mueller, N. (2017). Rapid, high-resolution detection of environmental change over continental scales from satellite data—The earth observation data cube. *International Journal of Digital Earth*, *9*, 106–111.

Lyzenga, DR. (1978). Passive remote sensing techniques for mapping water depth and bottom features. *Applied Optics*, *17*(3), 379–383. Available from https://doi.org/10.1364/AO.17.000379.

Mahrad El, B, Newton, A, Icely, JD, Kacimi, I, Abalansa, S, & Snoussi, M (2020). Contribution of remote sensing technologies to a Holistic coastal and marine environmental management framework: A review. *Remote Sensing*, *12*(14), 2313. Available from https://doi.org/10.3390/rs12142313.

Mavraeidopoulos, AK, Oikonomou, E, Palikaris, A, & Poulos, S. (2019). A hybrid bio-optical transformation for satellite Bathymetry modeling using Sentinel-2 imagery. *Remote Sensing*, *11*(23), 2746. Available from https://doi.org/10.3390/rs11232746.

Maxwell, A. E., Warner, T. A., & Fang, F. (2018). Implementation of machine-learning classification in remote sensing: An applied review. *International Journal of Remote Sensing*, *39*, 2784–2817.

Melet, A., Teatini, P., Le Cozannet, G., et al. (2020). Earth observations for monitoring marine coastal hazards and their drivers. *Surveys in Geophysics*, *41*, 1489–1534. Available from https://doi.org/10.1007/s10712-020-09594-5.

Mentaschi, L, Vousdoukas, MI, Pekel, JF, Voukouvalas, E, & Feyen, L. (2018). Global long-term observations of coastal erosion and accretion. *Scientific Reports*, *8*(1), 12876. Available from https://doi.org/10.1038/s41598-018-30904-w.

O'Carroll, AG, Armstrong, EM, Beggs, HM, Bouali, M, Casey, KS, Corlett, GK, Dash, P, Donlon, CJ, Gentemann, CL, Høyer, JL, Ignatov, A, Kabobah, K, Kachi, M, Kurihara, Y, Karagali, I, Maturi, E, Merchant, CJ, Marullo, S, Minnett, PJ, Pennybacker, M, Ramakrishnan, B, Ramsankaran, R, Santoleri, R, Sunder, S, Saux Picart, S, Vázquez-Cuervo, J, & Wimmer, W (2019). Observational needs of sea surface temperature. *Frontiers in Marine Science*, *6*, 420. Available from https://doi.org/10.3389/fmars.2019.00420.

Oppenheimer M., B. Glavovic,J. Hinkel, R. van de Wal, A.K. Magnan, A.Abd-Elgawad, R. Cai, M. Cifuentes Jara, R.M. Deconto, T. Ghosh, J. Hay, F. Isla, B. Marzeion, B. Meyssignac, Z. Sebesvari. 2019. Sea Level Rise and Implications for Low Lying Islands, Coasts and Communities.

Radočaj, D, Obhoaš, J, Jurišić, M, & Gašparović, M. (2020). Global open data remote sensing satellite missions for land monitoring and conservation: A review. *Land*, *9*(11), 402. Available from https://doi.org/10.3390/land9110402.

Schmitt, M., & Zhu, X. X. (2016). Data fusion and remote sensing: An ever-growing relationship. *IEEE Geoscience and Remote Sensing Magazine*, *4*(4), 6–23. Available from https://doi.org/10.1109/MGRS.2016.2561021.

Sentinel-2 User Handbook. Available online: https://earth.esa.int/documents/247904/685211/Sentinel- 2_User_Handbook (accessed on 21 September 2020).

Sishodia, RP, Ray, RL, & Singh, SK. (2020). Applications of remote sensing in precision agriculture: A review. *Remote Sensing*, *12*(19), 3136. Available from https://doi.org/10.3390/rs12193136.

Solari, L, Del Soldato, M., Bianchini, S., Ciampalini, A., Ezquerro, P., Montalti, R., Raspini, F., & Moretti, S. (2018). From ERS 1/2 to Sentinel-1: Subsidence monitoring in Italy in the last two decades. *Frontiers in Earth Science*, *6*.

Strozzi, T., Wegmuller, U., Tosi, L., Bitelli, G., & Spreckels, V. (2001). Land subsidence monitoring with differential sar interferometry. *Photogrammetric Engineering and Remote Sensing*, *67*(11), 1261–1270.

The Response of the Committee on Earth Observation Satellites (CEOS) to the Global Climate Observing System Implementation Plan 2010 (GCOS IP-10). 2012. https://unfccc.int/resource/docs/2012/smsn/igo/104.pdf.

Traganos, D., & Reinartz, P. (2018). Mapping Mediterranean seagrasses with Sentinel-2 imagery. *Marine Pollution Bulletin*, *134*, 197–209.

Yang, C, Yu, M, Hu, F, Jiang, Y, & Li, Y (2017). Utilizing cloud computing to address big geospatial data challenges. *Computers, Environment and Urban Systems*, *61*(Part B), 120–128.

Yao, X., Li, G., Xia, J., Ben, J., Cao, Q., Zhao, L., Ma, Y., Zhang, L., & Zhu, D. (2020). Enabling the big earth observation data via cloud computing and DGGS: Opportunities and challenges. *Remote Sensing*, *12*, 62. Available from https://doi.org/10.3390/rs12010062.

Yuan, H., Yang, G., Li, C., Wang, Y., Liu, J., Yu, H., Feng, H., Xu, B., Zhao, X., & Yang, X. (2017). Retrieving soybean leaf area index from unmanned aerial vehicle hyperspectral remote sensing: Analysis of RF, ANN, and SVM regression models. *Remote Sensing*, *9*, 309.

Zhang, J (2010). Multi-source remote sensing data fusion: Status and trends. *International Journal of Image and Data Fusion*, *1*(1), 5–24. Available from https://doi.org/10.1080/19479830903561035.

Zhao, Q., Yu, L., Du, Z., Peng, D., Hao, P., Zhang, Y., & Gong, P. (2022). An overview of the applications of earth observation satellite data: Impacts and future trends. *Remote Sensing*, *14*, 1863.

Zhu, X. X., Tuia, D., Mou, L., Xia, G.-S., Zhang, L., Xu, F., & Fraundorfer, F. (2017). Deep learning in remote sensing: A comprehensive review and list of resources. *IEEE Geoscience and Remote Sensing Magazine*, *5*, 8–36.

Zhu, Z. (2017). Change detection using Landsat time series: A review of frequencies, preprocessing, algorithms, and applications. *ISPRS Journal of Photogrammetry and Remote Sensing*, *130*, 370–384.

Chapter 12

Mineral exploration using multispectral and hyperspectral remote sensing data

Habes Ghrefat[1], Muheeb Awawdeh[1], Fares Howari[2] and Abdulla Al-Rawabdeh[1]

[1]Department of Earth and Environmental Sciences, Laboratory of Applied Geoinformatics, Yarmouk University, Irbid, Jordan, [2]College of Arts and Sciences, Fort Valley State University, Fort Valley, GA, United Arab Emirates

12.1 Introduction

Remote sensing (RS) is the science of acquiring, processing, and interpreting images and related data, acquired from aircraft and satellites, that record the interaction between matter and electromagnetic energy (Sabins, 1999). RS datasets are used for mineral exploration in two ways: (1) to map the geology and the faults and fractures of the region that localize ore deposits and (2) recognize hydrothermally altered rocks by their spectral signatures.

Increasing demands for minerals emphasize the need for replenishing depleting reserves by locating new prospective ore deposits. One of the most important applications of RS data for ore mineral exploration is to map and detect hydrothermal alteration minerals. Ore deposits such as orogenic gold, porphyry copper, massive sulfide, epithermal gold, podiform chromite, uranium, magnetite, and iron oxide copper-gold have been successfully prospected and discovered using RS data. Multispectral RS data such as the Advanced Spaceborne Thermal Emission and Reflection Radiometer (ASTER), Landsat data series, the Advanced Land Imager (ALI), and Worldview-3, and hyperspectral data such as Hyperion, HyMap, and the Airborne Visible/IR Image Spectrometer (AVIRIS) support cost-effective techniques for ore mineral exploration around the world. Synthetic Aperture Radar data have a high potential for structural analysis and mapping in metallogenic provinces.

Mapping of mineral resources usually covers geological mapping, structural mapping, spectral analysis, identification of hydrothermal alteration zones, mineral alteration mapping, and gold exploration. RS technology

Geoinformatics for Geosciences. DOI: https://doi.org/10.1016/B978-0-323-98983-1.00013-2

plays a vital role in mineral exploration. It is effectively decreasing initial investments and saving time by clearly targeting the most suitable locations for the occurrence of ore deposits. While remotely sensed images cannot replace direct ground observation or data derived from field and laboratory studies, they can form valuable supplements to more traditional methods and provide information and a perspective that is not otherwise available. Despite no replacement for fieldwork and other more traditional methods, RS can provide essential information from a truly new perspective. RS technology plays a vital role in the initial stages of ore mineral exploration.

Airborne and spaceborne satellites sensors used to provide informative data about a selected area or specific object for different applications, including mineral exploration. In many environments where traditional field surveys are difficult and time-consuming, the application of RS enables quick delineation of mineralized zones over wide areas in less time and cost (Goossens, 1993; Zeinalov, 2000; Carranza & Hale, 2002; Ciampalini, Garfagnoli, Antonielli, Moretti et al., 2013; Ciampalini, Garfagnoli, Antonielli, Ventisette et al., 2013; Daneshfar et al., 2006; Elsayed Zeinelabdein & El Nadi, 2014). Multispectral and hyperspectral RS data have been used in various environments for delineating structural elements that may have controlled mineralization (Moore, 1983; Unrug, 1988; van der Meer et al., 2012; Shahriari et al., 2013, 2014; Wang et al., 2017) as well as alteration zones (Abdelhamid & Rabba, 1994; Abdelsalam et al., 2000; Bennett et al., 1993; Elsayed Zeinelabdein & El Nadi, 2014; Madani et al., 2003; Madani, 2009; Kaufmann, 1988; Ramadan & Kontny, 2004). According to the technique used and the source of of the electromagnetic radiation, RS is divided into two major modes of sensing: active and passive. In active sensing such as radar, the target object is irradiated with an artificial source of electromagnetic energy and the reflected signals are collected and analyzed. The reflected natural electromagnetic radiations from the sun are collected and analyzed in passive sensing such as Landsat and ASTER data.

12.2 Spectral properties of minerals and rocks

Spectral features in the visible and near-infrared region (VNIR) ($\sim$0.4–1.0 μm) are part of the electromagnetic spectrum (EM) and are dominated by electronic processes in transition metals such as Fe, Mn, Cu, Ni, and Cr. Elements such as Si, Al, and various anion groups such as silicates, oxides, hydroxides, carbonates, and phosphates lack spectral features in the VNIR region. The presence of these elements leads to absorption bands at the appropriate wavelengths. For example, iron is the most important constituent having spectral properties in the VNIR region. The ferric-ion in Fe-O, exhibit its strong absorption of UV-blue wavelengths, due to the charge-transfer effect. This results in a steep fall-off in reflectance toward blue, and a general rise toward infrared, with the peak occurring in the 1.3–1.6 μm region.

The SWIR region (1–3 μm) is important as it is marked by spectral features of hydroxyls (OH) and carbonates (CO_3), which commonly occur in the Earth's crust. The hydroxyl ion is a widespread constituent in rock-forming minerals such as clays, micas, and chlorite. The hydroxyl ion shows a vibrational fundamental absorption band at about 2.74–2.77 μm and an overtone at 1.44 μm.

If both Mg-OH and AI-OH combinations are present, the absorption peak generally occurs at 2.3 μm and a weaker band at 2.2 μm, leading to a doublet in kaolinite minerals. On the other hand, montmorillonite and muscovite typically exhibit only one band at 2.3 μm due to Mg-OH. The broad-band absorption feature at ~2.1–2.4 μm is used to diagnose clay-rich areas.

Carbonates occur quite commonly in the Earth's crust in the form of calcite ($CaCO_3$), magnesite ($MgCO_3$), dolomite [(Ca-Mg) CO_3], and siderite ($FeCO_3$). The major carbonate absorption bands in the SWIR region occur at 1.9 μm, 2.35 μm, and 2.55 μm, produced due to combinations and overtones. The peak at 1.9 μm may interfere with that due to the water molecule at 2.35 μm with a similar feature in clays at around 2.3 μm. However, a combination of 1.9 μm and 2.35 μm and an extra feature at 2.5 μm is considered diagnostic of carbonates. Furthermore, the presence of siderite is accompanied by an electronic absorption band due to iron, occurring near 1.1 μm.

The thermal infrared (TIR) region, part of the electromagnetic spectrum, is characterized by spectral features exhibited by many rock-forming mineral groups such as silicates, carbonates, oxides, phosphates, sulfates, nitrates, nitrites, and hydroxyls. The carbonates show a weak absorption feature located around 11.3 μm. The sulfates display absorption bands near 9 μm and 16 μm. The phosphates also have fundamental features near 9.25 μm and 10.3 μm. The features in oxides usually occupy the same range (8–12 μm) as that of bands in Si-O. The nitrates have spectral features at 7.2 μm and at 8 μm and 11.8 μm. The hydroxyl ions found in the aluminum-bearing clays display fundamental vibration bands at 11 μm. The silicates (quartz, feldspars, muscovite, augite, hornblende, and olivine) which form the most abundant group of minerals in the Earth's crust display vibrational spectral features in the TIR region due to the presence of SiO_4-tetrahedron.

12.3 Characteristics of multispectral and hyperspectral remote sensing data

The RS data used in mineral exploration were acquired by various airborne and spaceborne sensors. Most multispectral sensors such as Landsat Enhanced Thematic Mapper Plus (ETM +), Landsat 8 OLI/TIR, ASTER, ALI, WordView, QuickBird, and Ikonos measure reflected energy at a few and separated wavelength bands (Table 12.1). These sensors are spaceborne and can provide global coverage with minimal cost. On the other hand, hyperspectral data such as AVIRIS and Hyperion collect reflected radiation

TABLE 12.1 Summary characteristics of selected remote sensing data used in mineral exploration.

Sensor	Subsystem	Spectral range (μm)	Spatial resolution (m)	Swath width (km)
AVIRIS	VNIR	0.38–0.900	18.5	10.5
	SWIR	0.900–2.500		
Hyperion	VNIR	0.400–0.900	30	7.7
	SWIR	0.900–2.400		
ALI	VNIR	0.480–0.690		
		0.433–0.453		
		0.450–0.515		
		0.525–0.605		
		0.633–0.690		36
		0.775–0.805	30	
		0.845–0.890		
	SWIR	1.200–1.300		
		1.550–1.750		
		2.080–2.350		
ASTER	VNIR	0.520–0.600		
		0.630–0.690		
		0.780–0.860	15	
		0.780–0.860		
	SWIR	1.600–1.700		60
		2.145–2.185		
		2.185–2.225	30	
		2.235–2.285		
		2.295–2.365		
		2.360–2.430		
	TIR	8.125–8.475	90	
		8.475–8.825		
		8.925–9.275		
		10.250–10.950		
		10.950–11.650		

(*Continued*)

TABLE 12.1 (Continued)

Sensor	Subsystem	Spectral range (μm)	Spatial resolution (m)	Swath width (km)
Landsat 7 ETM +	VNIR	0.520–0.900	15	
		0.450–0.515		
		0.525–0.605		
		0.630–0.690		185
		0.750–0.900		
	SWIR	1.550–1.750	30	
		2.090–2.350		
	Pan	0.520–0.900	15	
	TIR	10.45–12.50	60	
Landsat 8 OLI/ TIRS	VNIR	0.433–0.453	30	185
		0.450–0.515		
		0.525–0.600		
		0.630–0.680		
	SWIR	0.845–0.885	30	
		1.360–1.390		
		1.560–1.660		
		2.100–2.300		
	Pan	0.500–0.680	15	
	TIR	10.360–11.30		
		11.50–12.50	100	

at many narrow and continuous wavelength bands (Vane & Goetz, 1988). Hyperspectral instruments, also known as imaging spectrometers, are example of an airborne imaging spectrometer. According to Vane et al., the typical hyperspectral imaging spectrometers collect data in more than 200 bands with approximately 10 nm spectral resolution. These data can be used to distinguish between different targets that would otherwise appear spectrally similar in multispectral data because of their high information contents. The need for rigorous processing methods to uncover the wealth of information provided about mixtures of various surface materials is one of the disadvantages of these data. The spectral characteristics of selected multispectral and hyperspectral data are summarized below.

12.3.1 Airborne Visible/Infrared Imaging Spectrometer

The AVIRIS became operational in 1983 (Vane & Goetz 1988) and constitute of 224 bands covering the VNIR-SWIR spectrum between 0.38 and 2.5 μm. The bandwidths of AVIRIS are approximately 10 nm (Table 12.1). The instrument is flown aboard the ER-2 aircraft at an altitude of approximately 20 km above the sea level for high altitude imagery. The spatial resolution of AVIRIS data is 20 m, with swath width of 10.5 km.

12.3.2 Hyperion

Satellite-based hyperspectral imaging became a reality in November 2000 with the successful launch and operation of the Hyperion instrument aboard the EO-1 satellite (Pearlman et al., 2001). Hyperion has a spatial resolution of 30 m and spectral bandwidth of approximately 10 nm (Table 12.1). Hyperion has 242 overlapping channels (0.4–2.5 μm) with a low signal-to-noise ratio, with swath width of approximately 7.7 km by 180 km per image.

12.3.3 Advanced Land Imager

ALI is the first earth orbiting satellite of NASA's New Millennium Program and was launched on November 21, 2000 (Hearn et al., 2001; Pearlman et al., 2001). The Earth Observing 1 (EO-1) spacecraft is in a sun-synchronous orbit at 705 km altitude, flying one minute behind Landsat 7, passing over the equator in descending node at 10.01 am. ALI is designed to produce images comparable to those of Landsat 7 ETM + data. ALI measures the electromagnetic radiation through nine distinct spectral bands from 0.433 to 2.35 μm at 30-m spatial resolution. ALI has an additional panchromatic band with 10-m spatial resolution (Table 12.1) and swath width of 37 km.

12.3.4 Advanced Spaceborne Thermal Emission and Reflection Radiometer

ASTER is a joint project between the Ministry of International Trade and Industry of Japan and the US National Aeronautic and Space Administration (NASA) (Fujisada, 1995; Yamaguchi et al., 1998). ASTER was launched on the platform of Terra spacecraft of NASA in 1999 and has an extensive spectral coverage from visible to TIR region (Table 12.1). It is having three bands in the VNIR, six bands in the SWIR, and five bands in the TIR with 15, 30, and 60-m spatial resolutions, respectively. Since ASTER data are characterized by wide spectral coverage and relatively high spatial resolution, the identification of a variety of surface materials is possible. One of the major objectives of the ASTER mission is to obtain high-resolution image data in the 14 spectral bands over the entire land surface, including stereo images for orthorectification and Digital Elevation Module generation.

12.3.5 Landsat 7 Enhanced Thematic Mapper Plus

The Landsat program continuously collects multispectral optical data over the Earth's surface through a series of satellites. The program began in 1972 with the Landsat 1 and continue to acquire data till today with the Landsat 8 OLI and Landsat 7 ETM+ sensors. The Landsat 7 satellite was first launched in April 1999 as part of NASA's Earth Observing System system (Goward et al., 2001) and having eight spectral bands spanning the VNIR-SWIR and TIR spectrum with 15, 30, and 60 m spatial resolution, respectively (Table 12.1). Providing data continuity with Landsats 4 and 5 is the primary objective of the Landsat 7 mission.

12.3.6 Landsat 8 Operational Land Imager and Thermal Infrared Sensor

The Landsat 8 is an American Earth observation satellite launched on February 11, 2013, as a joint mission between NASA and the United States Geological Survey (USGS). Landsat 8 orbits Earth every 98.9 minutes at an altitude of 705 km, collects data over a 185 km swath, and covers the entire globe every 16 days. The Landsat 8 satellite carries a two-sensors payload, the Operational Land Imager (OLI) and the Thermal Infrared Sensor (TIRS) (Irons et al., 2012; Roy et al., 2014). As compared to the ETM+, the Landsat 8 OLI sensor has a blue coastal/aerosol band (band 1: 0.43–0.45 μm) and a new shortwave infrared Cirrus band (band 9: 1.36–1.39 μm) (Table 12.1) (Roy et al., 2014). According to Irons et al. (2012) and Roy et al. (2014), the optical bandwidths of the Landsat 8 data are narrower than those of the Landsat 7 ETM+. Consequently, detecting and mapping of clays and iron oxide minerals related to the hydrothermal alteration were enhanced and improved. Iron oxide minerals have strong absorption features in the near-infrared region.

12.4 Remote sensing methods used in mineral exploration

12.4.1 Minimum noise fraction transformation/principal component analysis

The minimum noise fraction (MNF) transformation process is a data reduction method designed to increase apparent signal-to-noise by estimating noise statistics from the data and segregating it to higher-order eigen channels while still retaining much of the original signal. According to Green et al. (1988), MNF is an algorithm that performs two consecutive data reduction operations. The first transformation is based on the estimation of the noise covariance matrix. This transformation decorrelates and rescales the noise in the data to unit variance with band-to-band correlation (Green et al., 1988). The second transformation is a standard principal components analysis

(PCA), which is used for noise-smoothed data. The MNF transformation is similar to PCA transformation but differs in the principal that MNF considers the noise separately from the noise, while PCA considers the overall data variation using a single covariance matrix (Richards, 1994; Smith et al., 1985). Generally, the signal (i.e., high information content) is concentrated in lower ordered eigenchannels, whereas the noisy bands are concentrated in higher ordered eigenchannels.

12.4.2 Spectral angle mapper

The spectral angle mapper (SAM) is a classification method that measures the similarity between image spectra and reference spectra collected from field or laboratory sources. SAM calculates the angle of similarity between pixel spectra and reference spectra by treating them as vectors in n-dimensional space, where n is the number of spectral bands used (Fig. 12.1). The angle is determined by measuring the arc cosine of the dot product of the two spectra. The smaller the angle between the two spectra, the higher their spectral similarity. The result of SAM classification is an image showing the best endmember match for each pixel. Additionally, SAM outputs rule images showing the actual angular distance (i.e., radians) between each spectrum in the image and each reference or endmember spectrum. One of the advantages of SAM is that it is not affected by gain factors (e.g., solar illumination) because the angle between two vectors is independent of the length or magnitude of the vector. According to Crosta et al., SAM classification results are highly dependent on the wavelength range used and on the selected thresholds governing the classification, which are arbitrary determined.

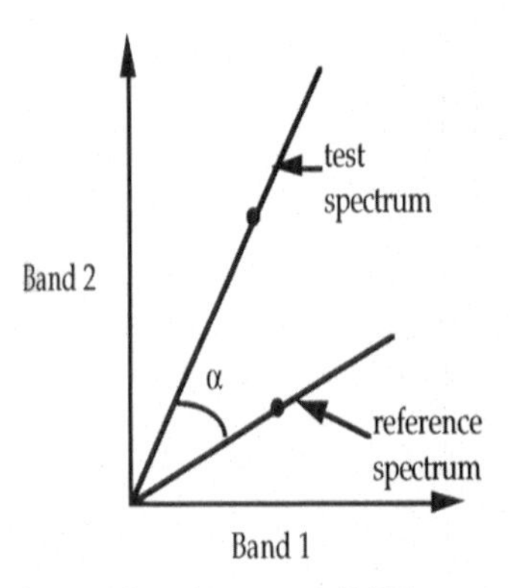

FIGURE 12.1 Illustration of spectral angle mapper (SAM) concept based on an ideal 2-band spectrum or 2-dimensional classification scenario.

12.4.3 Linear spectral unmixing

Spectral mixture analysis, or "unmixing," is an attempt to uncover the wealth of information provided by multi and hyperspectral sensors by assigning sub-pixel class membership to a wide range of cover types mixing within the scale of a pixel (Adams et al., 1993; Boardman, 1993). Spectral mixture analysis can be used to map the proportions of these cover types within each pixel of an image. For instance, the reflectance measured for each pixel is generally a linear combination of the reflectance of each individual endmember assuming they are aerially mixed present within the pixel. Also, because of unit-sum and positivity constrains, all mixing endmembers must be known prior to analysis. The number of endmembers must be no more than $n+1$ including shade, where n is the number of spectral bands. The linear mixing model can be described as (after Adams et al., 1993):

$$R_c = \sum_{i=1}^{n} F_i * R_{i,c} + E_c \tag{12.1}$$

where R_c is a pixel composite reflectance for channel "c," F_i is a fraction of endmember cover type "i," $R_{i,c}$ is a reflectance contributed by endmember "i" in channel "c," E_c is a channel "c" describing RMS fit on N spectral endmembers, c is hyperspectral or multispectral channels $>> > n$, and n is a number of spectral endmembers.

12.4.4 Matched filtering

Matched filtering (MF) is a spectral mapping procedure that maximizes the response of known endmembers and suppresses the response of the unknown composite background using least square regression methods (Boardman et al., 1995). It is partly based on foreground-background spectral unmixing and also provides a rapid means of identifying specific materials based on matches to library or image endmember spectra and does not require knowledge of all the endmembers within an image scene. MF can best be described as partial unmixing because it is not constrained by the unit sum rules of linear spectral unmixing. The results of MF are as gray-scale images with values ranging from ≤ 0 to ≥ 1, which provides a means of estimating the relative degree of match between a reference and image spectra, where ≥ 1 is the perfect fit.

12.4.5 Mixture tuned matched filtering

According to Boardman, mixture tuned matched filtering (MTMF) combines the strength of MF (i.e., no requirement for knowing all endmembers) (Boardman et al., 1995) with the unit sum and positivity constraints imposed by linear spectral unmixing (Adams et al., 1993; Boardman, 1993). The signature at any given pixel is a linear combination of the individual components contained in that pixel. Like linear spectral unmixing, it constrains the result to feasible mixtures to reduce false alarm rates. The outputs of MTMF are two sets of images: the MF score and infeasibility images. The MF scores are presented as gray-scale images with values ranging between 0 and 1. The image provides a means of estimating relative degree of match to reference spectrum, where 1 is a perfect match. The highly infeasible numbers in the infeasibility images indicates that mixing between the composite background and the target is not feasible. The best match to a target is obtained when the matched filter score is high (~ 1) and the infeasibility score is low (~ 0).

12.4.6 Band ratioing

Band ratioing is a technique used in RS to effectively display spectral variations (Goetz et al., 1985; Shalaby et al., 2010). Ratioing enhances the contrast between materials by dividing the brightness values (digital numbers) at peaks/maxima and troughs/minima in a reflectance curve after the additive atmospheric haze and additive sensor offset have been removed. Spectral band ratioing enhances the desired compositional information while suppressing other types of information, such as the terrain slope and grain size differences (Jensen, 1996; Vincent, 1997).

12.4.7 Pan sharpening

According to Johnson et al. (2012), the spectral sharpening technique is used to merge high spatial resolution panchromatic band with lower spatial resolution multispectral data to create a multispectral image with higher-resolution features.

12.5 Case studies

12.5.1 Case study 1: Mapping of gossan zones in the Eastern Arabian Shield, Saudi Arabia

The Khunayqiyah gossans (Fig. 12.2), Eastern Arabian Shield, Saudi Arabia, have been used to demonstrate the effectiveness of using the Landsat 8 OLI imagery to detect and delineate gossan zones in arid regions. The Khunayqiyah district ($\sim$900 m.a.s.l.) is located at the edge of the Paleozoic sedimentary cover. The basement of the Khunayqiyah forms an undulating physiographic relief of hills that has been transected by valleys. According to Testard et al. (1980) and Vaslet et al., the Khunayqiyah district is predominated by a sequence of felsic volcanic and volcano-sedimentary rocks interlayered with carbonates, comprising the Shalahib Formation ($\sim$1500 m thick). The Khunayqiyah

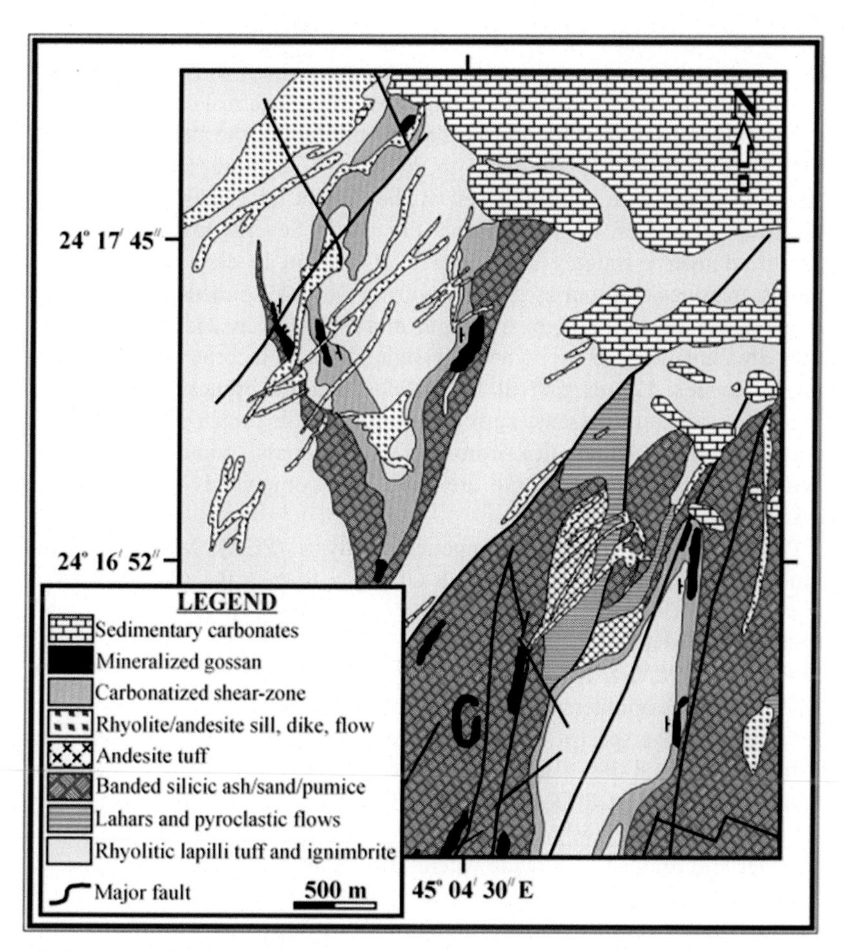

FIGURE 12.2 Geologic map of the study area (Gahlan & Ghrefat, 2018).

volcano-sedimentary rocks can be subdivided into three types of rhyolitic tuffs (BRGM Geoscientists, 1993): lapilli tuffs, rhyolitic tuffs, and vitroclastic siliceous ash-fall tuffs alternating with ignimbrite flow. The latter volcano-sedimentary succession is intruded by subvolcanic dikes, sills, and domes/stocks of rhyodacite or andesitic composition. Three types of carbonatic lithologies were identified in the Khunayqiyah area, that is, sedimentary, hydrothermal, and skarnoid by Testard (1983). Hydrothermal carbonates form discontinuous anastomosing bands along the N–S-trending shear zones. Two types of hydrothermal carbonates (Types I and II) were recognized in the study area (e.g., BRGM Geoscientists, 1993). Type I hydrothermal carbonates are foliated, large (up to 200 m in width and 3000 m in length), and contain little Mn-, Fe-, Zn-, and Cu-mineralization. In contrast, Type II hydrothermal carbonates are nonfoliated, chlorite-bearing, smaller (a few tens of meters in width and a few hundreds of meters in length), and highly Fe-, Zn-, and Cu-mineralized. Notably, the ancient workings are commonly restricted to the Type II hydrothermal carbonates.

The mineralization (Mn- and Fe-Zn-Cu-sulfides) of the study area is spatially associated with or restricted to the Type II hydrothermal carbonates. The mineralized zones are exposed at the surface as reddish brown, iron-rich, irregular bodies, termed gossans (Fig. 12.2). Several gossans have been recognized over a roughly circular area of ~4 km in diameter. Generally, the gossan zones crop out as discrete pods of different dimensions. The largest gossan is up to ~400 m in length and ~100 m in width. The gossan zones are characterized by Fe and Mn-stained lenses incorporated in crystalline carbonates. Lenses rich in Fe sulfides (mainly pyrite) were recorded within the siliceous gossans, surrounded by reddish brown (hematitic) and yellowish brown (limonitic) iron-staining materials. Limonite, goethite, hematite, malachite, and azurite are the major components of the massive gossans.

Pan sharpening, principal component analysis (PCA), MNF, and band ratio (BR) were applied to Landsat 8 OLI data to map the gossan zones in the study area (Figs. 12.3 and 12.4). The color RGB (red, green, blue) composites of pan-sharpened original bands (4, 3,2), PCA (PC3, PC2, PC1), and MNF (MNF2, MNF4, MNF3) (Fig. 12.4) were found to be the most useful to delineate gossan/alteration zones. The results revealed good correspondence between the spectra of the collected samples and image-derived spectra from Landsat 8 data (Fig. 12.5). The findings of this study demonstrate multispectral Landsat 8 OLI data, and the above used RS techniques were useful in exploring new gossan occurrences in the Arabian Nubian Shield and other arid regions worldwide where little in situ geological data exist.

12.5.2 Case study 2: Delineation of copper mineralization zones at Wadi Ham, Northern Oman Mountains, United Arab Emirates (UAE) using multispectral Landsat 8 (OLI) data

Wadi Ham which is located (Fig. 12.6) in the Northern Oman Mountains, United Arab Emirates (UAE), is an area of mainly Semail ophiolite exposures, having its base the metamorphic sole.

The Southern ophiolite zone extends south from Wadi Ham to Wadi Hatta (Fig. 12.7). The ophiolite complex is divided into two units: the Mantle sequence and Crustal sequence which consist of the cumulus and noncumulus. This zone comprises a sequence of complex north-trending belts whose composition becomes more basic from east to west.

The Layering of rocks which constitute the ophiolite suite in the southern zone shows a generally east and east northeast dip, although there are numerous local variations from this trend. The major fractures controlling many of

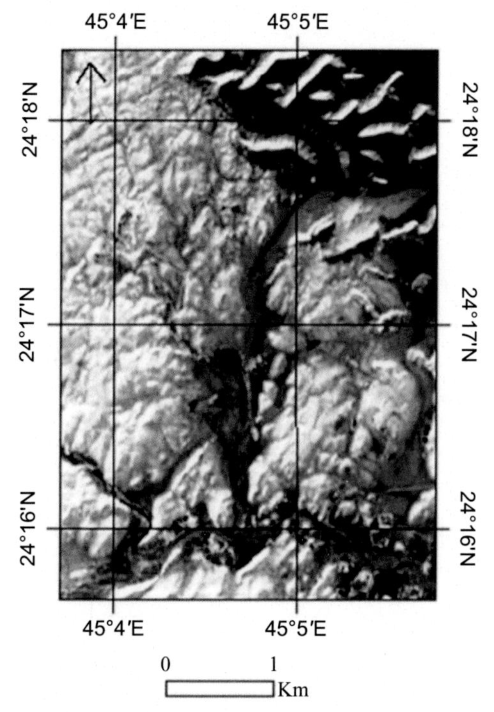

FIGURE 12.3 RGB composite of pan-sharpened band 4 (=red), band 3 (=green), and band 2 (=blue) images of Landsat 8 OLI data of the study area. Gossan zones and host rocks appear as green and whitish colors, respectively (Gahlan & Ghrefat, 2018).

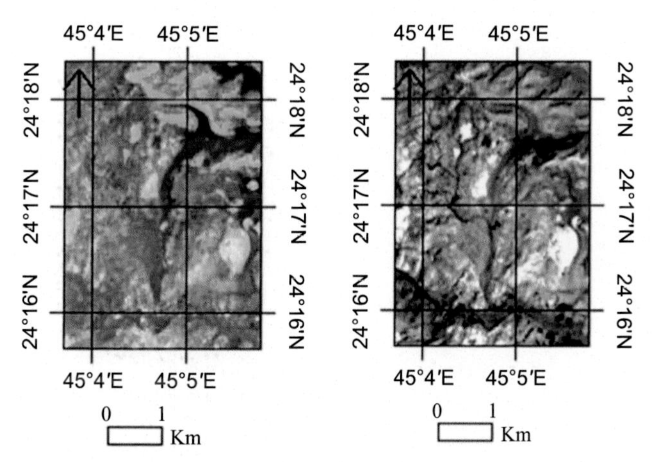

FIGURE 12.4 Color composite of MNF2 (=red), MNF4 (=green), MNF3 (=blue) (left). Gossan zones and host rocks appear as whitish blue and yellowish pink colors, respectively. Band ratio 6/5 image (right). Gossan zones appear as the bright pixels (Gahlan & Ghrefat, 2018).

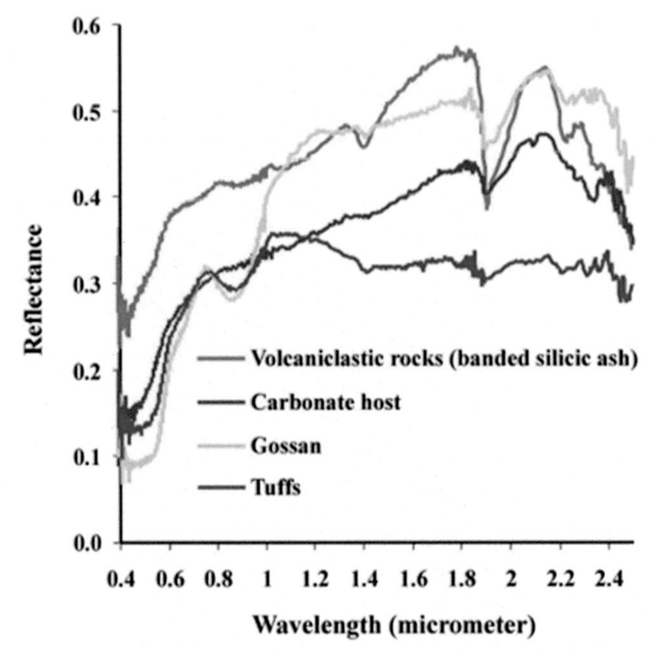

FIGURE 12.5 Measured laboratory spectra of the representative samples collected from the Khunayqiyah gossan zones and the host lithologies using the GER 3700 spectroradiometer (Gahlan & Ghrefat, 2018).

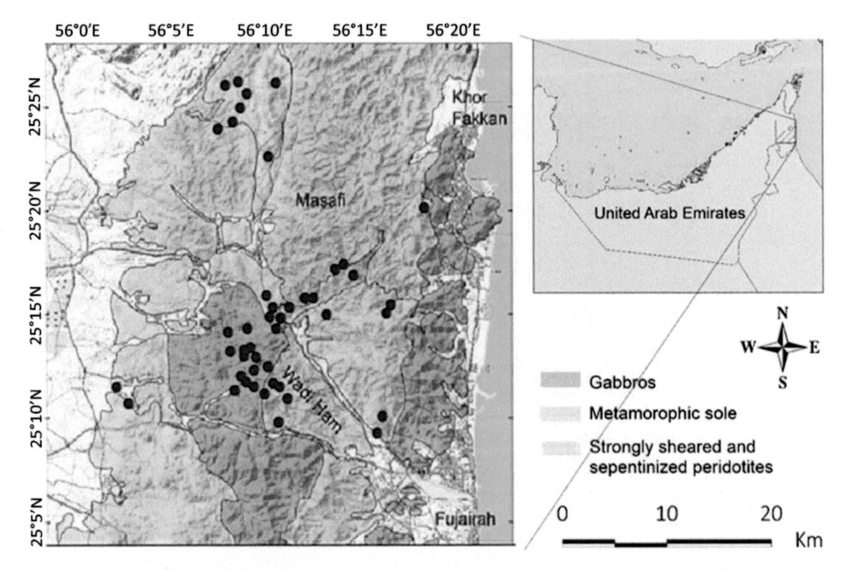

FIGURE 12.6 Geological and location map of the Wadi Ham area. The locations of copper occurrence are shown in red circles (Howari et al., 2022).

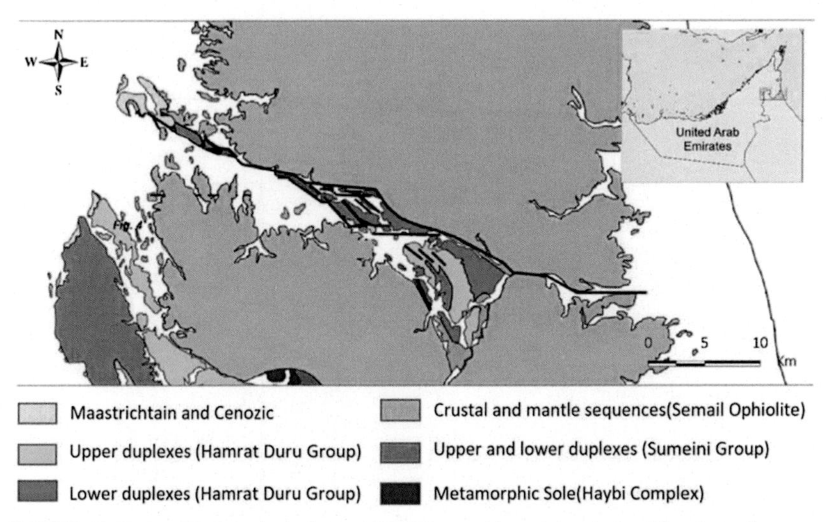

FIGURE 12.7 Modified geological map of the Hatta region, updated after Glennie et al.

the larger Wadies, for example Wadi Mudayq and Wadi Munayi, trend north west to north west, roughly parallel with the Wadi Ham zone. The principal copper mineralized zones occur in these wadies and have similar trend.

The alteration processes accompanying the introduction of copper mineralization at Wadi Ham vary widely in composition and type. Very close to mineralization, some features of sulfurization and epidotization predominates, then grading outwards into a zone of secondary Cu-mineralization (dominated by Malachite and Mn-minerals). At outer zone, other processes such as carbonization and development of secondary iron oxides. The zone of secondary copper-manganese oxides is characterized by the development of hydrated oxides of copper and manganese. The zone is remarkable by the drasting increase of Cu, Mg, Zn, and Ni abundances. The zone gossan is characterized by the remarkable reddish-brown colors of hematite and limonite. The gossan zone is particularly developed at the roof and also on the orebody's sides. The wide zone gossan is the most outer alteration zone of the Cu mineralization in the Wadi Ham area due to the high mobility and abundances of Fe in the secondary environment.

Copper deposits are widespread in the ultramafic rocks of Semail ophiolite massifs of the northern Oman Mountains, United Arab Emirates (UAE). Copper-bearing mineralized zones (Fig. 12.8) were detected and mapped using PCA, MNF, BR, and decorrelation stretch (DS) applied to Landsat 8 OLI data. The spectra of malachite and azurite samples are characterized by diagnostic absorption features in the VNIR region (0.6–1.0 μm) (Fig. 12.9). The results obtained from the PCA, MNF, BR, DS, spectral reflectance analyses, and mineralogical and chemical analyses were in good agreements.

FIGURE 12.8 False color composite images of principal components PC1, PC2, PC3 displayed in R-G-B for the study area. The dark purple colors pixels represent gabbro, peridotite is displayed in green colors, and metamorphic sole is shown in yellowish green. The locations of copper are shown in black circles.

12.5.3 Case study 3: Mapping Maqna Gypsum Deposits in the Midyan Area, Northwestern Saudi Arabia, using ASTER Thermal Data

The Maqna gypsum deposits (Fig. 12.10) have an outcrop area of approximately 270 km^2 in the Midyan area of northwestern Saudi Arabia. The area is part of the Maqna massif and is bounded west by steep cliffs along the Gulf of Aqaba and the Wadi Ifal plain to the east and south. The massif has a steep relief of several hundred meters and is incised by numerous wadis.

The deposit is part of the thick evaporite, carbonate, and clastic sedimentary Miocene Raghama Formation. The group occupies the flanks of a northeasterly oriented, 45 km long, south plunging pericline with a core of Proterozoic granitic rocks. The evaporites are found mainly in the bed formation (Middle to Late Miocene) in the uppermost part of the Raghama formation. The formation varies in thickness from approximately 150 to 300 m.

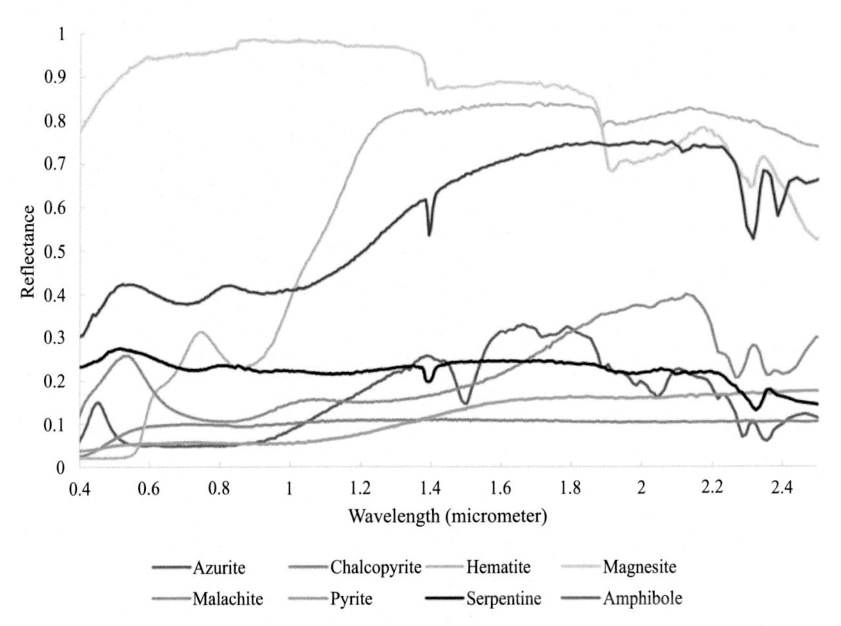

FIGURE 12.9 US Geological Survey spectral reflectance curves of selected iron, sulfur, and copper minerals (Clark, 1999).

The formation is overlain by poorly consolidated deposits of the clastic Pliocene Lisan Formation.

The GER 3700 spectroradiometer (0.3–2.5 μm) was used to measure the laboratory spectra of the samples (Fig. 12.11). ASTER (AST_05) data (8–12 μm) were used to detect and map the gypsum deposits in the Midyan area of northwestern Saudi Arabia. These data were processed using the MNF, pixel purity index, and nD visualization to derive image endmembers. The Gypsum spectrum showed diagnostic absorption features at 1.4, 1.9, 2.2, and 8.63 μm (Fig. 12.12). The spatial maps of gypsum deposits were determined using BR (band 12/11) (Fig. 12.13) and MF techniques. The results of BR and MF match well with the published geologic map and field data. The result of this study is of great economic importance in hyper-arid environments such as Saudi Arabia.

12.5.4 Case study 4: Mapping oolitic iron ore deposits in western Saudi Arabia using Landsat 7 Enhanced Thematic Mapper plus Data (ETM +)

The area of study (Fig. 12.14), a part of the Makkah Quadrangle (Moore and Al-Rehaili, 1989; Rehman et al., 2016), is located in the west-central part of

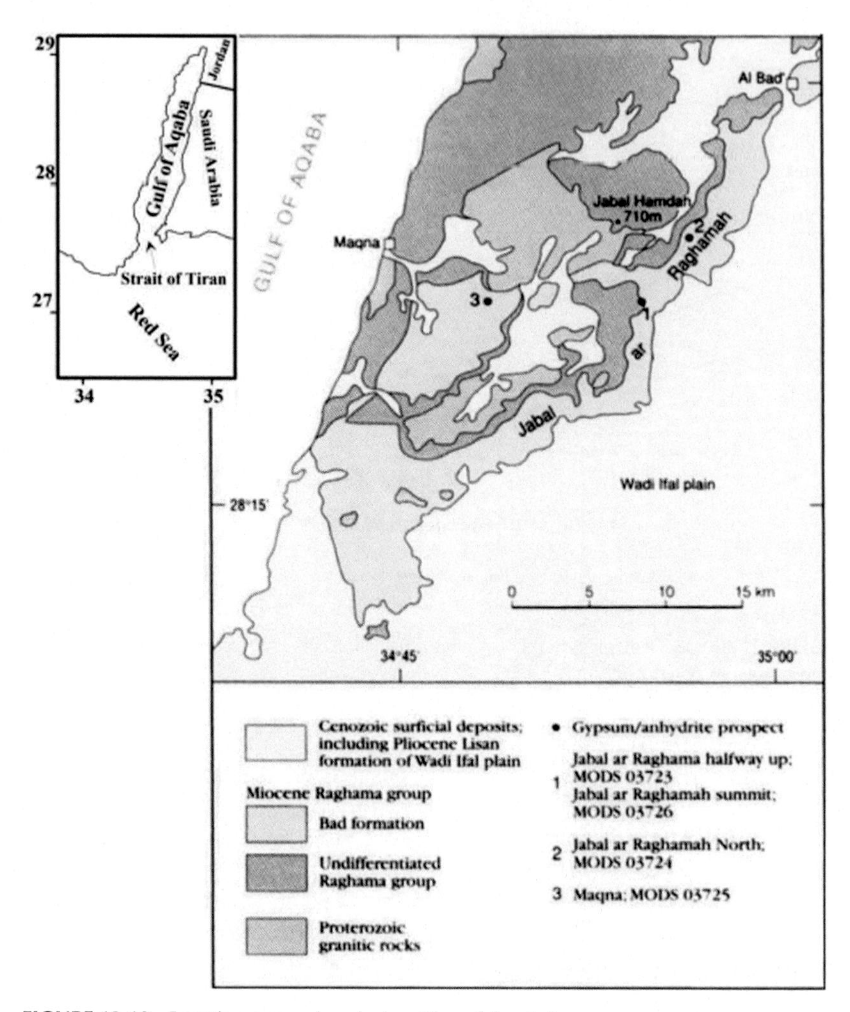

FIGURE 12.10 Location map and geologic setting of the study area.

the Arabian Shield bordering the Red Sea from the west. The study area, which comprises about 1265 km^2, lies from 21° 15′ to 21° 45′ N latitude and 39° 25′ to 39° 45′ E longitude and includes the three major cities of Makkah, Jeddah, and At Ta'if. It consists of late Precambrian layered rocks (10%) and plutonic rocks (90%) (Fig. 12.15). Tertiary sedimentary rocks dominate the western part of the study area, and the northern part is dominated by Miocene to Pliocene lavas.

In the study area, major economic mineral resources are abundant in Cenozoic sedimentary deposits. The Ashumaysi formation in the study area contains two beds of oolitic iron ore ranging in thickness from 7 to 10 feet.

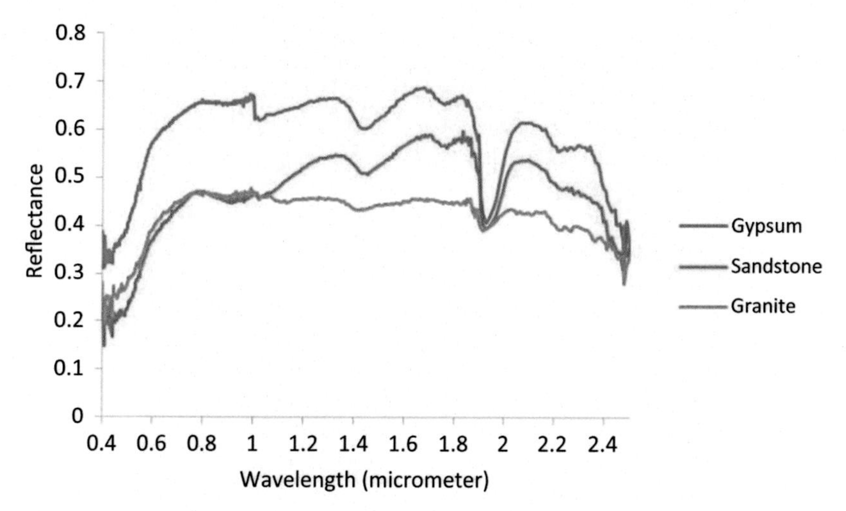

FIGURE 12.11 The laboratory measured spectral reflectance of gypsum, granite, and sandstone using the GER 3700 spectroradiometer.

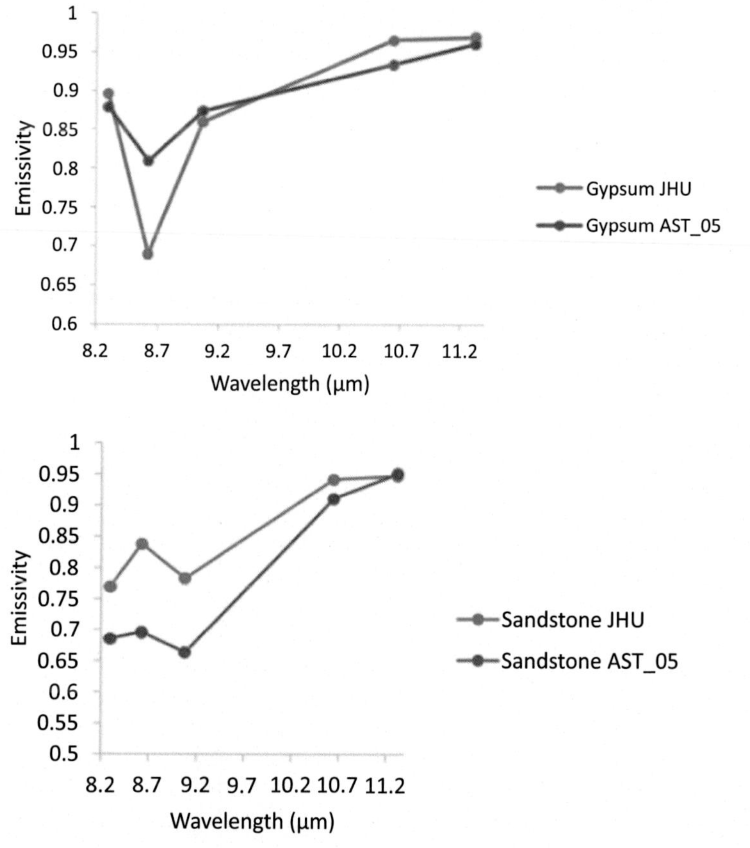

FIGURE 12.12 Emittance spectra of gypsum, granite, and sandstone.

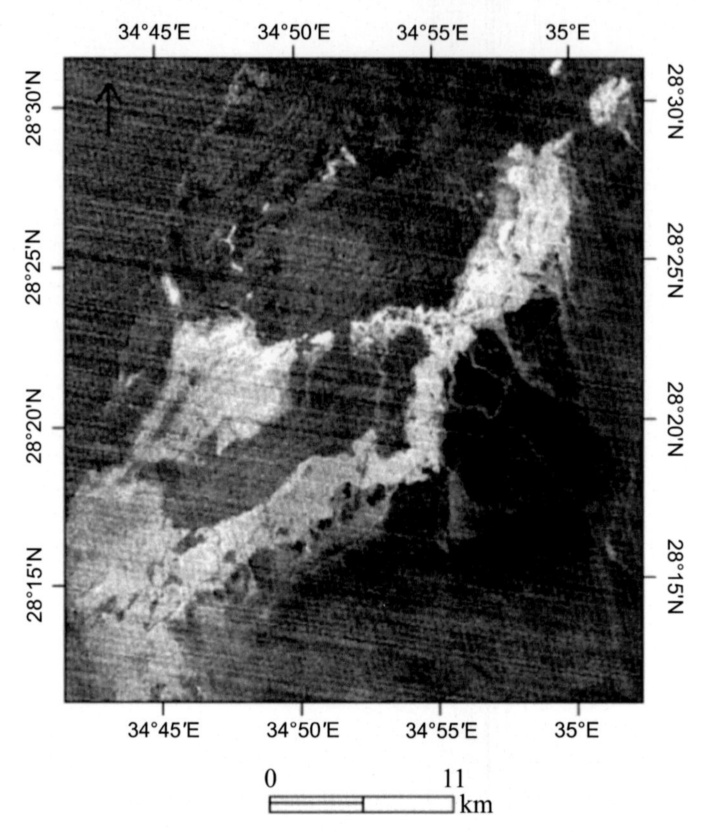

FIGURE 12.13 ASTER band ratio (12/11) of gypsum deposits (Ghrefat, 2020).

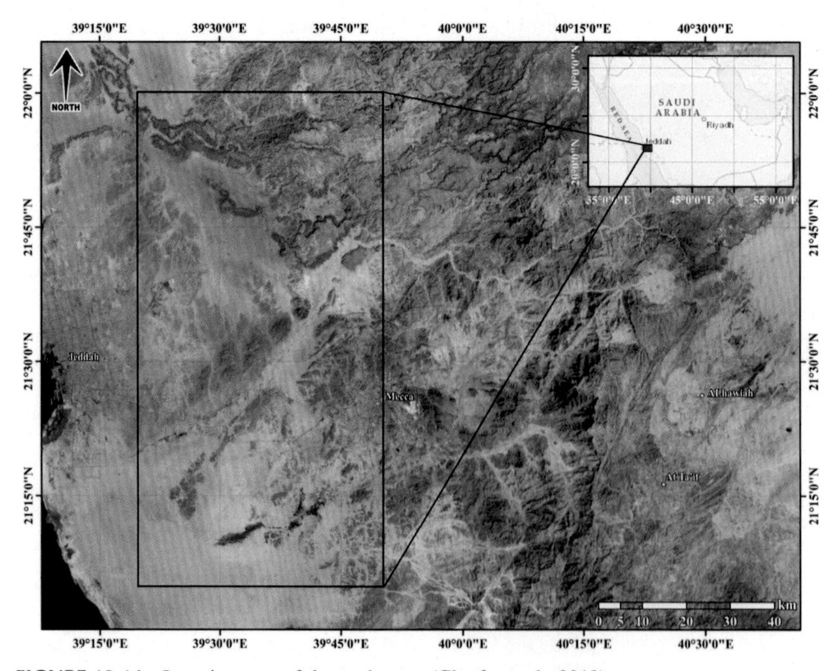

FIGURE 12.14 Location map of the study area (Ghrefat et al., 2018).

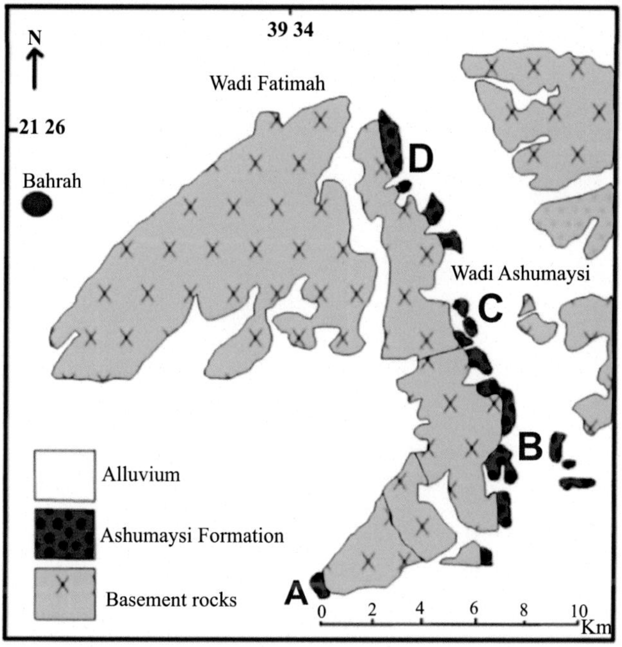

FIGURE 12.15 Geologic map of the study area (modified after Al-Shanti, 1966 and Moore & Al-Rehaili, 1989).

The total reserves in the study area are estimated to be about seventy million tons of ore. The iron contents range between 45% and 48%. The goals of the current study were to use Landsat 7 ETM + to locate iron oxide deposits in the Ashumaysi formation, western Saudi Arabia, and validate the results of different RS techniques, PCA, MNF, and supervised classification, applied to Landsat 7 ETM + data using 16 samples collected from different locations in the current study. The findings are of great importance for detecting other mineral resources in arid and semiarid environments.

Mapping of oolitic iron ore deposits hosted in the Oligo-Miocene sedimentary rocks of the Ashumaysi formation, western Saudi Arabia, was carried out using Landsat 7 ETM + data. Ore samples were collected from six various locations in the study area and were subjected to the laboratory using the GER 3700 Spectroradiometer (0.4–2.5 μm) (Fig. 12.16) and X-ray diffraction (XRD). PCA, MNF, and minimum distance classification were used and assessed to map mineralization zones in the study area (Fig. 12.17). Good correspondences were observed between the results obtained from PCA, MNF, supervised classification, spectral reflectance analyses, and XRD. Confusion matrix results revealed that mapping of iron ores using MNF is better and more accurate than using PCA. Good matching was observed between the spectral reflectance

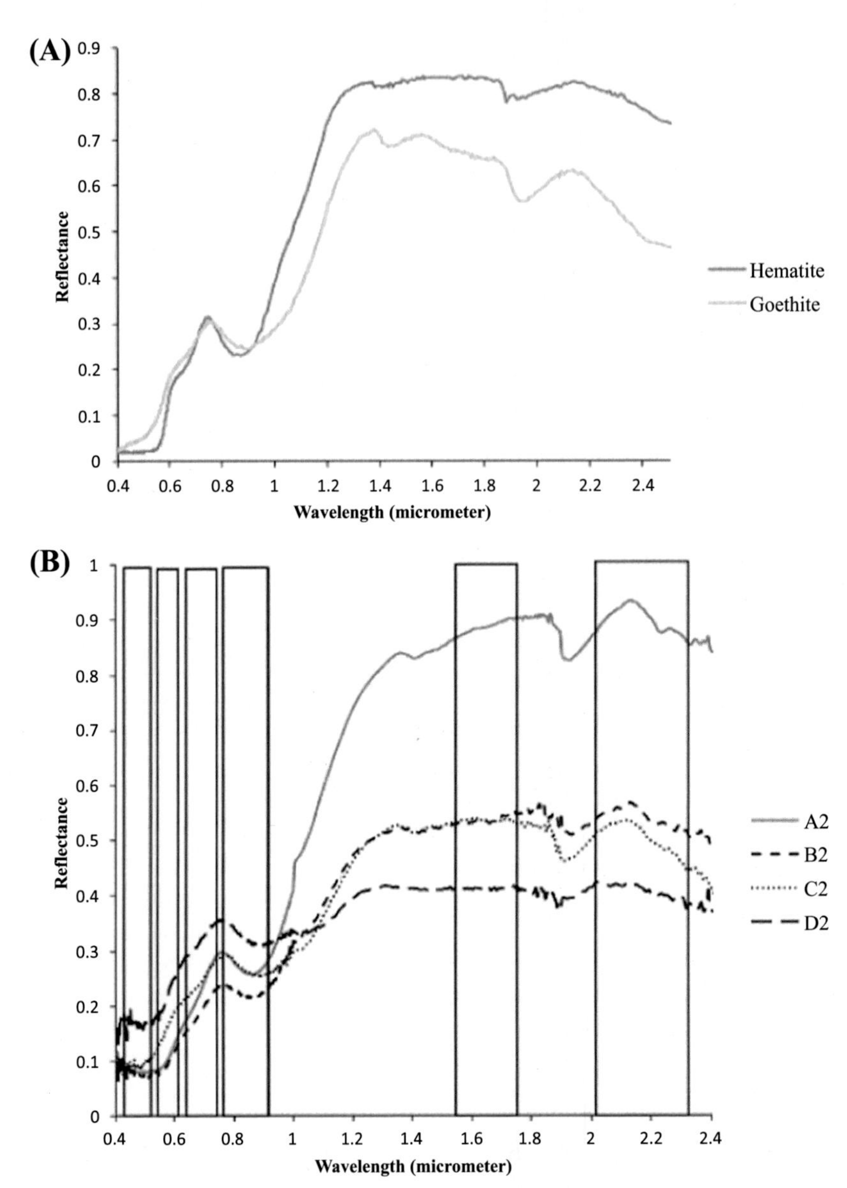

FIGURE 12.16 USGS reflectance spectra (from Clark et al., 1993) of iron oxide (hematite) and iron hydroxide (goethite) (A) and laboratory-measured spectra of the collected samples using the GER 3700 spectroradiometer (B).

curves of the collected samples and the corresponding pixels from Landsat 7 ETM + . Landsat 7 ETM + data are useful in the detection of iron ore deposits in arid environments.

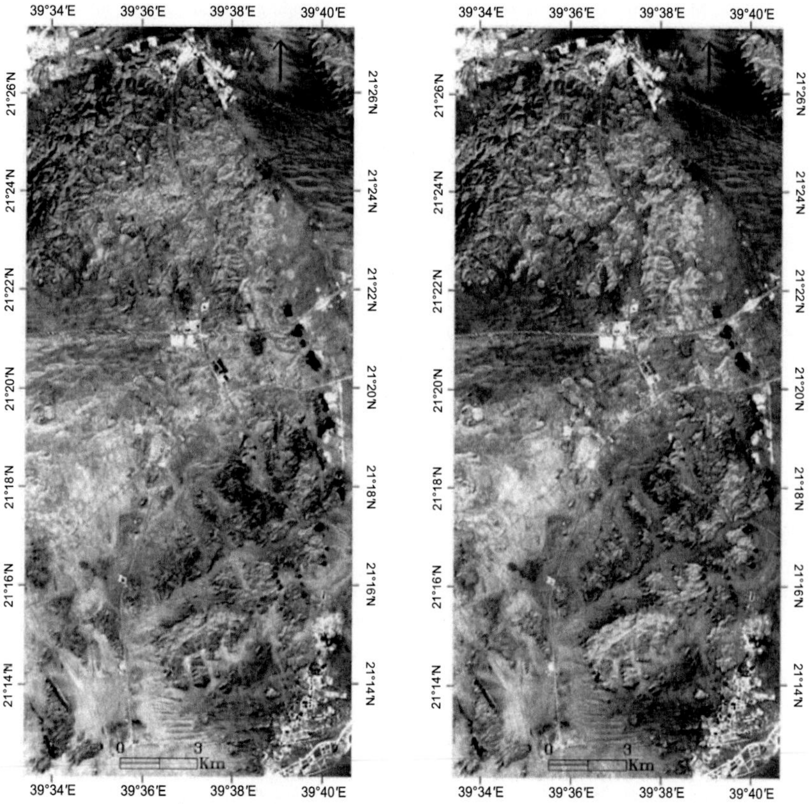

(A) Principal component band 2 (PC2)

(B) Minimum noise band 3 (MNF3)

FIGURE 12.17 Gray-scale images of PC2 (A) and MNF3 (B) derived from Landsat 7 ETM data. The presence of oolitic iron deposits corresponds to the darkest zone.

12.6 Conclusions

There are advantages and limitations to using RS data in mineral exploration. Multispectral and hyperspectral RS data (e.g., Landsat 8, ASTER, AVIRIS, and Hyperion) are important in the initial stages of ore deposits exploration especially in arid and semiarid regions. The various image processing techniques such as PCA, MNF, BR, and image classification effectively map various metallic and nonmetallic mineral resources.

References

Abdelhamid, G., & Rabba, I. (1994). An investigation of mineralized zones revealed during geologic mapping, Jabal Hamra Faddan-Wadi Araba, Jordan, using Landsat TM data. *International Journal of Remote Sensing*, *15*, 1495–1506.

Abdelsalam, M., Stern, R. J., & Berhane, W. G. (2000). Mapping gossans in arid environments with Landsat TM and SIR-C images: the Beddaho alteration zone in northern Eritrea. *Journal of African Earth Sciences*, *30*, 903–916.

Adams, J. B., Smith, M. O., & Giliespie, A. R. (1993). Imaging spectroscopy: Interpretation based on spectral mixture analysis. In Pieters C. M. & Englert P. A. J. (Eds.), *Remote geochemical analysis: Elemental and mineralogical composition* (pp. 145–166). New York, Cambridge University Press.

Al-Shanti, A. (1966). Oolitic iron ore deposits in Wadi Fatima between Jeddah and Mecca, Saudi Arabia. Saudi Arabian Directorate General of Mineral Resources. *Bulletin*, *2*.

Bennett, S. A., Atkinson, W. W., & Kruse, F. K. (1993). *Use of thematic mapper imagery to identify mineralization in the Santa Teresa District, Sonora, Mexico* Roterdam: Balkema *International Geology Review* (35, pp. 1009–1029).

Boardman, J. W. (1993). Automated spectral unmixing of AVIRIS data using convex geometry concepts: In Summaries. Fourth JPL Airborne Geoscience Workshop, JPL Publication 93-26, Vol. 1, pp. 11–14.

Boardman, J. W., Kruse, F. A., & Green, R. O. (1995). Mapping target signatures via partial unmixing of AVIRIS data: In Summaries. Fifth JPL Airborne Earth Science Workshop, JPL Publication 95-1, Vol. 1, pp. 23–26.

BRGM Geoscientists. (1993). Khnaiguiyah zinc–copper deposit prefeasibility study: Synopsis of geology and mineralization. Saudi Arabian Directorate General of Mineral Resources, Technical Report, BRGM-TR-13-4.

Carranza, E. J. M., & Hale, M. (2002). Mineral imaging with Landsat Thematic Mapper data for hydrothermal alteration mapping in heavily-vegetated terrane. *International Journal of Remote Sensing*, *23*, 4827–4852.

Ciampalini, A., Garfagnoli, F., Antonielli, B., Moretti, S., & Righini, G. (2013a). Remote sensing techniques using Landsat ETM + applied to the detection of iron ore deposits in Western Africa. *Arabian Journal of Geosciences*, *6*, 4529–4546.

Ciampalini, A., Garfagnoli, F., Del Ventisette, C., & Moretti, S. (2013b). Potential use of remote sensing techniques for exploration of iron deposits in Western Sahara and Southwest of Algeria. *Natural Resources Research*, *22*, 179–190.

Clark, R. N. (1999). Spectroscopy of rocks and minerals, and principles of spectroscopy. In A. N. Rencz (Ed.), *Manual of remote sensing* (pp. 3–58). New York: Wiley.

Daneshfar, B., Desrochers, A., & Budkewitsch, P. (2006). Mineral-potential mapping for MVT deposits with limited data sets using Landsat data and geological evidence in the Borden Basin, Northern Baffin Island. *Resources Research*, *15*, 129–149.

Elsayed Zeinelabdein, K. A., & El Nadi, A. H. (2014). The use of Landsat 8 OLI image for the delineation of gossanic ridges in the Red Sea Hills of NE Sudan. *American Journal of Earth Sciences*, *1*, 62–67.

Fujisada, H. (1995). Design and performance of ASTER instrument. *Proceedings of SPIE*, *2583*, 16–25.

Gahlan, H., & Ghrefat, H. (2018). Delineation of gossan zones in arid regions by Landsat 8 (OLI): An application to mineral exploration in the Eastern Arabian Shield, Saudi Arabia. *Natural Resources Research Journal*, *27*, 109–124.

Ghrefat, H. (2020). Mapping Maqna Gypsum Deposits in the Midyan Area, Northwestern Saudi Arabia, Using ASTER Thermal Data. *Journal of Geology and Geoscience*, *4*(1), 1–6.

Ghrefat, H. A., Al Zahrani, A. A., & Galmed, M. A. (2018). Mapping oolitic iron ore deposits in the Ashumaysi formation, Western Saudi Arabia, Using Different Remote Sensing Techniques Applied to Landsat 7 Enhanced Thematic Mapper Plus Data (ETM +). *Journal of the Indian Society of Remote Sensing*. Available from https://doi.org/10.1007/s12524-018-0786-y.

Goetz, A. F. H., Vane, G., Solomon, J. E., & Rock, B. N. (1985). Imaging spectrometry for earth remote sensing. *Science*, *228*, 1147–1153.

Goossens, M. A. (1993). Integrated analysis of Landsat TM, airborne magnetic, and radiometric data, as an exploration tool for granite-related mineralization, Salamanca province, Western Spain. *Nonrenewable Resources*, *2*, 14–30.

Goward, S. N., Masek, J. G., Williams, D. L., Irons, J. R., & Thompson, R. J. (2001). The Landsat 7 mission. *Remote Sensing of Environment*, *78*, 3–12.

Green, A. A., Berman, M., Switzer, P., & Craig, M. D. (1988). A transformation for ordering multispectral data in terms of image quality with implications for noise removal. *IEEE Transactions on Geoscience and Remote Sensing*, *26*(1), 65–74.

Hearn, D. r., Digenis, C. J., Lencioni, D. E., Mendenhall, J. A., Evans, J. B., & Welsh, R. D. (2001). EO-1 Advanced Land Imager overview and spatial performance. *Geosciences and Remote Sensing Symosium*, *6*, 3114–3117.

Howari, F. M., Ghrefat, H., Nazzal, Y., Galmed, M. A., Abdelghany, O., Fowler, A. R., Sharma, M., AlAydaroos, F., & Xavier, C. M. (2022). Delineation of copper mineralization zones at Wadi Ham, Northern Oman Mountains, United Arab Emirates Using Multispectral Landsat 8 (OLI) Data. *Frontiers in Earth Science*, *8*. Available from https://doi.org/10.3389/feart.2020.578075.

Irons, J. R., Dwyer, J. L., & Barsi, J. A. (2012). The next Landsat satellite: The Landsat data continuity mission. *Remote Sensing of Environment*, *122*, 11–21.

Jensen, J. R. (1996). Introductory digital image processing: A remote sensing perspective. Prentice Hall: Prentice Hall Series in Geographic Information Science.

Johnson, B. A., Tateishi, R., & Hoan, N. T. (2012). Satellite image pansharpening using a hybrid approach for object-based image analysis. *ISPRS International Journal of Geo-Information*, *1*(3), 228–241.

Kaufmann, H. (1988). Mineral exploration along the Aqaba-Levant structure by use of TM data: Concepts, processing and results. *International Journal of Remote Sensing*, *9*, 1639–1658.

Madani, A. M. (2009). Utilization of Landsat ETM + data for mapping gossans and iron rich zones exposed at Bahrah area, western Arabian Shield, Saudi Arabia. *Journal of King Abdulaziz University: Earth Sciences*, *20*, 35–49.

Madani, A., Abdel Rahman, E. M., Fawzy, K. M., & Emam, A. (2003). Mapping of the hydrothermal alteration zones at Haimur Gold Mine Area, South Eastern Desert, Egypt by using remote sensing techniques. *The Egyptian Journal of Remote Sensing & Space Sciences*, *6*, 47–60.

Moore, J. M. (1983). Tectonic Fabric and Structural Control of Mineralization in the Southern Arabian Shield: A Compilation Based on Satellite lmagery Interpretation. Saudi Arabian Deputy Ministry for Mineral Resources Open-File Report USGS-OF-03-105.

Moore, T. A. & Al-Rehaili, M. H. (1989). Geologic map of the Makkah Quadrangle, Sheet 21D, Kingdom of Saudi Arabia, Ministry of Petroleum and Mineral Resources, Directorate General of Mineral Resources Geoscience map GM-107C, 1:250,000 scale.

Pearlman, J., Carman, S., Segal, C., Jarecke, P., Clancy, P., & Browne, W. (2001). *Overview of the Hyperion Imaging Spectrometer for the NASA EO-1 mission*, *7*, 3036–3038.

Ramadan, T. M., & Kontny, A. (2004). Mineralogical and structural characterization of alteration zones detected by orbital remote sensing at Shalatein District area, SE Desert, Egypt. *Journal of African Earth Sciences*, *40*, 89–99.

Richards, J. A. (1994). *Remote sensing digital image analysis: An introduction.* (340p.). Berlin, Germany: Springer-Verlag.

Roy, D. P., Qin, Y., Kovalskyy, V., Vermote, E. F., Ju, J., Egorov, A., Hansen, M. C., Kommareddy, I., & Yan, L. (2014). Conterminous United States demonstration and characterization of MODIS-based Landsat ETM + atmospheric correction. *Remote Sensing of Environment, 140*, 433–449.

Sabins, F. (1999). Remote sensing for mineral exploration. *Ore Geology Reviews, 14*, 157–183.

Shahriari, H., Ranjbar, H., & Honarmand, M. (2013). Image segmentation for hydrothermal alteration mapping using PCA and concentration-area fractal model. *Natural Resources Research, 22*, 191–206.

Shahriari, H., Ranjbar, H., Honarmand, M., & Carranza, E. J. M. (2014). Selection of less biased threshold angles for SAM classification using the real value–area fractal technique. *Resource Geology, 64*, 301–315.

Shalaby, M. H., Bishta, A. Z., Roz, M. E., & Zalaky, M. A. (2010). Integration of geologic and remote sensing studies for the discovery of uranium mineralization in some granite plutons, Eastern Desert, Egypt. *Journal of King Abdulaziz University: Earth Sciences, 21*, 1–25.

Smith, M. O., Johnson, P. E., & Adams, J. B. (1985). Quantitative determination of mineral types and abundances from reflectance spectra using principal components analysis. *Journal of Geophysical Research, 90*, 797–804.

Testard, J. (1983). *Khnaiguiyah: A synsedimentary hydrothermal deposit comprising Cu–Zn–Fe sulfides and Fe-oxides in an ignimbritic setting*. Saudi Arabian Deputy Ministry for Mineral Resources, Open-File Report, BRGM-OF-03-9.

Testard, J., Tegyey, M., Picot, P., Naury, M., & Kosakevitch, M. (1980). Khnaiguiyah: Mineralization in an acid volcanosedimentary complex. Institute of Applied Geology Bulletin (vol. 3, pp. 79–98). King Abdulaziz University.

Unrug, R. (1988). Mineralization controls and source metals in the Lufilian Fold Belt, Shaba (Zaire), Zambia and Angola. *Economic Geology, 83*, 1247–1258.

van der Meer, F. D., van der Werff, H. M. A., van Ruitenbeek, F. J. A., Hecker, C. A., Bakker, W. H., Noomen, M. F., van der Meijde, M., Carranza, E. J. M., de Smeth, J. B., & Woldai, T. (2012). Multi- and hyperspectral geologic remote sensing: A review. *International Journal of Applied Earth Observation and Geoinformation, 14*, 112–128.

Vane, G., & Goetz, A. F. H. (1988). Terrestrial imaging spectroscopy. *Remote Sensing of Environment, 24*, 1–29.

Vincent, R. K. (1997). Fundamentals of geological and environmental remote sensing. Upper Saddle River: Prentice Hall.

Wang, G., Du, W., & Carranza, E. J. M. (2017). Remote sensing and GIS prospectivity mapping for magmatic-hydrothermal base- and precious-metal deposits in the Honghai district, China. *Journal of African Earth Sciences, 128*, 97–115.

Yamaguchi, Y., Anne, B. K., Hiroji, T., Kawakami, T., & Pniel, M. (1998). Overview of Advanced Spaceborne Thermal Emission and Reflection Radiometer (ASTER). *IEEE Transactions on Geosciences and Remote Sensing, 36*(4), 1062–1071.

Zeinalov, G. A. (2000). Importance of remote-sensing data in structural geologic analysis of oil- and gas-bearing regions of Azerbaijan. *Natural Resources Research, 9*, 307–313.

Chapter 13

Geographic information systems and remote sensing for local development. Reservoirs positioning

Kleomenis Kalogeropoulos[1], Andreas Tsatsaris[1], Nikolaos Stathopoulos[2], Demetrios E. Tsesmelis[3], Athanasios Psarogiannis[1] and Evangelos Pissias[1]

[1]Department of Surveying and Geoinformatics Engineering, University of West Attica (UniWA), Athens, Greece, [2]Operational Unit "BEYOND Centre for Earth Observation Research and Satellite Remote Sensing", Institute for Astronomy, Astrophysics, Space Applications and Remote Sensing, National Observatory of Athens, Athens, Greece, [3]Laboratory of Technology and Policy of Energy and Environment, School of Applied Arts and Sustainable Design, Hellenic Open University, Patras, Greece

13.1 Introduction

Water is a natural resource and often a scarce one (Kalogeropoulos & Chalkias, 2013; Tsesmelis et al., 2019; Kalogeropoulos et al., 2023). The importance of water has been recognized since the time of early societies. The absence or abundance of this resource could demonstrate the ideal area for permanent or temporary settlement (Perry, 2013). Today, water is seen as an extremely valuable resource that affects social well-being and stability, economic development, and ecosystem functioning. Although water is a renewable resource and occurs in relative abundance, only 0.41% (groundwater: 0.397% and surface water: 0.022%) of the total volume of water on earth is suitable and relatively accessible to humans (Tsesmelis, Karavitis et al., 2022). This reality, combined with the ever-increasing demand for water and the concurrent degradation of available water supplies, makes water an even more valuable resource. Therefore the concept of water resources management (WRM) becomes of great importance and is called upon to play a special role in human well-being and the maintenance of ecological structures and functions.

During the last decades, the pressure which arises on water resources has increased significantly due, mainly, to population growth, immigration to

Geoinformatics for Geosciences. DOI: https://doi.org/10.1016/B978-0-323-98983-1.00014-4

urban and coastal areas (Stamellou et al., 2021), climate change, and desertification (Gioti et al., 2013; Kalogeropoulos et al., 2013; Kalogeropoulos & Chalkias, 2013; Karalis et al., 2014; Chalkias et al., 2016; Stathopoulos et al., 2019; Stamellou et al., 2021; Tsanakas et al., 2016; Tsesmelis et al., 2021). In the future, this pressure is expected to be maximized. Therefore the problems will be even higher as a result of water shortages in most areas (Tsesmelis, Vasilakou et al., 2022). The biggest challenge in dryland urban and rural economic activities is low and erratic rainfall. Another challenge, which constitutes the specificity of WRM plans in insular clusters, is that of territorial-geographical discontinuity (Drouart & Vouillamoz, 1999; Kalogeropoulos & Chalkias, 2013; Kim & Newman, 2019).

Increasing demand for water, and even more so when it is not used properly, will accelerate the arrival of water scarcity conditions. The widespread misconception that there is an abundance of water and that the only problem is getting it to the right place at the right time, still remains, resulting in the orientation of water resource management toward supply. The reduction and management of water demand, the imposition of and the introduction of water-saving measures now require the use of water immediate policy and legislative interventions (Grigg, 1997; Stephenson, 1998).

To solve those exceptional WRM problems and meet these needs, the adoption of adequate socio-economic approaches, on top of them those assuring demand reduction, is required. It requires also deep and particular geographic analysis of spatial factors and constraints, as well as suitable technologies preventing the implementation of heavy—although vulnerable—and noncoherent—although expensive—water systems. The depletion of groundwater resources, related to the degradation of their quality due to intensive exploitation of aquifers as well as of fossil waters, shall also be prevented. The use of nonrenewable resources has to be minimized if not forbidden and the promotion of projects aiming at an optimal valuation of surface runoff has to be encouraged (Kalogeropoulos et al., 2020).

Projects that ensure the utilization of surface water through adequate, decentralized, small-scale harvesting systems (small dams and mountainous-hilly reservoirs) could be more environmentally friendly than big-scale ones or those over-exploiting groundwater. Those projects may also serve to several parallel tasks, among them the protection of the forester environment, and, in some cases, the implementation of local scale hydro-power energy systems. Those surface water management projects could become even more attractive if related to appropriate technologies aiming at the use of locally–regionally existing human and natural resources and produced materials. In that case, they strengthen local–regional development processes and increase the opportunity for the creation of new jobs. Considering the developmental constraints that arose during the actual economical crisis, the above low capital investment and running cost water management schemes become "*projects of the moment*" and not "*momentum projects*" challenging for local development.

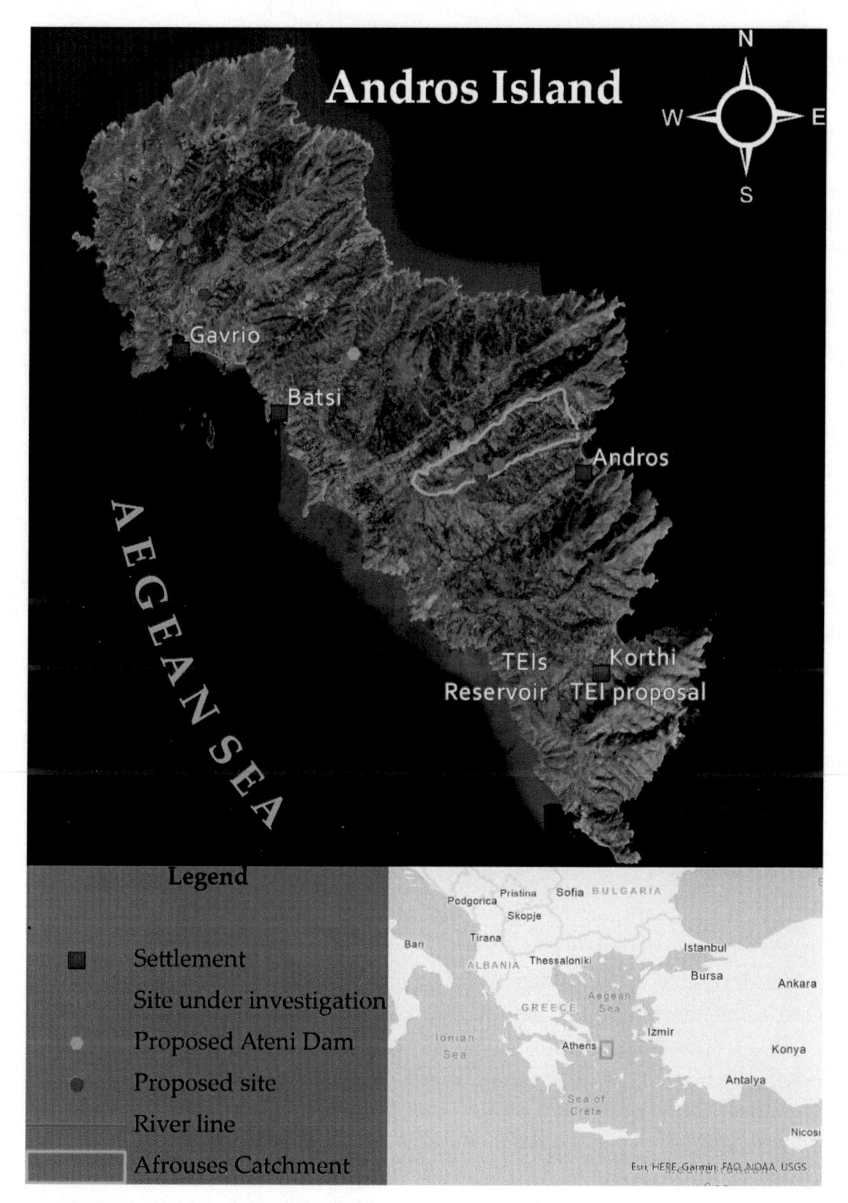

FIGURE 13.1 The proposed/under investigation sites for reservoir creation.

This chapter presents also a new technology package that can help build effective, low-cost systems for harvesting rainwater. This will deliver larger and more stable water supplies for peak season's urban uses, for crops and livestock, even in very dry and isolated areas (Drouart & Vouillamoz, 1999).

It is based on a methodology of choosing (selecting) the water basins and the suitable sites within, for the construction of small reservoirs. The production and use of primary data such as meteorological and spatial data inserted in a geographic information systems (GIS) context (Stathopoulos et al., 2017, 2019; Tsatsaris et al., 2021), using a suitable hydrological model can reveal sites that are appropriate for the construction of those reservoirs. Andros Island (Cyclades, Greece) is the study area for the proposed methodology. Several water basins and sites on the island were identified, among them, seven have been identified according to our principal criteria and, ultimately, three were studied to create small reservoirs (Fig. 13.1).

13.2 The historical background and the future of reservoirs in Greece

13.2.1 Complementarity, integrality, and adversariality within water resources management schemes

The concepts of complementarity as well as the concept of adversariality between different WRM schemes are reflecting advantageous but also inappropriate solutions. Analysis of recent technical practices in Greek insular clusters has proven that a combination of different types of water infrastructures could be successful since assuring a variety of alternative supply capacities. Nevertheless, practices that have been—or still are—implemented, considered either as the sole valuable practice, for example, water drillings, or as the main practice which has to be combined with other complementary practices, shall be, nowadays, reconsidered. The question then arises is to evaluate three main stream practices, referring to three following technologies: (1) water drillings, (2) dams and reservoirs, (3) seawater desalination units, within a different WRM context.

The design of an integrated WRM system in an inland region is subject to territorial discontinuity. This fact resulting two management models, the first is an open-trans territorial and the second is a local model. The first model, which is a typical model of WRM, is concerning mainly the so-called arid islands and relies on seawater transport from other areas that have a potential surplus. This model includes sometimes the extremely high-cost solution of water transport by ship and sometimes the solution of expensive water transfer projects with underwater pipelines. The second model is almost entirely local and seeks water self-sufficiency by exploiting the local natural hydro system of each island and possibly the installation of a desalination plant.

WRM was more empirical in the past, on a collective or individual level, and was based on well-known solutions from rural-residential island tradition like house cisterns (small open tanks) for drinking water or shallow irrigation wells. Today, these solutions are maintaining their usefulness. These solutions also have a positive environmental sign, as they are of low cost and probably the only problem lies in their low efficiency which makes them inadequate in modern times. The solution of drillings of reasonable depth has been for some time a satisfactory solution and has the potential to provide irrigation water also in reasonable quantities. During this period, which in some islands ended during the 1970s and in others during the 1980s, the water balance was equilibrated without any threat to the aquifers.

With the increasing need for the new high-consumption tourist-rural development model of that period, the drillings proliferate at high rates while overpumping of groundwater is observed from greater depths. The rise of this phenomenon is largely due to uncontrolled and, in most cases, to its expansion resulting in depletion of the normal underground aquifer and the rapid degradation of water quality. It was then the period when has been attempted unsuccessfully the implementation of a regulatory framework for water wells and when also the first dams and reservoirs program has been applied. It was then a period when the first desalination plant in Syros has been inaugurated, followed later (around 2000), by the provision, design, and installation of other units in most of the islands to partially cover the water supply needs exclusively.

This process is well known, and the description of each sector (or by case), as well as the physical inventory yield of main hydro systems, have, to some extent, taken place. But the critical assessment of public interest, at least, technical projects which make up the water management infrastructure, through a system of continuous monitoring and evaluation, did not take place.

However, explicitly or implicitly, the technical system of production, design, and implementation of projects in the country, which includes research groups, consultants, manufacturers—suppliers, public services, local government, and individuals that influence decision-making processes, especially the "subsystem" that operates in islands, dictates the shape and trends in terms of the basic categories of projects mentioned above. The first category, which relates to household infrastructure (mainly cisterns), was not enough supported as it should have, and the second, which is concerning groundwater, continues its excessive antienvironmental activity. The third one relates to desalination and is now gaining ground, and the fourth, known as small dams, is falling, despite its powerful advantages, one of which is energy autonomy in times of intense questioning of headlong energy-environmental crisis. Other methods, some of these environmentally friendly (reusing gray waters and/or wastewater), are not yet supported.

This chapter certainly cannot cover in any way the above failures; however, it can evaluate, to some extent, the findings of other research projects and studies mentioned in the area of Aegean. It can also draw conclusions from the research work which have been done on Andros Island and concerns to the local water system and the existent, or rather nonexistent projects of technical-hydraulic infrastructures.

The summary of these findings highlights two principles where the first put and encourages complementarity of different types of projects and the second one limit and/or exclude certain types of projects focusing on contrariety in terms of scale but also in their physical and anthropogenic/geographic factors that constitute each water system. The first principle serves the criterion of diversity and complementarity and interchangeability of the solutions which ensures the reduction of risk arising from a unidimensional development of a management plan. The second principle obeys the criterion of exclusion of inappropriate and abusive solutions which tend both to the depletion of water resources and to the destruction of the hydro system (Point, 1999).

13.2.2 The challenge of small reservoirs

It is known that reservoirs of the capacity of billions of m^3 have been built and probably, unfortunately, the same is going to be in the future (and of course reservoirs of hundreds or tens of millions of m^3 of capacity). These reservoirs, despite their huge differences between them, are classified in the category of large reservoirs-dams. The lower category, that of reservoirs of a few dozen or even of a few million m^3 (in single digits), is considered a medium category, while an even lower category, comprising less than 2–3 million m^3 capacity dams and reservoirs, is characterized as a class of small reservoirs. According to the research program, HYDROMED for small mountainous and upland reservoirs in North and Southeast Mediterranean characterizes as small reservoirs those classified between a few tens of thousands up to 1,000,000 m^3.

Nevertheless, it is less known, almost not known in our country, that more than 90% of reservoirs, actually under construction worldwide, are small reservoirs, that some countries have the privilege to be covered by extended and dense small reservoir "chains," among them many Mediterranean countries north and south. Greece is an exception, maybe the biggest exception.

In this current research project, small mountainous reservoirs are characterized as those which are serving purposes of local development (Helvetas, 1985) are absolutely safe (Z according to ICOLD classification < 20), and the relevant international experience is classified in a range of capacity from 40,000 to 100,000 m^3. Without of course excluding recommendations of the project for small deviations.

The purpose of this chapter is to support the idea of the creation of a network of small, really small reservoirs, with adequate technology, engineering, technique, and know-how, using local human resources and materials;

therefore, they are projects of low-cost and of high domestic-local added value (Forzieri et al., 2008; Kalogeropoulos et al., 2020). The methodological choice is primarily the comparative juxtaposition of specific concepts and related works belonging to the same technological "family," as in this case the family of reservoirs, and only secondarily the critical approach to issues and arguments concerning general WRM frames, referring to:

The superiority of the integrated design versus the piecemeal planning in WRM, for example, multidimensional versus unidimensional approach, complementarity versus exclusivity, etc.

The comparative evaluation of the general usefulness of a particular method of utilization of water resources in relation to another method, such as the use of runoff versus groundwater exploitation and vice versa.

The juxtaposition of a technical solution over other technical solutions, such as dams against drilling or desalination plants, and vice versa.

The chosen methodology does not focus on the rivalry of a technical choice to another, of a different kind, option, but between similar technical choices, of radically different scale projects, such as, for example, the rivalry between reservoirs of suitable small scale over others, allegedly as small (however, for our particular insular reality), rather medium and bigger, absolutely of improper scale. This methodology deals with the philosophy of the small reservoir, the prioritizing of goals, objectives, and priorities, the socio economic and environmental considerations, the application of a specific technology and technique, etc.

The result of our research is the result of a thorough multicriteria analysis, where the criteria have been preprocessed to get rid of established rationalities such as:

The superiority of economic developmental rationality against socio-economic rationality.

The superiority of the short-term versus the medium-term rationality.

The superiority of the rationality of lowering investment-functional finance costs arising from the under-evaluation and/or indeterminacy of high external, social and environmental, costs.

And beliefs, such as:

The overall effectiveness of the supposedly technologically advanced—new generation—solutions to previous technological generation solutions.

The general superiority of the scientific-technological advanced solution against empirical embedded and trusted technical solution.

The superiority of the capital-intensive model to any labor-intensive model (even that of specialized labor).

The continuous progress in methods and materials.

The scale economy is secured through large projects versus small projects.

Other established rationalities or beliefs can of course be mentioned. The above, however, suffice for a first practical approach to the problem of the utilization of surface runoff in Andros Island, through three exemplary solutions. The first two are derived from studies commissioned by the Ministry of Agriculture under a program for the construction of dams and reservoirs in the Northern Cyclades islands in 1991. The third solution (Fig. 13.2) is resulting from a research project commissioned by the Water Resources Laboratory (Department of Surveying Engineering, ex Technological Educational Institute of Athens).

The first solution proposed the construction of dams or alternatively the river reservoirs of a capacity (five of them) between 300,000 to 500,000 m^3. In one of the sites, in Ateni, a dam reservoir of 1,100,000 m^3 was proposed.

This solution of the six reservoirs has gradually abandoned and simultaneously a second solution matured, which arrived at the final design stage and involved the construction of only one of the above six reservoirs, that which was projected to post Ateni with the above capacity.

The third solution, which came "after 20 years," having investigated the feasibility of reservoirs in many places in a sufficient number of watersheds, opted initially for seven projects. In four of these basins a reservoir, by the above-mentioned Ministry of Agriculture (MA) study, was also predicted. The dimensions of the proposed—river—reservoirs in the context of this solution were ranging between 40,000 and 80,000 m^3 but one position in Ateni was of a 132,000 m^3 capacity (where the proposed MA dam's capacity was 1.1 million m^3 and 27 m height, type RCC, solution 2). During the last phase of the current research program, three of the proposed seven—out of river—reservoirs were selected with a capacity of 45,000–50,000 m^3. Those reservoirs are already at the stage of the prefinal study.

FIGURE 13.2 (A) Current meter measurement, (B) cross-section, (C) spill, and (D) meteorological station.

The reasons that led to the abandonment of the first two solutions are many and relate to a wide range of design weaknesses, both in the field to investigate the socio-economic feasibility and the scope of the technical-economic study.

The reasons, however, revived the issue of the small reservoirs to the forefront and allowed investigation of a third solution, are also numerous. Below is an attempt to document in a certain way the main reasons. However, a focal point for understanding the problem of acceptance or rejection of a reservoir or of a reservoir network is to consider solutions such as those mentioned above (many medium reservoirs, or one large or many small?) not in a frame of variance but in a context of contrarity. In a context where the quantitative difference indicated by the above sizes of reservoirs on three solutions is a qualitative difference and the ranking does not indicate simple graduation but technological- a technique in-depth change.

13.2.3 Socio-economic expediency and techno-economic feasibility

This current research study as an object has at first the water basin and then the reservoir. Where the purpose of a project is the creation of small reservoirs then the referential natural-geographic and human-geographic entity is, in general, the basin within which they will be constructed. Unlike other previous studies that resulted in recommendations to construct reservoirs for the full exploitation of the water resources of a river (to cover mainly regional than local demand), the aim of this project is the partial exploitation of the water resources of the basin to meet justifiable local demand to ensure sustainable development. This objective defines—and finally limits—the scale of the proposed engineering projects.

For this reason, the proposed small reservoirs are intended to cover only the critical, real, potable, and irrigation water demand of studded settlements and rural perimeters, the total area of which does not exceed hundreds of hectares in Andros. But they also intended to secure and improve environmental harmony, create natural reserves, and produce micro-climatic effects. As it is, almost exclusively, for mountainous and upland reservoirs of this class, the position within the basin where the reservoir will be constructed is selected following the criteria of its harmonization with the human-geographical and physical-geographical conditions of its area of influence.

More specifically, this research project starts from the initial assumption that a network of properly located small reservoirs of total capacity $V_{\text{all}} = (V_1 + V_2 + \ldots + V_n)$, in selected watersheds, is more beneficial than a larger reservoir capacity V, even more of V_{all}. The fact that the typical cost of a large reservoir will be minor than the cumulative cost of many small reservoirs, of smaller total capacity, and the fact that the unit cost of water in small reservoirs (cost per cubic meter) will be higher than the unit cost of water in a large reservoir, is neither confirmed nor an obvious, self-understood advantage. This seemingly paradoxical case is revealed, in many cases, by the accurate budgeting of heavy and expensive complementary works. It is related mainly to principal water supply and irrigation networks, essentially when geomorphology obliges. That same paradox is also reversed in many other cases if the cost-benefit analysis is not trapped by the highly controversial argument of economies of scale or by the criterion of hypothetical economic rationality. Instead, that paradox includes additional important criteria such as local development, the rational spatial distribution of activities, low environmental impact, and socio-economic regional isometry (commensurability).

13.2.4 The specifics of the technical design

The first attempts to address the problem of the design of small reservoirs (in our case water reservoirs of usable capacity $V < 100{,}000\ m^3$) adopting all the assumptions required in the design of large dams and applying their technical specifications. In this case, the resulting technical work is often a replica of a small scale, a miniature of a large project.

The second approach addresses the problem of underestimating the basic-fundamental design requirements and the difficulties of making a small reservoir. Therefore ignoring (perhaps unconsciously) the scientific data and the technical manufacturing requirements ensure durability, sustainability, and most importantly the safety of the project.

There are three major issues for a reservoir to be solved in terms of positioning and technical study. The first problem is related to the capacity of the reservoir, the second problem is related to the choice of the location of its construction, and the third with its material and technical characteristics.

13.2.5 The capacity issues

The capacity of the reservoir must only meet the reasonable needs of users. It is therefore not the offer resulting from the demand (the not always justifiable), but the demand has to be overdetermined and contingent-limited by the sustainable offer. In the case therefore of the capacity of the reservoir, its determination depends on demand's control, such as the modern understanding of sustainable management of water resources requires.

On this issue, the international perception is constantly strengthened and is looking for partial only withdrawal and storage of the annual runoff of a catchment. Withdrawer quantity should be then between 10% and 25% not to disturb the ecosystem of the river (or small stream) subject to unregulated flow decrease due to reservoir construction (FRAPNA, 2005). This principle was not taken into account either by the first solution (six small and medium reservoirs) or by the second solution (one medium reservoir capacity $> 1,100,000 \text{ m}^3$ in Ateni).

The losses due to evaporation (which is often referred to as a counterargument related to small reservoir construction in areas of high evaporation) are generally proportional to the free surface (mirror) of the water in the reservoir. It is therefore sufficient to obtain an estimate of the column of water that evaporates from the time that the spillway of the reservoir stops draining the excess water that flows to it (spillway crest level) until the end of the period of use of the reservoir. Typically, this period covers the period from April-May to October within the climatic conditions prevailing in the Greek islands.

A part of this research project was the fullest possible investigation of the problem. Earlier studies were examined, but many conclusions were made from the systematic study of the phenomenon in most Mediterranean countries. The calculations of the research project are referred to the worst possible evaporation which, however, will not affect the inflow and stored amount of water by more than 6%. The measurements of many years in the Mediterranean areas of France during the months of nonreplacement reach up to 220 $\text{m}^3/1000 \text{ m}^2$ of the water mirror of the reservoir. Consequently, they are also close to the estimates of our research.

To compare the results of these investigations to the results of systematic measurements in dams of Southwestern Mediterranean countries such as, mainly, Tunisia, Algeria, and Morocco (Mahe et al., 2018; Sadaoui et al., 2018) is useful to be reported that in a sample of 39 dams, the rate of average annual loss due to evaporation was measured at 6.5% of the total volume of average annual inflows to these reservoirs (Remini et al., 2009).

13.2.6 The problem of material-technical requirements

This research project is concerned with cases of reservoirs (outside the bed of the stream) and focuses on exploring relatively simple and low-cost solutions and proposes specific test methods to successfully implement them. With respect to the body of the technical work (of the configuration of the basin of the reservoir) and particularly in the perimeter of the embankment is proposing the construction to be done using earthen material (where is possible) of the excavation according to the type of flexible low earthen dam gravity aluminate impermeable core.

Furthermore, is proposing the sealing of the basin, as a whole, to be held with clay products of suitable quality, with also suitable screed condensation, and to ensure the desired minimum permeability. The main reason why this study considers the possible adoption of solutions that do not seek the theoretically complete sealing of the reservoir with geomembranes and especially the perimeter embankment of the reservoir is associated with some of the fundamental assumptions of this study. This case is concerning many manufacturing faults in reservoirs where the seal has been pursued by geomembranes without securing the conditions of properly implemented protection.

The second relates to the current occupancy of watercourses by beneficiaries which provides that the maximum period of possible nonreplacement of stored water of the reservoir will not exceed 2–3 summer months. In line with the above, the crucial importance of the geotechnical study, the laboratory tests, and the continuous field investigation shall be confirmed by strengthening the "earthy" character of the project. If the results will permit (this potential is highly likely to island sites in the country) as in the case of Andros, the sealing methods which are discussed are:

The proper concentration of the bottom surface using clay, creating in this way a restricted permeability layer thickness of about 50 cm. This method can be applied to positions of reservoirs that are located near soils and allow land extraction of high concentrations of clay. The implementation of this method requires the work of removing the surface vegetation, creating of the impermeable layer, and then protecting it using rock fills.

This latter ensures the consistency of the nonpermeable layer of the bottom, preventing the wear (corrosion-cracking) due to high-temperature changes in this period; for some reasons, it can happen that not be covered by water (e.g., maintenance). With respect to the wetted surface of the embankment, the protection by using clayey soils overlapping with filter, geotextile, and rock fills without using geomembranes is used successfully in many countries where the height does not exceed 10 m and the slope of the wetted sidewall is defined as 3:1.

In this way, it is possible to address erosion, due to the continuous change of moisture in the wetted wall of the embankment mainly due to a sharp change in the water level of the reservoir. The body of the low dike, with a maximum height of 8.00 m (5.00 m in Natura 2000 areas) will consist of the clay core, on a positioning cam, positioned closer to the wetting wall, which will fall below the level of the bearing of earth materials embankment that is placed on either side.

The wetted wall would also protect with a filter with a thickness of about 30 cm, and the rock fills which on its crest will approach 50 cm and on the base is 1.50 m (Quesnel, 1966). Concerning the geometric characteristics of the embankment is proposed:.

Coronation $b = 4.00$ m, $J_{in} = 3{:}1$, $J_{out} = 2.5{:}1$.

For the kernel is provided $b_{crest} = 2.00$ m and $B_{base} = 5.00$ m.

The geometric elements that are listed in the clay core will be reviewed until the last stage (preconstruction) of the geotechnical investigation and after considering the excavated material (and possible materials that will be sought in adjacent borrow).

The use of other technical materials for enhancing tightness such as bentonite etc. will be examined experimentally. The solution of the geomembranes is not excluded although is the last alternative.

13.3 The geographic information systems–based methodology for reservoir positioning

13.3.1 General

The simulation of the hydrological cycle with SWAT (Zheng et al., 2010), on the one hand, requires a large amount of meteorological and spatial data (vector and raster data); on the other hand, its execution (running) is quite easy, since many of the problems of the past have been overcome—compared to earlier versions—and of course, the fact that the successive steps of the model are executed serially (serial execution of the program's submenus) is a convenience. This means that the program does not allow the user to proceed to the next step without having finished the previous one. This procedure has taken place within the ArcGIS environment.

The methodology followed in this chapter involves the application of the hydrological model SWAT (Neitsch et al., 2005; Kalogeropoulos et al., 2011, 2020; Pissias et al., 2013) in three different sites within Andros Island. The basic idea is to estimate the annual runoff for a catchment based on meteorological and spatial data using the above hydrological model. The simulated runoff is used as an entry data in the Reservoir Simulation software, and the monthly failure is examined to qualify the optimal positioning of the reservoir and the annual volume of water release based on the failure rate of extraction of the required water volume.

The introduction of GIS technology led researchers to develop data processing automation and produce reliable simulation models. The hydrological model used in this work is SWAT, an acronym for Soil and Water Assessment Tool. The simulation of the hydrologic cycle in SWAT involves a large number of meteorological and spatial data (Zheng et al., 2010). SWAT uses the standard equations of hydrology to simulate the parameters of the hydrological cycle.

13.3.2 Data

The main entry data to run SWAT are (1) a Digital Elevation Model (DEM)—the DEM of the study area was created within an ArcGIS environment from digitized topographical data—(2) a land cover map, (3) a soil map, and (4) meteorological data (Stathopoulos et al., 2018). SWAT provides the opportunity either to enter historical data of rainfall, temperature, wind speed, and solar radiation or to create a statistical weather station, in the absence of previous data.

The input data required for the execution of Reservoir Simulation software are: (1) level-area curve, (2) hydrologic data (precipitation, temperature, and input per day), (3) latitude of the area under investigation, (4) annual extraction volume, (5) monthly outflow rates, (6) volume's function (m^3) based on lower and upper levels operation (overflow and lack of energy), and (7) deviation.

13.3.3 Methodology

The methodology will be analyzed for the Afrouses catchment on the island of Andros. The same methodology was followed for all the catchments under consideration. The next step was dedicated to the implementation of the simulation scenario. The SWAT simulation was performed for two selected locations within the Afrouses catchment to estimate the runoff for each upstream catchment. The simulation was also performed for the Afrouses catchment as a whole and showed an average annual rainfall of 565.2 mm and surface runoff of approximately 291.1 mm. This figure is almost in line with those given within the General Management Plan (due to the Framework Directive 2000/60) of the Ministry of Development, where the surface runoff is 293 mm. The results of the SWAT simulation are the input data for the reservoir simulation software, which simulates single or multifeasibility reservoir operation and can also process virtually unlimited points describing the change in reservoir volume in terms of level-surface point pairs.

The simulated values of precipitation and discharge (surface runoff) that SWAT exports in each of the two selected sites (for 1200 months of simulation-100 years) are the input data for the Reservoir Simulation. For the purposes of this study, the simulation period for the two dam sites and for the Afrouses catchment as a whole is from 1/1/2010 to 31/12/2109 (100 years).

The main steps of SWAT simulation are as follows:

1. Creating the project
2. Creation of the project
3. DEM—Watershed modeling (DEM)
4. Establishment of hydrological response units (HRUs)
5. Introduction of the synthetic artificial meteorological station
6. Run model
7. Display results.

13.4 Results

The proposed methodology attempts to specify suitable places to position small-scale reservoirs. As is already mentioned, SWAT model showed an average annual rainfall of almost 570 mm and surface runoff of about 290 mm (for the selected catchment), which is in line with the Master Management Plan of the Ministry of Development. Thus the simulation results are in principle judged acceptable based on the assumptions made and the data used (Kalogeropoulos et al., 2020).

Table 13.1 summarizes the SWAT results for the case during the first year of the simulation.

To evaluate the model, field measurements have been made using current meters for the water velocity/volume and a meteorological gauge station for the meteorological parameters.

TABLE 13.1 ArcSWAT results for the first year of simulation.

Unit					PERCO	Tile				Water
Time	PREC (mm)	SURQ (mm)	LATQ (mm)	GWQ (mm)	LATE (mm)	Q (mm)	SW (mm)	ET (mm)	PET (mm)	Yield (mm)
1	77.13	0.00	26.12	0.00	0.00	0.00	52.10	17.98	49.20	26.12
2	90.58	0.15	53.06	0.43	7.00	0.00	71.32	11.05	17.99	53.63
3	73.81	0.00	32.78	4.13	6.53	0.00	67.74	38.32	70.83	36.91
4	7.50	0.00	1.64	4.21	0.00	0.00	49.59	24.03	140.32	5.85
5	7.53	0.00	1.61	2.08	0.00	0.00	21.04	34.47	106.23	3.70
6	2.88	0.00	0.00	0.43	0.00	0.00	2.76	21.16	132.64	0.43
7	2.58	0.00	0.00	0.17	0.00	0.00	0.74	4.60	128.73	0.17
8	3.14	0.00	0.10	0.06	0.00	0.00	0.35	3.43	111.97	0.16
9	10.87	0.00	3.12	0.02	0.00	0.00	0.62	7.48	84.68	3.14
10	23.17	0.00	7.06	0.01	0.00	0.00	1.99	14.74	86.29	7.07
11	89.68	0.00	41.57	0.00	0.00	0.00	25.73	24.37	75.65	41.58
12	218.60	0.81	119.78	1.55	22.51	0.00	65.47	35.44	117.11	122.13
2020	607.46	0.95	286.84	13.09	36.05	0.00	65.47	237.071	121.64	300.87

For this study, several different scenarios of the annual volume of water extraction were considered for each of the two sites and for different reservoir heights. A certain scenario is adopted in terms of monthly outflow rates. These coefficients are smaller during months with low consumption (periods without large demands for water for domestic use and irrigation), such as the period from October to March, and increase during the summer period (Fig. 13.3).

After the selection of the suitable site for each reservoir, the technical design of the reservoir is followed. An example of this technical design is presented in Fig. 13.4.

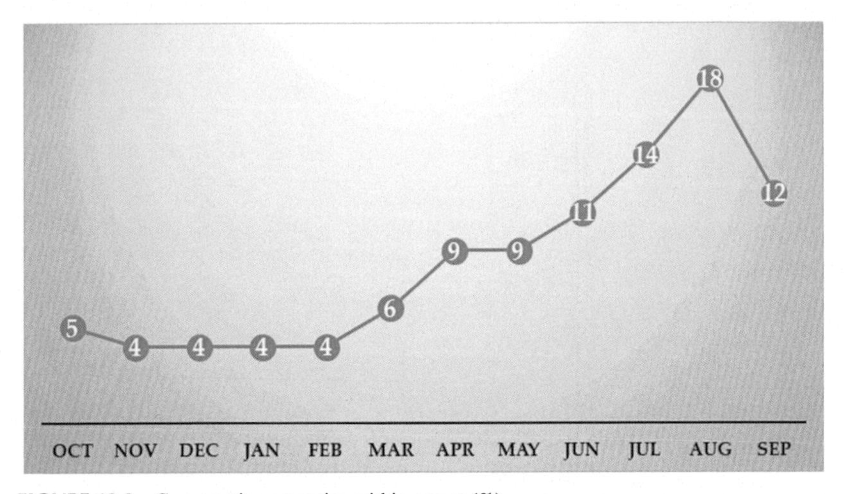

FIGURE 13.3 Consumption scenarios within a year (%).

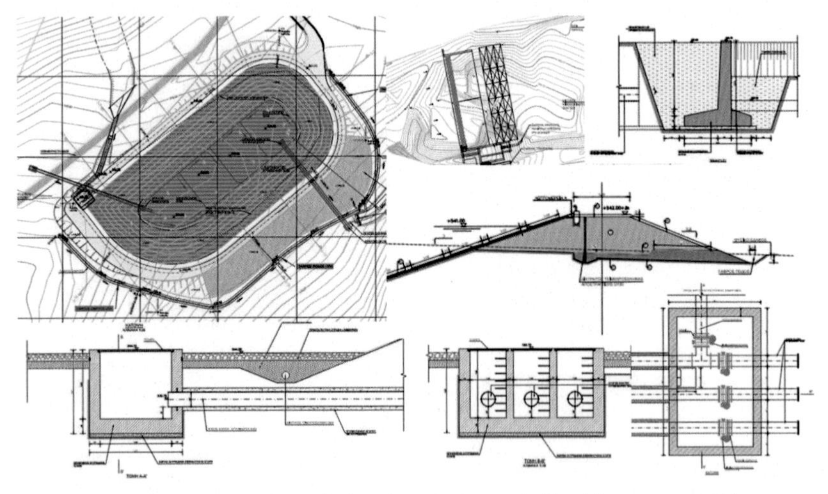

FIGURE 13.4 An example of technical design of a reservoir (the Afrouses reservoir).

13.5 Conclusions

This chapter describes the development of a methodology for the exploitation of water resources through the creation of small mountain and semi-mountain water reservoirs, which are based on Geoinformatics.

The discipline of Geoinformatics can be harmoniously combined with the science of hydrology, hydrogeology, hydraulics, and meteorology and contribute in general to the management of water resources. This coupling makes the best use of descriptive and spatial information to provide reliable results that will lead to the adoption of the best and most sustainable practices. The potential of geoinformatics and models is a tool at the disposal of the researcher–researcher to better simulate the natural environment. Perhaps of greater value than the results of this particular scenario is the fact that the proposed modeling can be used to obtain automated quantitative results for a number of different scenarios to simulate current and expected future conditions. In this particular case, the SWAT user has the possibility to parameterize elements of the model database to adapt it to their needs, as the elements and data used can be continuously refined and revised. This makes SWAT a reliable solution for investigating various phenomena (runoff, sediment, climate change, etc.).

Also, this chapter presents the long-run international and regional (Mediterranean essentially) experience in small dams and reservoirs as well as the results of an ongoing research program which is entitled "Exploitation of the surface runoff on the Island of Andros for the creation of mountain water reservoirs" and is an undertaking project by the Technological Educational Institute of Athens, Department of Surveying Engineering, Laboratory of Water Resources Management.

As it has been already mentioned, the purpose of this chapter is to support the idea of the creation of a network of small, really small reservoirs, with adequate technology, engineering, technique, and know-how, using local human resources and materials, therefore creating projects of low cost and high socio-economic benefits due to high domestic-local added value.

The methodological choice is primarily the comparative juxtaposition of specific concepts and related works belonging to the same technological "family," as in this case the family of reservoirs, and only secondarily the critical approach to issues and arguments concerning other technologies or different WRM frames.

This framework of WRM, considering small hydro systems as dynamic, four-dimensional entities, has the ability to provide solutions on a local scale. It ensures, in this way, sustainable development at a local scale. It is an element of microscale, which in its implementation through the network of scattered small reservoirs aspires to cover the entire island. This idea can provide solutions for drinking and irrigable water to the stakeholders, the derivatives, and to the growers versus the solutions of one big reservoir (the above-mentioned solution 2—a dam in Ateni).

Furthermore, due to the fact that there the retention of runoff is less than 20% (maximum) of the annual runoff (which is withdrawn from the river mainly during flood discharges), this solution is providing protection for the biodiversity, of the riparian zone and of the downstream catchment. This solution can create upland habitats and influence positively the microclimate of each area in which is applied. Also, in times of economic crises, it is crucial that the solution of small dams is an energy saver solution compared to any other solution for drinking and irrigable water.

For all the above-mentioned reasons, the solutions of a network of small reservoirs are taking under consideration additional important criteria such as local development, the rational spatial distribution of activities, low environmental impact, and socio-economic regional isometry (commensurability). Thus this is not a solution that is based only on economic criteria even if implementing all the proposed network of reservoirs in Andros Island (which all together have a capacity of about half of the capacity of the Ateni dam-reservoir), by using techniques in a local scale and earth materials from the region, the final unit cost (to the consumer) is the half (compares to solution 2 which covers only the 1/3 of the island and mostly Batsi and Gavrio settlements, while the solution of the reservoirs network covers the 100% of the island) in terms of the construction cost and the cost of the distribution water network.

WRM requires, beyond the adoption of policies preventing water wastage and depletion, the implementation of a water framework based on surface runoff development techniques to meet needs at the regional/local level. It is, therefore, necessary, in many cases, to conceive, plan, and build water projects founded on adequate/adapted technology, by using locally–regionally existing materials and techniques, aiming at the development of local socio-economical dynamics, creation of jobs, protection of the environment, etc. There is no contradiction in modern environments if combining the above-mentioned type of technical paradigms with high technical standards and, especially, adequate techniques.

References

Chalkias, C., Stathopoulos, N., Kalogeropoulos, K., & Karymbalis, E. (2016). *Applied hydrological modeling with the use of geoinformatics: Theory and practice. Empirical modeling and its applications*. InTech. https://doi.org/10.5772/62824.

Drouart, E., & Vouillamoz, J.-M. (1999). *Alimentation en eau des populations menacées*. Paris: Hermann.

Forzieri, G., Gardenti, M., Caparrini, F., & Castelli, F. (2008). A methodology for the pre-selection of suitable sites for surface and underground small dams in arid areas: A case study in the region of Kidal, Mali. *Physics and Chemistry of the Earth, Parts A/B/C, 33*, 74–85. https://doi.org/10.1016/j.pce.2007.04.014.

FRAPNA. (2005). Retenues Collinaires.

Gioti, E., Riga, C., Kalogeropoulos, K., & Chalkias, C. (2013). A GIS-based flash flood runoff model using high resolution DEM and meteorological data. *EARSeL eProceedings*. https://doi.org/10.12760/01-2013-1-04.

Grigg, N. S. (1997). Systemic analysis of urban water supply and growth management. *Journal of Urban Planning and Development, 123*, 23–33. Available from https://doi.org/10.1061/(ASCE)0733-9488(1997)123:2(23).

Helvetas, L. (1985). *Manuel technique pour l'approvisionnement en eau des zones rurales: avec de nombreux dessins detailles*. St. Gall, Leuven: SKAT (Swiss Centre for Development Cooperation in Technology and Management).

Kalogeropoulos, K., & Chalkias, C. (2013). Modelling the impacts of climate change on surface runoff in small Mediterranean catchments: Empirical evidence from Greece. *Water and Environment Journal, 27*, 505–513. Available from https://doi.org/10.1111/j.1747-6593.2012.00369.x.

Kalogeropoulos, K., Chalkias, C., Pissias, E., & Karalis, S. (2011). *Application of the SWAT model for the investigation of reservoirs creation. Advances in the research of aquatic environment* (pp. 71–79). Berlin, Heidelberg: Springer. https://doi.org/10.1007/978-3-642-24076-8_9.

Kalogeropoulos, K., Karalis, S., Karymbalis, E., Chalkias, C., Chalkias, G., & Katsafados, P. (2013). Modeling flash floods in Vouraikos River mouth, Greece. *Proceedings of the 10th Global Congress on ICM: Lessons Learned to Address New Challenges, EMECS 2013—MEDCOAST 2013 Joint Conference*, 1135–1146.

Kalogeropoulos, K., Stathopoulos, N., Psarogiannis, A., Pissias, E., Louka, P., Petropoulos, G. P., & Chalkias, C. (2020). An integrated GIS-hydro modeling methodology for surface runoff exploitation via small-scale reservoirs. *Water (Switzerland), 12*, 1–18. Available from https://doi.org/10.3390/w12113182.

Kalogeropoulos, K., Tsanakas, K., Stathopoulos, N., Tsesmelis, D. E., & Tsatsaris, A. (2023). Cultural heritage in the light of flood hazard: The case of the "Ancient" Olympia, Greece. *Hydrology, 10*(3), 61. Available from https://doi.org/10.3390/hydrology10030061.

Karalis, S., Karymbalis, E., Valkanou, K., Chalkias, C., Katsafados, P., Kalogeropoulos, K., Batzakis, V., & Bofilios, A. (2014). Assessment of the relationships among catchments' morphometric parameters and hydrologic indices. *International Journal of Geosciences, 05*, 1571–1583. Available from https://doi.org/10.4236/ijg.2014.513128.

Kim, Y., & Newman, G. (2019). Climate change preparedness: comparing future urban growth and flood risk in Amsterdam and Houston. *Sustainability, 11*, 1048. Available from https://doi.org/10.3390/su11041048.

Mahe, G., Amrouni, O., Moussa, T. B., Dezileau, L., El Aoula, R., Habaieb, H., Hzami, A., Kacimi, I., Khedimallah, A., Kotti, F., Meddi, M., Mhammdi, N., & Abdeljaouad, S. (2018). *Assessment of the impact of dams on river regimes, sediment transports to the sea, and coastal changes. Recent advances in environmental science from the Euro-Mediterranean and surrounding regions, advances in science, technology & innovation* (pp. 31–32). Cham: Springer International Publishing. Available from https://doi.org/10.1007/978-3-319-70548-4_12.

Neitsch, S. L., Arnold, J. G., Kiniry, J. R., Srinivasan, R., & Williams, J. R. (2005). *Soil and water assessment tool input/output file documentation version* (p. 2005).

Perry, C. (2013). ABCDE + F: A framework for thinking about water resources management. *Water International, 38*, 95–107. Available from https://doi.org/10.1080/02508060.2013.754618.

Pissias, E., Psarogiannis, A., & Kalogeropoulos, K. (2013). *Water savings—A necessity in a changing environment. The case of small reservoirs*. Tinos, Greece: WIN4life.

Point, P. (Ed.), (1999). *La valeur économique des hydrosystèmes: Méthodes et modèles d'évaluation des services délivrés*. Paris: Economica.

Quesnel, B. (1966). *Hydraulique fluviale appliqué*. E.N.G.R.E.F., ed. RIBER.

Remini, B., Leduc, C., & Hallouche, W. (2009). *Évolution des grands barrages en régions arides: Quelques exemples algériens*, *20*, 96–103.

Sadaoui, M., Ludwig, W., Bourrin, F., Bissonnais, Y., & Romero, E. (2018). Anthropogenic reservoirs of various sizes trap most of the sediment in the Mediterranean Maghreb Basin. *Water*, *10*, 927. Available from https://doi.org/10.3390/w10070927.

Stamellou, E., Kalogeropoulos, K., Stathopoulos, N., Tsesmelis, D. E., Louka, P., Apostolidis, V., & Tsatsaris, A. (2021). A GIS-cellular automata-based model for coupling urban sprawl and flood susceptibility assessment. *Hydrology*, *8*, 159. Available from https://doi.org/10.3390/hydrology8040159.

Stathopoulos, N., Kalogeropoulos, K., Dimitriou, E., Skrimizeas, P., Louka, P., Papadias, V., & Chalkias, C. (2019). *A robust remote sensing—spatial modeling—remote sensing (R-M-R) approach for flood hazard assessment. Spatial modeling in GIS and R for Earth and environmental sciences* (pp. 391–410). Elsevier. Available from https://doi.org/10.1016/B978-0-12-815226-3.00017-X.

Stathopoulos, N., Kalogeropoulos, K., Polykretis, C., Skrimizeas, P., Louka, P., Karymbalis, E., & Chalkias, C. (2017). *Introducing flood susceptibility index using remote-sensing data and geographic information systems empirical analysis in Sperchios River Basin, Greece. Remote sensing of hydrometeorological hazards* (pp. 381–400). CRC Press.

Stathopoulos, N., Skrimizeas, P., Kalogeropoulos, K., Louka, P., & Tragaki, A. (2018). *Statistical analysis and spatial correlation of rainfall in Greece for a 20-year time period.*

Stephenson, D. (1998). *Water supply management.*

Tsanakas, K., Gaki-Papanastassiou, K., Kalogeropoulos, K., Chalkias, C., Katsafados, P., & Karymbalis, E. (2016). Investigation of flash flood natural causes of Xirolaki Torrent, Northern Greece based on GIS modeling and geomorphological analysis. *Natural Hazards*, *84*, 1015–1033. Available from https://doi.org/10.1007/s11069-016-2471-1.

Tsatsaris, A., Kalogeropoulos, K., Stathopoulos, N., Louka, P., Tsanakas, K., Tsesmelis, D. E., Krassanakis, V., Petropoulos, G. P., Pappas, V., & Chalkias, C. (2021). Geoinformation technologies in support of environmental hazards monitoring under climate change: An extensive review. *ISPRS International Journal of Geo-Information*, *10*, 94. Available from https://doi.org/10.3390/ijgi10020094.

Tsesmelis, D. E., Karavitis, C. A., Oikonomou, P. D., Alexandris, S., & Kosmas, C. (2019). Assessment of the vulnerability to drought and desertification characteristics using the standardized drought vulnerability index (SDVI) and the environmentally sensitive areas index (ESAI). *Resources*, *8*, 6. Available from https://doi.org/10.3390/resources8010006.

Tsesmelis, D. E., Vasilakou, C. G., Kalogeropoulos, K., Stathopoulos, N., Alexandris, S. G., Zervas, E., Oikonomou, P. D., & Karavitis, C. A. (2022b). Drought assessment using the standardized precipitation index (SPI) in GIS environment in Greece. *Computers in Earth and Environmental Sciences*, 619–633, https://doi.org/10.1016/B978-0-323-89861-4.00025-7.

Tsesmelis, D. E., Karavitis, C. A., Kalogeropoulos, K., Tsatsaris, A., Zervas, E., Vasilakou, C. G., Stathopoulos, N., Skondras, N. A., Alexandris, S. G., Chalkias, C., & Kosmas, C. (2021).

Development and application of water and land resources degradation index (WLDI). *Earth*, *2*, 515–531. Available from https://doi.org/10.3390/earth2030030.

Tsesmelis, D. E., Karavitis, C. A., Kalogeropoulos, K., Zervas, E., Vasilakou, C. G., Skondras, N. A., Oikonomou, P. D., Stathopoulos, N., Alexandris, S. G., Tsatsaris, A., & Kosmas, C. (2022a). Evaluating the degradation of natural resources in the Mediterranean environment using the water and land resources degradation index, the case of Crete Island. *Atmosphere*, *13*, 135. Available from https://doi.org/10.3390/atmos13010135.

Zheng, J., Li, G., Han, Z., & Meng, G. (2010). Hydrological cycle simulation of an irrigation district based on a SWAT model. *Mathematical and Computer Modelling*, *51*, 1312–1318. Available from https://doi.org/10.1016/j.mcm.2009.10.036.

Section 4

Geospatial modeling and analysis

Chapter 14

An integrated approach for a flood impact assessment on land uses/cover based on synthetic aperture radar images and spatial analytics. The case of an extreme event in Sperchios River Basin, Greece

Nikolaos Stathopoulos[1], Kleomenis Kalogeropoulos[2], Melpomeni Zoka[1], Panagiota Louka[3,4], Demetrios E. Tsesmelis[5] and Andreas Tsatsaris[2]

[1]Operational Unit "BEYOND Centre for Earth Observation Research and Satellite Remote Sensing", Institute for Astronomy, Astrophysics, Space Applications and Remote Sensing, National Observatory of Athens, Athens, Greece, [2]Department of Surveying and Geoinformatics Engineering, University of West Attica (UniWA), Athens, Greece, [3]Department of Natural Resources Development and Agricultural Engineering, Agricultural University of Athens, Athens, Greece, [4]Department of EU Projects, NEUROPUBLIC S.A., Piraeus, Greece, [5]Laboratory of Technology and Policy of Energy and Environment, School of Applied Arts and Sustainable Design, Hellenic Open University, Patras, Greece

14.1 Introduction

Floods are considered one of the most destructive natural phenomena worldwide that affect humans (e.g., lives, population transfer) and animal life, properties, crops, infrastructure (e.g., communication networks), as well as the environment (e.g., sedimentation, soil loss, and pollution) (Somasundaram et al., 2003; Tsesmelis, Karavitis et al., 2022; Brivio et al., 2002; Tsesmelis et al., 2021; Stathopoulos et al., 2017; Chalkias et al., 2016; Zoka et al., 2018; Diakakis et al., 2017; Maantay et al., 2009; Jonkman et al., 2005;

Geoinformatics for Geosciences. DOI: https://doi.org/10.1016/B978-0-323-98983-1.00015-6

Psomiadis et al., 2019). River flooding is caused by intense rainfall that leads to the accelerated accumulation and discharge of runoff from the upstream to the downstream (Stamellou et al., 2021). These phenomena have increased in several regions in the last decades due to urbanization and morphological changes. Furthermore, an aftereffect of climate change is the increment in the frequency of these events on Earth (Stamellou et al., 2021; Han et al., 2022; Chalkias et al., 2016; Stathopoulos et al., 2018; Deng et al., 2022). Considering the aforementioned environmental, economic, and social challenges that are associated with floods, a cornucopia of directives, actions, and plans has been established at a European and International level. A characteristic example of this at the European Union Directive "2007/60/EC (EUR-Lex, 2007)," which aims to obligate the Member States to draw up flood mitigation plans by adopting contemporary approaches, like "*Risk Management*" and "*Hazard Mapping*" (Tsatsaris et al., 2021; Kalogeropoulos et al., 2023; Handmer, 1980; Pelletier et al., 2005; Shakun et al., 2013; Pulvirenti et al., 2011; Webster et al., 2004; Hubert-Moy et al., 1992).

Greece is characterized by a northern Mediterranean climatic type, with a rainy season between November and April and hot, arid summers (Tsesmelis, Vasilakou et al., 2022). The country is mostly affected by flash floods, which are caused by intense rainfall in a short time period due to the fact that the majority of Greek drainage basins have a relatively small extent and steep slopes (Karalis et al., 2014; Gioti et al., 2013). In general, a flash flood is related to several factors such as rain intensity and duration, topography, soil conditions, plant coverage (or cover), forests' destruction, and urbanization (Elekkas, 2009).

The technological evolution of recent decades resulted in the generation of several approaches, aiming to support European and International flood prevention and mitigation plans. Among these approaches, the exploitation of geographical information systems (GIS) and remote sensing (RS) is highlighted and frequently used in various aspects of floods. In specific, GIS and RS are widely used in hydrological and hydraulic modeling (Gul et al., 2010; Chalkias et al., 2016; Kalogeropoulos & Chalkias, 2013; Stathopoulos et al., 2019; Tsanakas et al., 2016; Kourgialas et al., 2010; Fortin et al., 2001), surface runoff modeling (Gioti et al., 2013; Fugura et al., 2011), identifying the most optimum location for the creation of water reservoirs (Kalogeropoulos et al., 2011; Kalogeropoulos et al., 2020; Pissias et al., 2013; Luís and Cabral, 2021), assessing susceptible areas to flood events (Stamellou et al., 2021; Stathopoulos et al., 2017; Kalogeropoulos et al., 2013; Popescu et al., 2010), flood mapping (Zoka et al., 2018; Li et al., 2018; Psomiadis et al., 2019; Zhang et al., 2021; Konapala et al., 2021; Martinis et al., 2022; Tan et al., 2022; McCormack et al., 2022; Bekele et al., 2022), etc.

In this work, Sentinel-1 satellite images and high-resolution land use/cover data were used for the estimation of the destructive effects of a flood event that took place in the Sperchios river valley from the end of January until the beginning of February 2015. The investigated event caused damage to the infrastructure and crops of the region that covers the greatest part of the Sperchios plain. This chapter aims to present an easy-to-use methodological approach for fast assessment of flood effects with a particular emphasis on agricultural regions.

14.2 Study area and rainfall event

Sperchios river catchment (Fig. 14.1) is located in the northern part of East Central Greece and is delimited west from Timfristos (2315 m), north from Orthis, south from Vardousia (2285 m), the Oiti (2141 m), and Kallidromo (1419 m), and in the east part by the sea and specifically the Gulf of Maliakos. The total surface area of the catchment is 2039 km^2, and the average altitude is 607 m. The route of Spechios River is approximately 80 km.

Sperchios river catchment is occupied by dense, native-leaved, coniferous, and mixed forest, transitional (between forest and bushes) regions and areas with sclerophyllous vegetation that covers approximately 60%–65% of the total area of the catchment. One-third of the catchment (30%–35%), which lies mainly in the valley of the river, is used for agricultural activities (cultivated land, olive trees, vineyards, etc.). The smallest area of the catchment (1%–2%) refers to urban areas.

The phenomenon under study concerns the period between the 23rd of January and the 25th of February, 2015. This period was marked by extreme rainfall events in terms of duration and volume, especially at its peak. Although the Sperchios river basin has frequent flood runoffs, the investigated event is one of the rarest and most severe events that occurred in the study site in the last 100 years.

The studied phenomenon had a significant impact on the region's crops and infrastructure, as large areas were inundated for a long time, along with recorded damages to roads and bridges. According to the meteorological data of the National Observatory of Athens (NOA), the peak of the phenomenon took place on the 31st of January 2015 and is characterized by the maximum 3 hours accumulated rainfall of 7 mm (9–12 am) and 5.4 mm (6–9 pm) in the meteorological stations of Makrakomi and Lamia, respectively (Fig. 14.1). The total recorded rainfall for the whole duration of the phenomenon was 101.6 mm and 45.2 mm, accordingly.

14.3 Materials and methods

14.3.1 Data

The delineation of the flooded regions was implemented by analyzing images of the satellite mission of SENTINEL-1 (S1) (Table 14.1) (Biggin & Blyth, 1996).

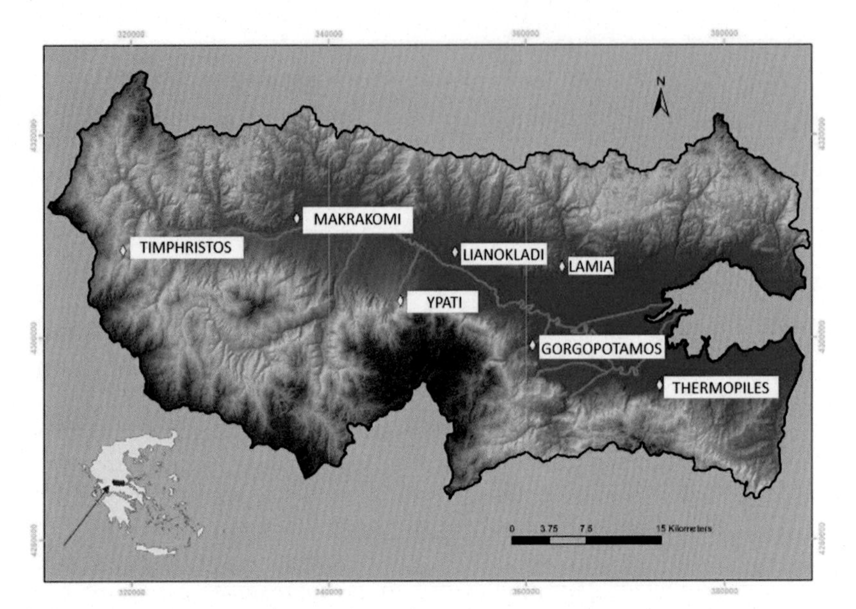

FIGURE 14.1 Water catchment of Sperchios River.

TABLE 14.1 Input datasets.

Datasets	Type	Characteristics	Sources
Satellite SAR images	SENTINEL-1A, Level 1 GRD, IW, HR, Dual Polarization, zip file	Spatial resolution of pixel 10 × 10 m	ESA – Copernicus Database (https://scihub.copernicus.eu)
Land use/ cover	Vector shapefile (polygon)	Polygons ilot	O.P.E.K.E.P.E.

In particular, the S1 satellite constellation provided images for the following dates: 2nd, 3rd, 8th, 9th, 14th, and 15th of February, 2015. The satellite's sensor belongs to the type of synthetic aperture radar (SAR) sensor, which is the most effective type for the study of such events, due to the fact that it is not affected by atmospheric conditions (e.g., cloud coverage) compared to optical satellites.

The utilized land use/cover background layer was provided by the Ilot data provided by the Greek Payment & Control Agency—OPEKEPE (Table 14.1). It is worth mentioning, that this dataset has a higher spatial resolution that the land use/cover data provided by CORINE (Copernicus Project).

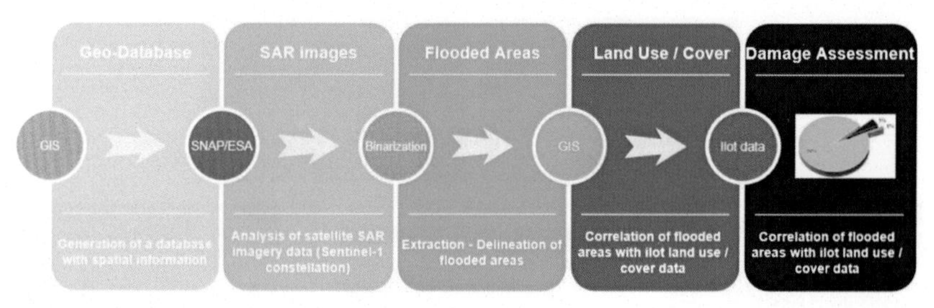

FIGURE 14.2 Flowchart.

The proposed methodology refers to the analysis of six S1 satellite images for the extraction of the continuously and periodically inundated areas throughout the duration of the flood event. Thence, the impact on the flooded areas was assessed by exploiting the detailed Ilot-based land use/cover data. The steps of the methodological approach are displayed in Fig. 14.2.

The analysis of each SENTINEL image was implemented by using the SNAP (Sentinel Application Platform) software which is provided at no charge by the European Space Agency (ESA) by following the next steps:

1. Choose a subset of the image—As the image covers a relatively broader region than the area of the study site, a subset of the image was used to facilitate and speed up the following processing steps [Fig. 14.3(1) and 14.3(2)].
2. Radiometric calibration—The result of this process is an output product that is characterized by a radiometrically calibrated backscatter value (the signal that is reflected by the surface and received by the sensor) [Fig. 14.3(3)].
3. Noise removal—Correct the "salt and pepper" effect, or in other words the multiple random spots of the imagery data [Fig. 14.3(4)].
4. Binarization—To separate the areas covered by water from the other areas (with no water), a threshold value should be applied. Thence, the pixels with the value of 1 corresponded to the class of water and shadows while the value of 0 represented the other classes [Fig. 14.3(5)–14.3(7)].
5. Geometrical correction—The aforementioned product was geometrically projected [Fig. 14.3(8)].
6. Extraction of flooded polygons—The water and shadow areas were converted into a polygon of information. The shadow areas were excluded by considering the orography, morphology (slope, aspect, and altitude) of the study area, and the expert knowledge of the researchers on the area and the specific event.

All the flood vector layers (polygons) that were generated by the analysis of the six satellite images were merged into one final polygon. This polygon refers to the maximum flooded area (throughout the whole duration of the event) displayed by combining satellite images (Fig. 14.3).

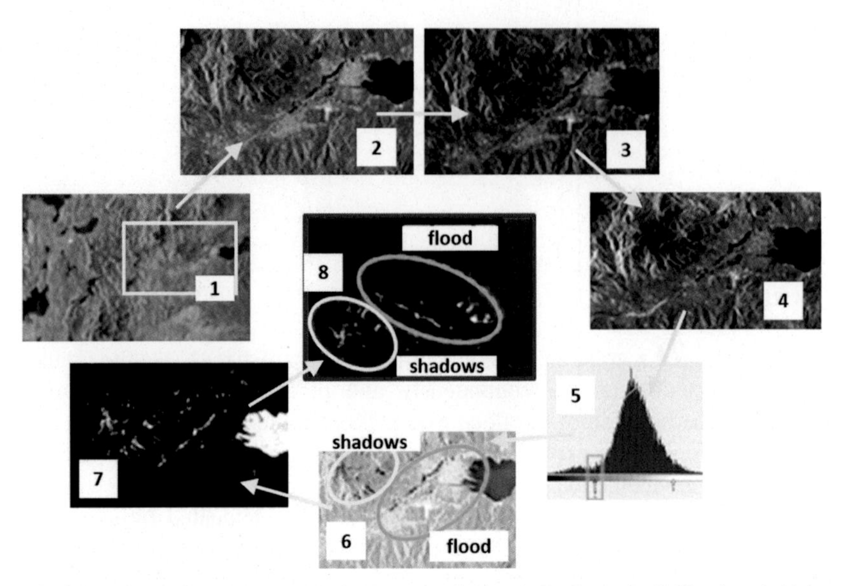

FIGURE 14.3 Satellite image analysis: 1. Image Subset, 2. Radiometric Calibration, 3. Noise Removal, 4. Binarization, 5. Geometrical Correction, 6. Extraction of Flooded polygons, 7. Initial Flood Map and 8. Differentiation between Flood and Shadows.

Finally, as previously mentioned, the flooded polygons were combined with the land use/cover data (Ilot polygons) in a GIS environment to assess the total impact of the flood in terms of land use/cover type. It aims to the categorization of the image by masking out the Ilot polygons information from the layer of the total extent of the inundated areas. Lastly, the results were statistically processed for the extraction of the conclusions and the assessment of the possible flood-related effects.

14.4 Results and discussion

The following map of Fig. 14.4 displays the categorization of the land use/cover data (Ilot) for the Sperchios river basin. The reddish hue portrays the flooded areas that were extracted by processing the SENTINEL-1 images.

It is observed that flooding was more extensive in the plain areas and close to the river bed and estuaries. The results of the implemented analysis are presented in Table 14.2. In particular, Table 14.2 includes the distribution of the land use/cover for the whole study site and for the regions where the flood was recorded.

The statistical analysis of the inundated area for each land use/cover category was generated by accounting for the data in Table 14.2, and the grouped results are displayed in Fig. 14.5A. According to the analysis of the imagery data, it was recorded that 2.3% of the cultivated areas were flooded.

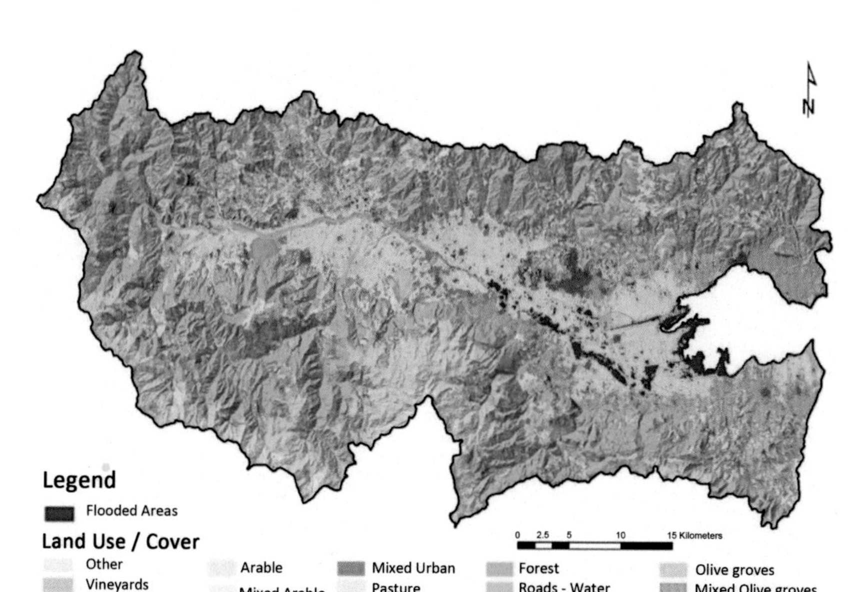

FIGURE 14.4 Land use/land cover and flooded areas map.

TABLE 14.2 Land use/land cover and flooded areas and percentages.

Code	Land use/cover class	Total area (km^2)	Flooded area (km^2)	Per total land use/land cover class extent – flood percentage (%)	Per total flood extent – flood percentage (%)
10	Forest	996.085	0.361	0.036	1.864
12	Forest Pasture	234.285	0.051	0.022	0.264
30	Natural Grassland	174.543	0.23	0.132	1.186
(10,12, 30)	Natural vegetation	1.404912	0.643	0.046	3.315
40	Arable	263.685	11.163	4.234	57.571
41	Mixed arable	76.499	0.56	0.732	2.889
50	Permanent crops	9.914	0.003	0.035	0.018

(Continued)

TABLE 14.2 (Continued)

Code	Land use/cover class	Total area (km²)	Flooded area (km²)	Per total land use/land cover class extent – flood percentage (%)	Per total flood extent – flood percentage (%)
51	Mixed permanent	13.287	0.014	0.103	0.071
60	Olive groves	115.627	0.081	0.07	0.42
61	Mixed olive groves	34.76	0.056	0.161	0.288
70	Vineyards	0.859	0	0	0
71	Mixed vineyards	2.171	0.002	0.071	0.008
(40,41, 50,51, 60,61, 70,71)	Cultivated areas	516.802	11.879	2.299	61.264
20	Urban	25.054	0.007	0.029	0.038
21	Mixed urban	4.356	0.001	0.028	0.006
91	Roads—water	29.111	0.66	2.268	3.404
92	Abandoned areas	8.813	0.001	0.009	0.004
90	Other	38.525	6.199	16.091	31.969
(20,21, 91,92, 90)	Artificial surfaces, water, etc.	105.859	6.868	6.488	35.422
Total		2.027573	19.39	0.956*	100

*Percentage of the total flooded area divided by the total area of the river basin.

More precisely, 4.23% of the total area of the arable lands was inundated, 0.73% of the mixed arable, and 0.44% of the permanent crops, olive crops, and vineyards were flooded, respectively. The percentage of flooded infrastructure surfaces and other land uses was the highest (6.49%), while the

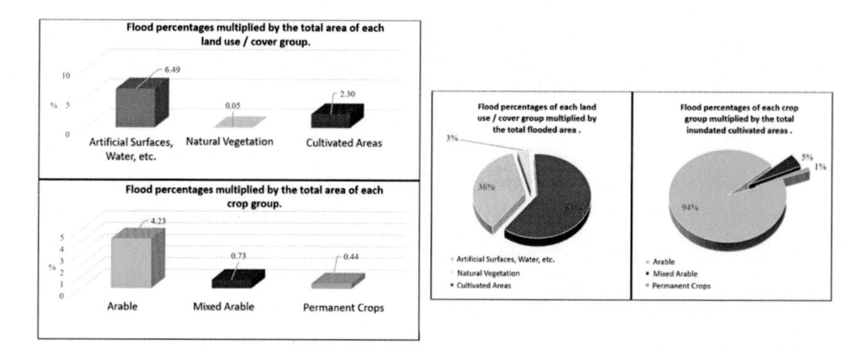

(A) Percentage of flood per land use – land cover. **(B) Apportionment of land use – land cover in the flooded areas.**

FIGURE 14.5 Allocation of the land use/land cover category in the flooded areas. (A) Percentage of flood per land use—land cover and (B) Apportionment of land use—land cover in the flooded areas.

lowest percentage of the flood was recorded in forest areas and natural grasslands (0.05%). The apportionment of the land use/cover of the flooded areas was calculated based on the information provided in Table 14.2, and it is illustrated in Fig. 14.5B. It is observed that 61% of the inundated regions referred to crops, 36% to infrastructure surfaces and other land uses, such as urban areas, roads, and surfaces covered by water (also before the flood event) while solely the 3% refer to forest areas and natural grasslands. Further analysis of the flooded cultivated areas shows that 94% of the affected areas belong to arable crops, 5% to mixed arable crops, and 1% to permanent crops, vineyards, and olive groves.

According to the above, it can be concluded that arable crops were more exposed in terms of a flood risk compared to the permanent and arboreal crops for the event under investigation. Further analysis of the flooding conditions of the Sperchios river basin and the generation of a flood time series of the region could provide the necessary information so as to deduce essential conclusions for the agricultural policies of the region as far as cultivation practices and flood-related prevention and mitigation measures are concerned.

14.5 Conclusions

In the present work, a methodology was presented for the utilization and processing of SAR and land use/cover Ilot data for the extraction of the inundated areas, the estimation of the allocation of the flood in the several land use/cover categories, and the assessment of the probable damages due to floods. The satellite SAR data, such as the SENTINEL's products, are suitable for the study of intense rainfall and flood events as they are not affected by atmospheric conditions in contrast to the optical satellite systems (e.g., Landsat). Furthermore, the high-resolution land use/cover Ilot data represent a fundamental basis for a detailed evaluation and appraisal of the possible damages and losses provoked by extreme events such as floods. The methodology was deemed efficient and fast for the analysis and assessment of the catastrophic effects stimulated by floods. A paramount advantage of the method is its low application cost as both the input data (SAR images, land use/cover) and software (SNAP and open-source GIS [e.g., QGIS]) are open and free of charge to everyone.

The proposed methodology represents an essential tool that stakeholders and decision-makers can utilize for the management of the negative effects of floods. The application of this methodology supports the procedure of taking the appropriate measures immediately as well as the related policy-making for restoring the damages that took place. Moreover, this methodology assists the process of taking protection/prevention actions for future phenomena, for example, the construction of antiflood structures.

It is noteworthy, that this methodological approach can be also utilized for the estimation of damages to crops, and thence support the estimation of the corresponding compensations from the competent bodies and the design of agricultural policies aiming to choose the appropriate and the most enduring crops in regions that are prone to flood events. Lastly, yet importantly, this methodology can be reclaimed and further optimized for research purposes by research institutes and universities.

References

Bekele, T. W., Haile, A. T., Trigg, M. A., & Walsh, C. L. (2022). Evaluating a new method of remote sensing for flood mapping in the urban and peri-urban areas: Applied to Addis Ababa and the Akaki catchment in Ethiopia. *Natural Hazards Research*, *2*(2), 97–110. Available from https://doi.org/10.1016/j.nhres.2022.03.001.

Biggin, D. S., & Blyth, K. (1996). A comparison of ERS-1 satellite radar and aerial photography for river flood mapping. *Water & Environment Journal, 10*, 59–64.

Brivio, P. A., Colombo, R., Maggi, M., & Tomasoni, R. (2002). Integration of remote sensing data and GIS for accurate mapping of flooded areas. *International Journal of Remote Sensing*, *23*(3), 429–441. Available from https://doi.org/10.1080/01431160010014729.

Chalkias, C., Stathopoulos, N., Kalogeropoulos, K., & Karymbalis, E. (2016). *Applied hydrological modeling with the use of geoinformatics: Theory and practice. Empirical modeling and its applications*. InTech. Available from https://doi.org/10.5772/62824.

Deng, Z., Wang, Z., Wu, X., Lai, C., & Zeng, Z. (2022). Strengthened tropical cyclones and higher flood risk under compound effect of climate change and urbanization across China's Greater Bay Area. *Urban Climate*, *44*, 101224. Available from https://doi.org/10.1016/j.uclim.2022.101224.

Diakakis, M., Deligiannakis, G., Katsetsiadou, K., Lekkas, E., Melaki, M., & Antoniadis, Z. (2017). Mapping and classification of direct effects of the flood of October 2014 in Athens. *Bulletin of the Geological Society of Greece*, *50*(2), 681. Available from https://doi.org/10.12681/bgsg.11774.

EUR-Lex. Directive 2007/60/EC of the European Parliament and of the Council of 23 October 2007 on the assessment and management of flood risks. Retrieved https://eur-lex.europa.eu/legal-content/EN/TXT/?uri = CELEX:32007L0060. Accessed October 26, 2022.

Elekkas. (2009) (in Greek). Retrieved from http://www.elekkas.gr/images/stories/pdfdocs/books/FUSIKES_KATASTROFES_09.pdf. Accessed October 2022.

Fortin, J., Turcotte, R., Massicotte, S., Moussa, R., Fitzback, J., & Villeneuve, J. (2001). Distributed watershed model compatible with remote sensing and GIS data I: Description of model. *Journal of Hydrologic Engineering*, *6*(2), 91–99. Available from https://doi.org/10.1061/(ASCE)1084-0699(2001)6:2(91).

Fugura, A. A., Billa, L., Pradhan, B., Mohamed, T. A., & Rawashdeh, S. (2011). Coupling of hydrodynamic modeling and aerial photogrammetry-derived digital surface model for flood simulation scenarios using GIS: Kuala Lumpur flood, Malaysia. *Disaster Advances*, *4*, 20–28.

Gioti, E., Riga, C., Kalogeropoulos, K., & Chalkias, C. (2013). A GIS-based flash flood runoff model using high resolution DEM and meteorological data. *EARSeL eProceedings*, 10.12760/01-2013-1-04.

Gul, G. O., Harmancioglu, N., & Gul, A (2010). A combined hydrologic and hydraulic modeling approach for testing efficiency of structural flood control measures. *Natural Hazards*, *54*(2), 245–260. Available from https://doi.org/10.1007/s11069-009-9464-2.

Han, X., Mehrotra, R., Sharma, A., & Rahman, A. (2022). Incorporating non-stationarity in regional flood frequency analysis procedures to account for climate change impact. *Journal of Hydrology*, *612*, 128235. Available from https://doi.org/10.1016/j.jhydrol.2022.128235.

Handmer, J. W (1980). Flood hazard maps as public information: An assessment within the context of the Canadian flood damage reduction program. *Canadian Water Resources Journal*, *5*, 82–110.

Hubert-Moy, L., Ganzetti, I., Bariou, R., & Mounier, J. (1992). Maps of flooded areas in Ille-et-Vilaine through remote sensing [Une cartographie des zones inondables en Ille-et-Vilaine par teledetection]. *Norois*, *155*, 337–347.

Jonkman, S. N., & Kelman, I. (2005). An analysis of the causes and circumstances of flood disaster deaths. *Disasters*, *29*(1), 75–97. Available from https://doi.org/10.1111/j.0361-3666.2005.00275.x.

Kalogeropoulos, K., & Chalkias, C. (2013). Modelling the impacts of climate change on surface runoff in small Mediterranean catchments: Empirical evidence from Greece. *Water and Environment Journal*, *27*, 505–513. Available from https://doi.org/10.1111/j.1747-6593.2012.00369.x.

Kalogeropoulos, K., Chalkias, C., Pissias, E., & Karalis, S. (2011). *Application of the SWAT model for the investigation of reservoirs creation. Advances in the research of aquatic environment* (pp. 71–79). Berlin, Heidelberg: Springer. Available from https://doi.org/10.1007/978-3-642-24076-8_9.

Kalogeropoulos, K., Karalis, S., Karymbalis, E., Chalkias, C., Chalkias, G., & Katsafados, P. (2013). Modeling flash floods in Vouraikos River mouth, Greece. *Proceedings of the 10th Global Congress on ICM: Lessons learned to address new challenges, EMECS 2013—MEDCOAST 2013 Joint Conference*, 1135–1146.

Kalogeropoulos, K., Stathopoulos, N., Psarogiannis, A., Pissias, E., Louka, P., Petropoulos, G. P., & Chalkias, C. (2020). An integrated GIS-hydro modeling methodology for surface runoff exploitation via small-scale reservoirs. *Water (Switzerland)*, *12*, 1–18. Available from https://doi.org/10.3390/w12113182.

Kalogeropoulos, K., Tsanakas, K., Stathopoulos, N., Tsesmelis, D. E., & Tsatsaris, A. (2023). Cultural heritage in the light of flood hazard: The case of the "Ancient" Olympia, Greece. *Hydrology*, *10*(3), 61. Available from https://doi.org/10.3390/hydrology10030061.

Karalis, S., Karymbalis, E., Valkanou, K., Chalkias, C., Katsafados, P., Kalogeropoulos, K., Batzakis, V., & Bofilios, A. (2014). Assessment of the relationships among catchments' morphometric parameters and hydrologic indices. *International Journal of Geosciences, 05*, 1571–1583. Available from https://doi.org/10.4236/ijg.2014.513128.

Konapala, G., Kumar, S. V., & Khalique Ahmad, S. (2021). Exploring Sentinel-1 and Sentinel-2 diversity for flood inundation mapping using deep learning. *ISPRS Journal of Photogrammetry and Remote Sensing, 180*, 163–173. Available from https://doi.org/10.1016/j.isprsjprs.2021.08.016.

Kourgialas, N. N., Karatzas, G. P., & Nikolaidis, N. P. (2010). An integrated framework for the hydrologic simulation of a complex geomorphological river basin. *Journal of Hydrology, 381*(3-4), 308–321. Available from https://doi.org/10.1016/j.jhydrol.2009.12.003.

Li, Y., Martinis, S., Plank, S., & Ludwig, R. (2018). An automatic change detection approach for rapid flood mapping in Sentinel-1 SAR data. *International Journal of Applied Earth Observation and Geoinformation, 73*, 123–135. Available from https://doi.org/10.1016/j.jag.2018.05.023.

Luís, A. D. A., & Cabral, P. (2021). Small dams/reservoirs site location analysis in a semi-arid region of Mozambique. *International Soil and Water Conservation Research, 9*(3), 381–393. Available from https://doi.org/10.1016/j.iswcr.2021.02.002.

Maantay, J., & Maroko, A. (2009). Mapping urban risk: Flood hazards, race, and environmental justice in New York. *Applied Geography, 29*, 111–124. Available from https://doi.org/10.1016/j.apgeog.2008.08.002.

Martinis, S., Groth, S., Wieland, M., Knopp, L., & Rättich, M. (2022). Towards a global seasonal and permanent reference water product from Sentinel-1/2 data for improved flood mapping. *Remote Sensing of Environment, 278*, 113077. Available from https://doi.org/10.1016/j.rse.2022.113077.

McCormack, T., Campanyà, J., & Naughton, O. (2022). A methodology for mapping annual flood extent using multi-temporal Sentinel-1 imagery. *Remote Sensing of Environment, 282*, 113273. Available from https://doi.org/10.1016/j.rse.2022.113273.

Pelletier, J. D., Mayer, L., Pearthree, P. A., House, P. K., Demsey, K. A., Klawon, J. E., & Vincent, K. R. (2005). An integrated approach to flood hazard assessment on alluvial fans using numerical modeling, field mapping, and remote sensing. *Bulletin of the Geological Society of America, 117*, 1167–1180.

Pissias, E., Psarogiannis, A., & Kalogeropoulos, K. (2013). *Water savings—A necessity in a changing environment. The case of small reservoirs*. Tinos, Greece: WIN4life.

Popescu, I., Jonoski, A., Van Andel, S. J., Onyari, E., & Moya Quiroga, V. G. (2010). Integrated modelling for flood risk mitigation in Romania: Case study of the Timis–Bega river basin. *International Journal of River Basin Management, 8*, 269–280.

Psomiadis, E., Soulis, K. X., Zoka, M., & Dercas, N. (2019). Synergistic approach of remote sensing and gis techniques for flash-flood monitoring and damage assessment in Thessaly plain area, Greece. *Water, 11*(3), 448. Available from https://doi.org/10.3390/w11030448.

Pulvirenti, L., Pierdicca, N., Chini, M., & Guerriero, L. (2011). An algorithm for operational flood mapping from synthetic aperture radar (SAR) data using fuzzy logic. *Natural Hazards and Earth Systems, 11*(2), 529–540. Available from https://doi.org/10.5194/nhess-11-529-2011, Sciences.

Skakun, S., Kussul, N., Shelestov, A., & Kussul, O. (2013). Flood hazard and flood risk assessment using a time series of satellite images: A case study in Namibia. *Risk Analysis, 34*(8), 1521–1537. Available from https://doi.org/10.1111/risa.12156.

Somasundaram, G., Norris, F. H., Asukai, N., & Murthy, R. S. (2003). Natural and technological disasters. *Trauma Interventions in War and Peace*, 291–318. Available from https://doi.org/10.1007/978-0-306-47968-7_13.

Stamellou, E., Kalogeropoulos, K., Stathopoulos, N., Tsesmelis, D. E., Louka, P., Apostolidis, V., & Tsatsaris, A. (2021). A GIS-cellular automata-based model for coupling urban sprawl and flood susceptibility assessment. *Hydrology*, *8*, 159. Available from https://doi.org/10.3390/hydrology8040159.

Stathopoulos, N., Kalogeropoulos, K., Dimitriou, E., Skrimizeas, P., Louka, P., Papadias, V., & Chalkias, C. (2019). A Robust remote sensing–spatial modeling–remote sensing (R-M-R) approach for flood hazard assessment. *Spatial Modeling in GIS and R for Earth and Environmental Sciences*, 391–410. Available from https://doi.org/10.1016/B978-0-12-815226-3.00017-X, Elsevier.

Stathopoulos, N., Kalogeropoulos, K., Polykretis, C., Skrimizeas, P., Louka, P., Karymbalis, E., & Chalkias, C. (2017). Introducing flood susceptibility index using remote-sensing data and geographic information systems empirical analysis in Sperchios River Basin, Greece. *Remote Sensing of Hydrometeorological Hazards*, 381–400. Available from https://doi.org/10.1201/9781315154947-18, CRC Press.

Stathopoulos, N., Skrimizeas, P., Kalogeropoulos, K., Louka, P., & Tragaki, A. (2018). *Statistical analysis and spatial correlation of rainfall in Greece for a 20-year time period.*

Tan, J., Chen, M., Ao, C., Zhao, G., Lei, G., Tang, Y., Wang, B., & Li, A. (2022). Inducing flooding index for vegetation mapping in water-land ecotone with Sentinel-1 & Sentinel-2 images: A case study in Dongting Lake, China. *Ecological Indicators*, *144*, 109448. Available from https://doi.org/10.1016/j.ecolind.2022.109448.

Tsanakas, K., Gaki-Papanastassiou, K., Kalogeropoulos, K., Chalkias, C., Katsafados, P., & Karymbalis, E. (2016). Investigation of flash flood natural causes of Xirolaki Torrent, Northern Greece based on GIS modeling and geomorphological analysis. *Nat. Hazards*, *84*, 1015–1033. Available from https://doi.org/10.1007/s11069-016-2471-1.

Tsatsaris, A., Kalogeropoulos, K., Stathopoulos, N., Louka, P., Tsanakas, K., Tsesmelis, D. E., Krassanakis, V., Petropoulos, G. P., Pappas, V., & Chalkias, C. (2021). Geoinformation technologies in support of environmental hazards monitoring under climate change: An extensive review. *ISPRS International Journal of Geo-Information*, *10*, 94. Available from https://doi.org/10.3390/ijgi10020094.

Tsesmelis, D. E., Karavitis, C. A., Kalogeropoulos, K., Tsatsaris, A., Zervas, E., Vasilakou, C. G., Stathopoulos, N., Skondras, N. A., Alexandris, S. G., Chalkias, C., & Kosmas, C. (2021). Development and application of water and land resources degradation index (WLDI). *Earth*, *2*, 515–531. Available from https://doi.org/10.3390/earth2030030.

Tsesmelis, D. E., Karavitis, C. A., Kalogeropoulos, K., Zervas, E., Vasilakou, C. G., Skondras, N. A., Oikonomou, P. D., Stathopoulos, N., Alexandris, S. G., Tsatsaris, A., & Kosmas, C. (2022). Evaluating the degradation of natural resources in the Mediterranean environment using the water and land resources degradation index, the case of Crete Island. *Atmosphere*, *13*, 135. Available from https://doi.org/10.3390/atmos13010135.

Tsesmelis, D. E., Vasilakou, C. G., Kalogeropoulos, K., Stathopoulos, N., Alexandris, S. G., Zervas, E., Oikonomou, P. D., & Karavitis, C. A. (2022). Drought assessment using the standardized precipitation index (SPI) in GIS environment in Greece. *Computers in Earth and Environmental Sciences*, 619–633. Available from https://doi.org/10.1016/B978-0-323-89861-4.00025-7, Elsevier.

Webster, T. L., Forbes, D. L., Dickie, S., & Shreenan, R. (2004). Using topographic lidar to map flood risk from storm-surge events for Charlottetown, Prince Edward Island, Canada.

Canadian Journal of Remote Sensing, *30*(1), 64–76. Available from https://doi.org/10.5589/m03-053.

Zhang, X., Chan, N. W., Pan, B., Ge, X., & Yang, H. (2021). Mapping flood by the object-based method using backscattering coefficient and interference coherence of Sentinel-1 time series. *Science of the Total Environment*, *794*, 148388. Available from https://doi.org/10.1016/j.scitotenv.2021.148388.

Zoka, M., Psomiadis, E., & Dercas, N. (2018). The complementary use of optical and SAR data in monitoring flood events and their effects. *EWaS3 2018*, *2*(11). Available from https://doi.org/10.3390/proceedings2110644.

Chapter 15

Quantitative comparison of geostatistical analysis of interpolation techniques and semiveriogram spatial dependency parameters for soil atrazine contamination attribute

Aqil Tariq[1,2]
[1]Department of Wildlife, Fisheries and Aquaculture, Mississippi State University, MS, United States, [2]State Key Laboratory of Information Engineering in Surveying, Mapping and Remote Sensing (LIESMARS), Wuhan University, Wuhan, P.R. China

15.1 Introduction

Discrimination of agrochemicals has resulted in contamination of agricultural lands and food crops. Agrochemicals, especially pesticides, harm the ecosystem and health of humans. It is the most widely used herbicide globally, and it is the most stable and dangerous of all pesticides. It is used to control weeds in crops such as corn, sugarcane, berries, sorghum, and other agricultural products (Mamián et al., 2009). It is also an essential pollutant due to its poor biodegradability and a high potential for degradation of surface and groundwater soil and water bodies (Mamián et al., 2009; Satsuma, 2009) through precipitation and leaching of soil (Fang et al., 2015; Mamián et al., 2009) and then reduce the quality of the water (Byer et al., 2011; Panshin et al., 2000).

There is increasing awareness regarding the dangers of atrazine to human safety, including acute eye and skin irritation and persistent impact on central nervous systems and immunity (Banat et al., 2010; Vonberg et al., 2014). It has long-term reproductive and endocrine-disrupting effects in living organisms, including humans(Hayes et al., 2002; Mukhtar et al., 2006).

Geoinformatics for Geosciences. DOI: https://doi.org/10.1016/B978-0-323-98983-1.00016-8

Among the regulated environmental degradation products (EDPs) listed in several countries, atrazine is the most contaminated, and due to this reason, European Union (EU) outlawed its usage in 2004. Atrazine has been classified as an herbicide with limited usage in the United States to prevent the chemical's potential environmental consequences. The United States of America Environmental Protection Agency (USA-EPA) initiated a comprehensive re-evaluation of atrazine in 2013 to determine whether or not additional limits are required to safeguard human health and safety.

The remaining study focused on the absorption, biotransformation, toxicity, and atrazine remediation aspects of the problem (Colborn et al., 1994; Lasserre et al., 2009; Mukhtar et al., 2006; Shah et al., 2021). Valuable findings have been obtained for recognizing the environmental destiny of atrazine, promoting improved herbicide control, and avoiding its unnecessary application. There is, however, limited information for characterizing the atrazine toxicity situation in soils. Furthermore, there is no proper monitoring technique that enables identifying potentially hazardous areas of atrazine contamination. This calls for an intervention of a framework where geographic information systems (GIS) modeling tools could help determine potential hazardous areas for pesticide residues.

GIS is a cost-effective and reliable alternative method for processing large amounts of data and can support quantitative statistical analysis to consider soil dynamics. Geostatistics is generally regarded as an excellent method for the spatial interpolation of soil parameters and their incoherence when measured in terms of the frequency of measurement errors and the cost of execution, among other things (Byer et al., 2011; Gokalp et al., 2010).

Previously, researchers tested the regional changeability of soil characteristics to determine spatial interaction of soil utilizing geospatial interpolation techniques (Hayes et al., 2002; Liu et al., 2014; Wills, 2005) and analyzed the spatial variability of biological and physio-chemical properties of soil (Bhunia et al., 2018; Hayes et al., 2002; Bureau of Pest Management Pesticide Product Registration Section, 2015; Zaya et al., 2011). However, spatial interpolation of the pollution position of atrazine in soils is usually limited, and there is a lack of knowledge on the most appropriate method of interpolation in soil atrazine prediction. Precision agriculture is consequently dependent on accurate calculations of the geographical distribution of the soil's atrazine status, which serves as a foundation for developing plans and strategies for decision-makers and policymakers. Thus successful environmental monitoring, modeling, and precision farming applications require excellent quality and economical soil data. Specifically, this chapter aims to conduct a detailed analysis of GIS-based interpolation techniques to estimate the spatial distribution of soil atrazine status in three tehsils (Rawalpindi, Taxila, and Gujar Khan) of the Rawalpindi district, which will provide fundamental data for risk management and atrazine monitoring.

15.2 Materials and methods

15.2.1 Study area

This study was carried out in three tehsils (Rawalpindi, Gujar Khan, and Taxila) of district Rawalpindi located in the Barani region of Punjab, Pakistan. In addition to having undulating surface topography and average annual precipitation of around 1200 mm, the region has undulating surface topography and undulating surface topography (Fig. 15.1). Aside from that, June and August are the wettest months of the year in this region. Heavy rainfall occurs in these months resulting in widespread eroding (Tariq et al., 2020).

15.2.2 Soil sampling and measurement

Using random sampling from the whole research region, 98 soil samples were obtained in the field in 2014 (36 from Rawalpindi, 32 from Gujar Khan, and 30 from Taxila soil samples, respectively). The coordinates of each sample site were obtained using a portable global positioning system (GPS). To create a hybrid soil sample, the undisturbed soil samples were collected at depths of 0–15 and 15–30 cm and mixed well together. Soil samples were dried at room temperature and passed through a filter with a 2 mm opening for laboratory analysis of atrazine residues in the soil.

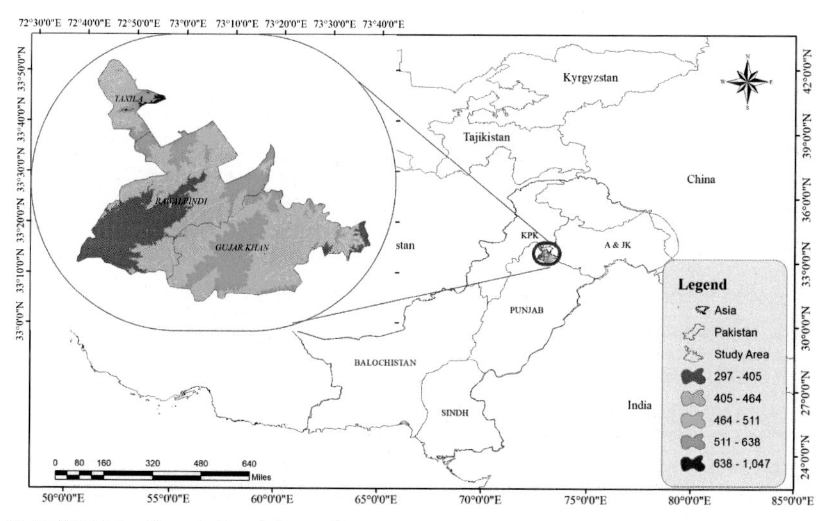

FIGURE 15.1 Description of the study area map.

15.2.3 Soil atrazine analysis

Whatman filter paper ($CHCl_3$) is used to extract atrazine with the use of therapeutic grade chloroform. The extracts were condensed to near dryness in a rotary evaporator at 40°C under reduced pressure after collecting chloroform from the samples. The final volume of the extract was 1 mL after the chloroform was recovered from the samples. Using high-performance liquid chromatography in the reverse phase, the extracted atrazine was dissolved in 30 mL of methanol and evaluated using the technique provided in this section by an expert in the field (Hutta et al., 2009).

15.2.4 Interpolation methods

A variety of deterministic interpolation techniques were used to generate. The spatial distribution of soil includes inverse distance weighted (IDW), local polynomial interpolation (LPI), smoothing degree, that is, radial base functions (RBF), and Geostatistical interpolation techniques, including ordinary kriging (OK) and Empirical Bayes (EBK), as described by Fang et al. (2015). The results were divided into two groups: validation samples, chosen from the homogeneously distributed samples to check the accuracy and efficiency of each interpolation method, and projection maps, chosen from the samples that were not chosen for validation samples.

15.2.4.1 Inverse distance weighted method

The IDW is one of the most often used deterministic interpolation algorithms in soil research, and it is focused on the areas immediately around known sites. The inversion of their width from the point of interpolation is used to determine the weights assigned to the interpolation locations. It is consequently formed so that the close points have more weights (and so have more significant influence) than the distant points and vice versa. The fact that the system is self-governing is obvious as it creates sample sites (Mousavifard et al., 2013; Shahbeik et al., 2014).

$$Z(x_o) = 1 + \frac{nx}{1!} + \frac{\sum_{i=1}^{n} \frac{x_i}{h_u^u}}{\sum_{i=1}^{n} \frac{1}{h_u^u}} \tag{15.1}$$

where $Z(x_o)$ represents the interpolated value, n represents the overall sample data value, x_i represents the ith data value, h_{ij} represents the difference between the interpolated value and the sample data value, and β represents the weighting force.

15.2.4.2 Local polynomial interpolation method

Instead of using all of the data, LPI is more suitable for local polynomials since it uses just points in the defined neighborhood (Seyedmohammadi et al., 2016). The surface value at the neighborhood's center is calculated to derive the expected value so that the communities can overlap. LPI is capable of producing surfaces that are capable of absorbing variations over a short time (Johnston et al., 2001).

15.2.4.3 Ordinary kriging method

When estimating unmeasured site values (un-sample locations) x^0, the OK interpolation strategy was employed on the assumption that $Z^*(x^0)$ is identical to the known, computed value-line total, the OK interpolation approach was used (field measured value). The Kriging technique is measured with the use of the following (15.2) (Shahbeik et al., 2014; Xie et al., 2008).

$$Z^*(x_o) = \sum_{i=0}^{n} \lambda_i Z(x_i) \tag{15.2}$$

where $Z^*(x_o)$ is the expected value at the location, $Z(x_i)$ is the actual value at the sampling site x_i, k_i is the weighting coefficient of the calculated site, and n is the number of sites scanned for interpolation within the neighborhood.

15.2.4.4 Radial base function method

When using the RBF, you may forecast values that are close to those measured at the same spot while also allowing each computed point to traverse through the surface you have generated. The predicted values will differ from the observed values due to a factor of two or three below or above the smallest of the stately values (Johnston et al., 2001). Using the RBF method, you may combine five deterministic, accurate interpolation methods: multi-quadratic (MQ), tension spline (SPT), thin-plate spline (TPS), inverse multi-quadratic (IMQ), and complete regularized spline (CRS). The RBF fits a surface based on the estimated sample values, resulting in a decrease in the overall curvature of the surface (Li et al., 2011). In situations when surface values fluctuate substantially within a short distance, RBF is rendered ineffective (Li & Heap, 2011; Li et al., 2011). CRS was chosen as the RBF for this research since it is the most often seen.

15.2.4.5 Empirical Bayes method

The most intricate features of a problem may be reduced to an automated process of subsetting and simulation with empirical Bayesian kriging. It is implicitly assumed by the EBK phase that the approximate semivariograms are the actual interpolation area semivariograms and that it is a linear approximation involving varying spatial damping in the approximate semivariograms. For geophysical corrections that interpolate spatially, the effect is a dependable nonstationary usable approach. When data coverage is high, this algorithm distributes local patterns, and when the data coverage is low, this algorithm makes bending to the prior means a priority (Bhunia et al., 2018; De Lange et al., 2010; Yu et al., 2009).

15.2.5 Semivariograms

When describing the spatial structure of variables, the semivariogram is particularly useful since it provides some insight into the processes that could be impacting the distribution of data (Marchetti et al., 2012). Based on the definition of the regionalized component and the underlying assumptions (Timm et al., 2006), a semivariogram is expressed as:

$$\gamma(h) = \frac{1}{2N(h)} \sum_{i=1}^{N(h)} [Z(x_i) - Z(x_i + h)]^2 \tag{15.3}$$

The semivariance $\gamma(h)$, the lag distance h, Z, the soil field parameter, $N(h)$, the number of pairs of sites split by the lag distance h, and the Z values at x_i and $x_i + h$ places are given by $Z(x_i + h)$ and $Z(x_i)$, respectively (Wang et al., 2015).

Because of the empirical semivariograms produced from the data, theoretical semivariogram models were used to construct geostatistical constraints, including nugget variance (C_0), sill variance ($C_0 + C_1$), distance parameter range (k) and organized variance (C_1). The nugget/sill ratio, $C_0/(C_0 + C_1)$, was determined better to understand the geographical dependency of values in the data. An average nugget/sill ratio of less than 25% implies significant geographical reliance, whereas an average nugget/sill ratio of more than 75% suggests low spatial dependence (Seyedmohammadi et al., 2016). The Gaussian, exponential, and spherical functions are the three often used geostatistics models, with the exponential function being the most popular (Wills, 2005). For features that are more susceptible to short-range fluctuation, the exponential and spherical models demonstrate linear behavior in origin, which makes them suitable models to use (Nixon, 2001).

$$\gamma(h) = \left\{ \begin{array}{c} c\left\{\frac{3h}{2a} - \frac{1}{2}\left(\frac{h}{a}\right)^3\right\} \text{for } h \leq a \\ c \text{ for } h > a \end{array} \right. \tag{15.4}$$

$$\gamma(h) = c\left\{1\text{-exp}\left(-\frac{\mathrm{h}}{a}\right)\right\} \tag{15.5}$$

$$\gamma(h) = c\left\{1 - \exp\left(-\frac{h^2}{a^2}\right)\right\} \tag{15.6}$$

where c is the sill, the gap from the lag, and a is the effective range (Bhunia et al., 2018; Wills, 2005). Arc GIS_v10.8 software used for construction of semivariogram parameters for modeling.

15.2.6 Cross-validation

As defined by Li & Heap (2011), the cross-validation approach was developed to assess and evaluate the effectiveness of various interpolation methods. The sample points were divided into two databases at random, with one database being used to train a process and the second database being used to test the model. During each cycle, the preparation and testing sets will be switched over to reduce variability and ensure that each data point is tested against the other. Several error measurement metrics, including root mean square error (RMSE), mean absolute error (MAE), and the determination coefficient (R^2) for error measurement, were calculated to assess the accuracy of the interpolation techniques. The RMSE is a set of error-based stages that define the accuracy of the interpolation techniques that are being employed.

$$\text{MAE} = \sum_{i=1}^{N} \left|Z^*(x_i)\text{-}Z(x_i)\right|/N \tag{15.7}$$

$$\text{RMSE} = \sqrt{\frac{\sum_{i=1}^{N} (Z(x_i)\text{-}Z^*(x_i))^2}{N}} \tag{15.8}$$

where N is the number of samples, $Z(x_i)$ is observed value at point x_i, and $\left(Z^*(x_i)\right)$ is the predicted value at point xi.

15.2.7 Delineation of base map

GPS coordinate reading was taken with Etrex Gemini, and base maps were then created using the Arc GIS (v.10.8) GIS program based on the result obtained from the atrazine residue study of collected samples.

15.3 Results and discussion

15.3.1 Descriptive statistics

The spatial variations in the soil atrazine were observed from the measured descriptive statistical values among all three tehsils of the studied area, as presented in Table 15.1. Measures of the tendency of soil atrazine were determined by mean and median, while its dispersion was computed in standard deviation, variance, and coefficient of skewness and kurtosis. The results showed significant differences between the mean values across the studied area. The mean highest atrazine content value (0.356) was observed in tehsil Rawalpindi, whereas the mean lowest value (0.262) was found in tehsil Taxila. The variance was ranged from 0.030 to 0.076, which showed differential degrees of heterogeneity among the atrazine contents of the

TABLE 15.1 Descriptive statistics of the soil atrazine status in the studied area.

Tehsil	Mean	Median	SD*	Variance	Skewness	Kurtosis
Rawalpindi	0.356	0.295	0.276	0.076	0.535	0.517
Gujar Khan	0.347	0.361	0.259	0.067	−1.411	−0.001
Taxila	0.262	0.245	0.174	0.03	0.687	0.455

SD*, standard deviation.

studied area. In the region, the number of variants might be attributed to varying degrees of variability in field management, such as land-use techniques, irrigation, and fertilization, as well as intrinsic factors, such as deforestation, flooding, and drainage conditions (Mousavifard et al., 2013). The coefficients of skewness and kurtosis were conducted to learn the effect on spatial analysis of the distribution structure of the atrazine products. In the cases of Rawalpindi and Taxila tehsils, positive values of skewness and kurtosis coefficients were found, but negative values of skewness and kurtosis coefficients were found in the case of Gujar Khan. The data within −1 to +1 one skewness and kurtosis, however, as defined (Di Virgilio et al., 2007; Timm et al., 2006).

15.3.2 Interpolation techniques comparison

The comparative analysis of interpolation strategies was achieved using cross-validation statistics as shown in Table 15.2, and their interpolated maps in Figs. 15.2–15.6. Interpolation was used in deterministic methods (IDW, LPI, and RBF) to evaluate spatial distributions of soil atrazine within the areas tested. The tests of the analysis revealed that LPI is more reliable than the other two processes. In Rawalpindi, Gujar Khan, and Taxila tehsils, the coefficient of determination (R^2) for LPI ranged from 0.863, 0.912, and 0.928 with RMSE values of 0.082, 0.072, and 0.036, respectively. Nevertheless, the RBF value showed less precision in the estimate system where R^2 was reported in Rawalpindi, Gujar Khan, and Taxila tehsils at 0.728, 0.746, and 0.847 with RMSE values of 0.091, 0.090, and 0.046, respectively.

In geostatistical approaches (OK and EBK), the spatial variation of the soil atrazine was interpolated. Analysis of data showed that OK gives more accuracy than the EBK method. The coefficient of determination (R^2) value for OK varied from 0.855, 0.697, and 0.846 with a smaller RMSE value of 0.103, 0.101, and 0.043 in Rawalpindi, Gujar Khan, and Taxila tehsils,

TABLE 15.2 Spatial distribution of soil atrazine through comparison interpolation techniques.

Interpolation	Rawalpindi			Gujar Khan			Taxila		
	MAE	RMSE	R^2	MAE	RMSE	R^2	MAE	RMSE	R^2
Deterministic methods									
IDW	0.064	0.072	0.824	0.056	0.065	0.845	0.036	0.046	0.832
LPI	0.060	0.082	0.863	0.060	0.072	0.912	0.031	0.036	0.928
RBF	0.074	0.091	0.728	0.068	0.091	0.746	0.031	0.046	0.847
Geostatistical methods									
OK	0.089	0.103	0.655	0.076	0.101	0.607	0.036	0.043	0.846
EBK	0.098	0.117	0.656	0.075	0.010	0.647	0.057	0.069	0.761

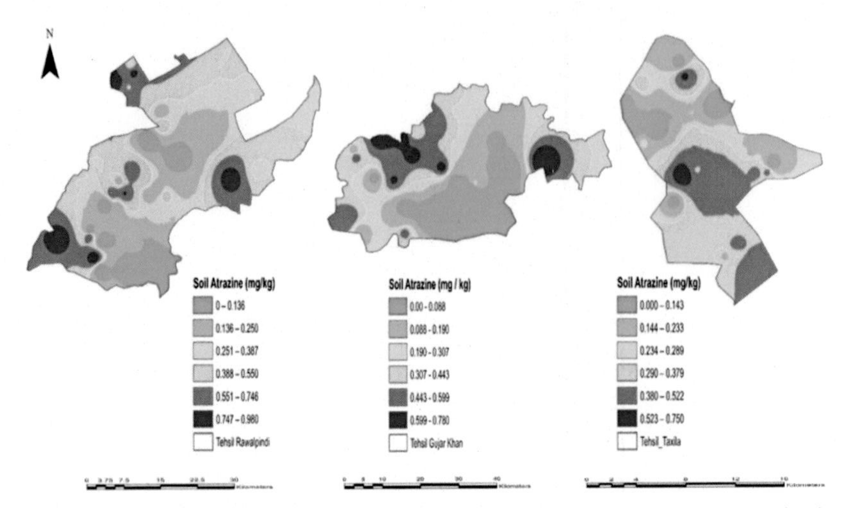

FIGURE 15.2 IDW deterministic interpolation map of spatial distribution of soil atrazine in studied area. *IDW*, inverse distance weighted.

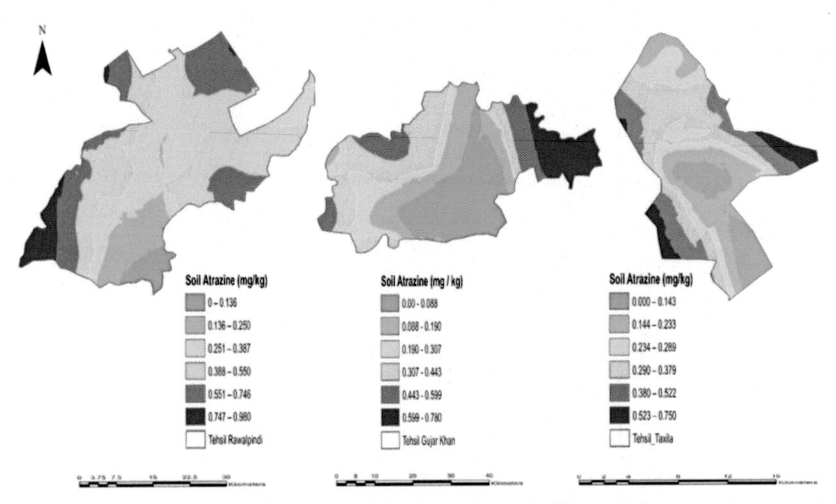

FIGURE 15.3 LPI deterministic interpolation map of spatial dispersal of soil atrazine in studied area. *LPI*, local polynomial interpolation.

respectively. OK gives an approximation around a calculated sample for the whole field (Mousavifard et al., 2013; Parrón et al., 2011).

Overall, the efficiency of deterministic methods was contrasted comprehensively with that of geostatistical methods. LPI showed high determination coefficients and low RMSE value, and MAE suggested a strong fit between

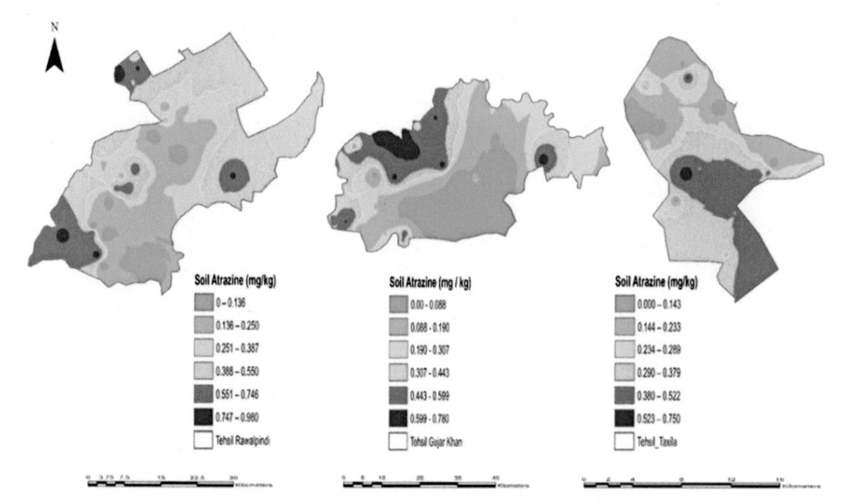

FIGURE 15.4 RBF deterministic interpolation map of spatial dispersal of soil atrazine in studied area. *RBF,* radial base functions.

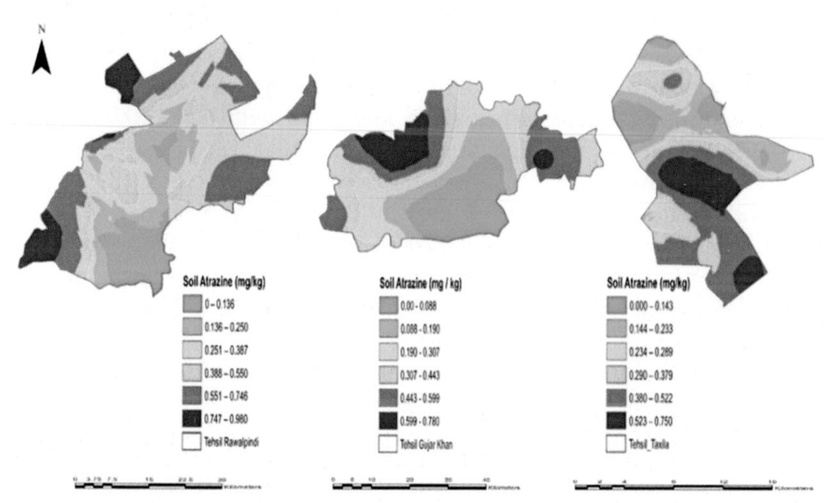

FIGURE 15.5 OK geostatistical interpolation map of spatial dispersal of soil atrazine in studied area. *OK*, ordinary kriging.

the observed and expected concentration of soil atrazine in Rawalpindi, Gujar Khan, and Rawalpindi Taxila tehsils. Consequently, LPI approaches often provide improved interpolation to approximate values at unknown positions.

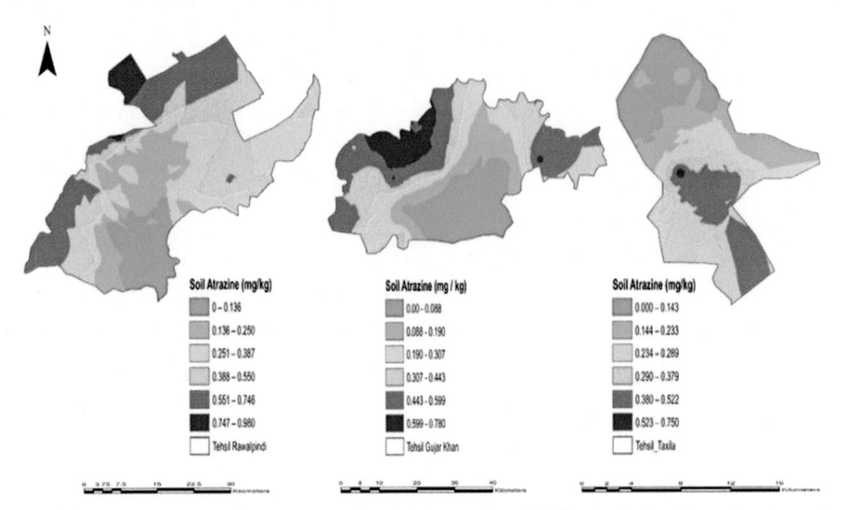

FIGURE 15.6 EBK geostatistical interpolation map of spatial dispersal of soil atrazine in studied area. *EBK*, Empirical Bayes.

15.3.3 Semivariograms spatial dependence of modeling

Table 15.3 shows the soil atrazine where variable characteristics were generated from semivariograms with the best-fitted model. Gaussian model showed the highest nugget effect (24% and 23%) with range (0.05, 0.06, and 0.039 km) tehsil Rawalpindi, Gujar Khan, and Taxila, respectively, while spherical model demonstrated the lowest nugget effect (89% and 83%) with range (0.06 and 0.264 km) in Tehsil Gujar Khan and Taxila, respectively. The Gaussian model showed strong spatial dependency with $C0/C0 + C1$ of value 23% in Tehsil Gujar Khan, while the linear model exhibited moderate spatial dependency with $C0/C0 + C1$ of value 43% in tehsil Rawalpindi.

The details obtained from ISO semivariograms Figs. 15.7–15.9 demonstrated the reality of specific spatial dependency for field vector samples collected soil atrazine. The nugget effect is correlated with spatial variation over distances shorter than the shortest gap between measurements (Lopez-Granados, 2002). Data analysis has shown in this research that the overall nugget effect means that an additional extensive number of sampling of soil atrazine at smaller distances may be required to identify spatial dependency of the parameter of the model semivariogram to produce more accurate results. A wide range suggests that other values determine the measured values of variable soil atrazine for this variable over more considerable distances than soil variables with narrower ranges (Gokalp et al., 2010). The nugget-to-sill ratio relates to the spatial dependency of soil products (Hayes et al., 2002; Lopez-Granados, 2002; Seyedmohammadi et al.,

TABLE 15.3 Semivariograms parameters with models for spatial dependence of soil atrazine.

Tehsil	Model	Nugget variance (C_0)	Sill variance ($C_0 + C_1$)	Range (km)	Nugget/sill(%) $C_0/(C_0 + C_1)$	Spatial dependence
Rawalpindi	Linear	0.053	0.092	0.413	43	Moderate
	Spherical	0.035	0.067	0.039	48	Moderate
	Gaussian	0.060	0.079	0.050	24	Strong
	Exponential	0.051	0.127	1.770	60	Moderate
Gujar Khan	Linear	0.039	0.081	0.230	52	Moderate
	Spherical	0.007	0.065	0.060	89	Weak
	Gaussian	0.048	0.063	0.067	24	Strong
	Exponential	0.031	0.085	0.390	63	Moderate
Taxila	Linear	0.013	0.045	0.114	71	Moderate
	Spherical	0.009	0.041	0.112	77	Weak
	Gaussian	0.022	0.029	0.039	23	Strong
	Exponential	0.009	0.055	0.264	83	Weak

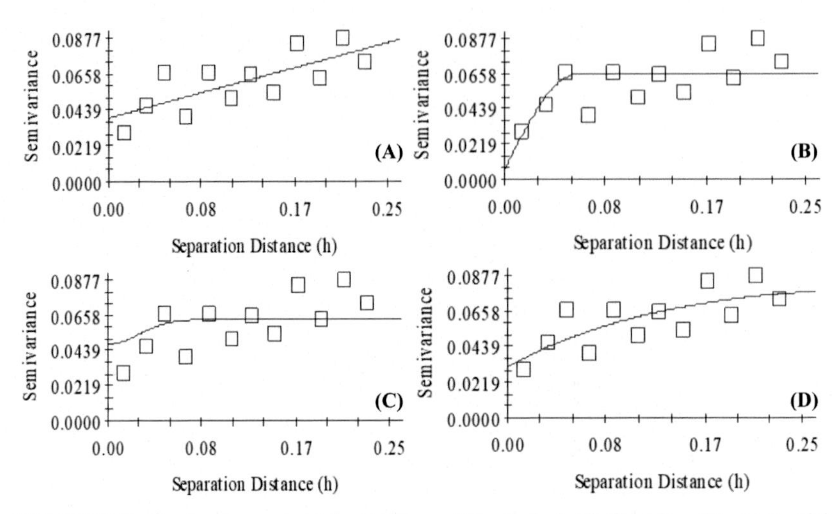

FIGURE 15.7 Semivariogram models: (A) linear, (B) spherical, (C) Gaussian, (D) exponential for the spatial dependency of soil atrazine variable in tehsil Gujar Khan.

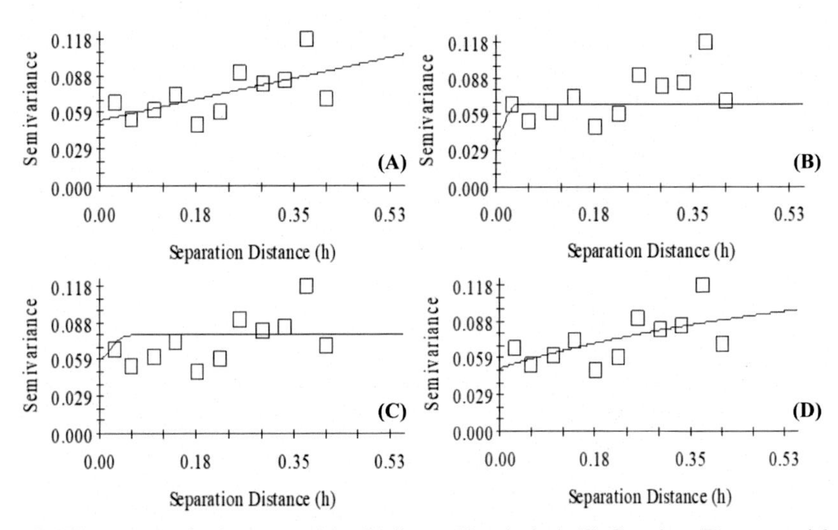

FIGURE 15.8 Semivariogram models: (A) linear, (B) spherical, (C) Gaussian, (D) exponential for the spatial dependency of soil atrazine variable in tehsil Rawalpindi.

2016). Similar parameters in this analysis to those (Timm et al., 2006) were used to describe the data presented in Table 15.3. The higher ratio suggests that the spatial heterogeneity is primarily due to stochastic influences such as agricultural steps such as the application of pesticides and

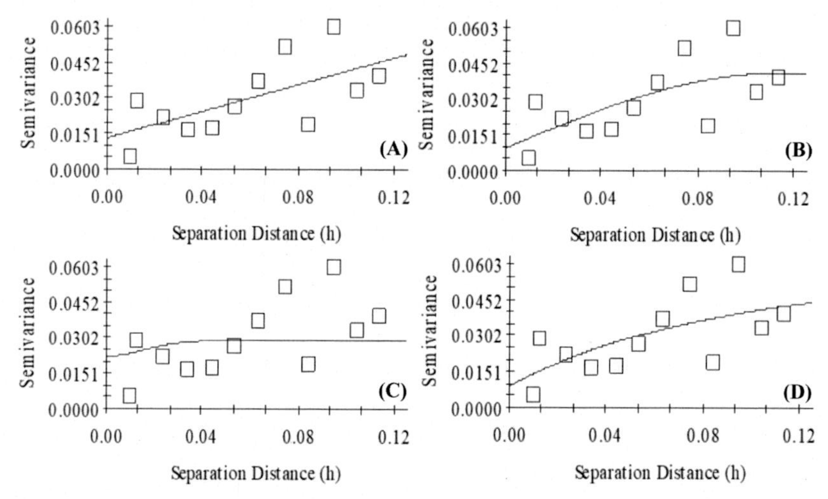

FIGURE 15.9 Semivariogram models: (A) linear, (B) spherical, (C) Gaussian, (D) exponential for the spatial dependency of soil atrazine variable in tehsil Taxila.

fertilizers, and other human activities in various cropping systems. The lower ratio indicates that structural factors such as parent material, topography, surface products, atmosphere, and other natural factors play an essential role in spatial variability (Goovaerts, 1997).

15.4 Conclusion

Because of its poor biodegradability and high potential to contaminate soil and water bodies, atrazine is one of the most commonly used, effective, poisonous herbicides, and important environmental pollutants. It has long-term reproductive and endocrine-disrupting effects in living organisms, including humans. There is limited information regarding characterization with proper monitoring techniques, which enables the identification of a potentially hazardous area of atrazine contamination. This calls for an intervention of a framework where GIS modeling tools in determining potential hazardous areas for soil atrazine contamination. A variety of deterministic interpolation (IDW, LPI, and RBF) or geostatistical interpolation (OK and EBK) techniques have been used in the present study to generate the spatial distribution of soil atrazine contamination attribute in three tehsils, namely Rawalpindi, Gujar Khan, and Taxila of Rawalpindi district. Descriptive statistical analysis of data showed that tehsil Rawalpindi had the mean highest atrazine content value (0.356 mg/kg), whereas tehsil Taxila had the mean lowest value (0.262 mg/kg). Among deterministic and geostatistical interpolation, the LPI reported a high value of the coefficients of determination (0.863) with a low value of RMSE (0.082) and MAE (0.060), suggesting a strong fit between the observed and expected soil atrazine pollution attributes. The semivariogram provided a clear explanation of the spatial structure of variables and some insight into potential mechanisms influencing the distribution of results. The Gaussian model demonstrated heavy spatial dependence (23%), whereas the spherical model displayed low spatial dependence (89%). Hence, the GIS-based LPI interpolation technique and semivariograms Gaussian modeling parameters might be an accurate alternative tool for providing basic data for risk management and control of soil atrazine contamination through spatial statistical analysis.

Acknowledgments

We would like to pay special and heart-whelming thanks to Prof. Muhammad Imran of the University of Arid Agriculture, Rawalpindi, and Saqib Abu Bakar, Soil Department, Govt. of Punjab, Pakistan for their facilitation at various stages of the field campaign.

Authors contribution

Aqil Tariq conducted the overall analysis and led the writing of the manuscript, design, data analysis, technical inputs, overall supervision for the research, and reviewed the paper.

Conflict of interests

The author declares that there is no conflict of interest in this manuscript's publication. Moreover, the writers have thoroughly addressed ethical issues, including plagiarism, informed consent, fraud, data manufacturing and/or falsification, dual publication and/or submission, and redundancy.

References

Banat, I. M., Franzetti, A., Gandolfi, I., Bestetti, G., Martinotti, M. G., Fracchia, L., Smyth, T. J., & Marchant, R. (2010). Microbial biosurfactants production, applications and future potential. *Applied Microbiology and Biotechnology*, *87*(2), 427–444, 01757598. Available from https://doi.org/10.1007/s00253-010-2589-0.

Bhunia, GS, Shit, PK, & Chattopadhyay, R (2018). Assessment of spatial variability of soil properties using geostatistical approach of lateritic soil (West Bengal, India). *Annals of Agrarian Science*, *16*(4), 436–443, 15121887. Available from https://doi.org/10.1016/j.aasci.2018.06.003.

Byer, J. D., Struger, J., Sverko, E., Klawunn, P., & Todd, A. (2011). Spatial and seasonal variations in atrazine and metolachlor surface water concentrations in Ontario (Canada) using ELISA. *Chemosphere*, *82*(8), 1155–1160. 00456535. https://doi.org/10.1016/j.chemosphere.2010.12.054. http://www.elsevier.com/locate/chemosphere.

Bureau of Pest Management Pesticide Product Registration Section. (2015). Active ingredient data package. Atrazine.

Colborn, T., vom Saal, F. S., & Soto, A. M. (1994). Developmental effects of endocrine-disrupting chemicals in wildlife and humans. *Environmental Impact Assessment Review, 14* (5-6), 469–489, 01959255. Available from https://doi.org/10.1016/0195-9255(94)90014-0.

De Lange, H. J., Sala, S., Vighi, M., & Faber, J. H. (2010). Ecological vulnerability in risk assessment – A review and perspectives. *Science of the Total Environment, 408*, 3871–3879. Available from https://doi.org/10.1016/j.scitotenv.2009.11.009.

Di Virgilio, N., Monti, A., & Venturi, G. (2007). Spatial variability of switchgrass (Panicum virgatum L.) yield as related to soil parameters in a small field. *Field Crops Research, 101*(2), 232–239, 03784902. Available from https://doi.org/10.1016/j.fcr.2006.11.009.

Fang, H., Lian, J., Wang, H., Cai, L., & Yu, Y. (2015). Exploring bacterial community structure and function associated with atrazine biodegradation in repeatedly treated soils. *Journal of Hazardous Materials, 286*, 457–465. 18733336. https://doi.org/10.1016/j.jhazmat.2015.01.006. http://www.elsevier.com/locate/jhazmat.

Gokalp, Z., Basaran, M., Uzun, O., & Serin, Y. (2010). Spatial analysis of some physical soil properties in a saline and alkaline grassland soil of Kayseri, Turkey. *African Journal of Agricultural Research*, 1991637X*5*(10), 1127–1137.

Goovaerts, P. (1997). Geostatistics for Natural Resources and Evaluation, Geostatistics for natural resources evaluation.

Hayes, T. B., Collins, A., Lee, M., Mendoza, M., Noriega, N., Stuart, A. A., & Vonk, A. (2002). Hermaphroditic, demasculinized frogs after exposure to the herbicide atrazine at low ecologically relevant doses. *Proceedings of the National Academy of Sciences of the United States of America, 99*(8), 5476–5480. 00274248. https://doi.org/10.1073/pnas.082121499. http://www.pnas.org.

Hutta, M., Chalányová, M., Halko, R., Góra, R., Dokupilová, S., & Rybár, I. (2009). Reversed phase liquid chromatography trace analysis of pesticides in soil by on-column sample pumping large volume injection and UV detection. *Journal of Separation Science, 32*(12), 2034–2042, 16159314. Available from https://doi.org/10.1002/jssc.200900036.

Johnston, K., Hoef., Krivoruchko, K., & Lucas, N. (2001). Using ArcGIS geostatistical analyst. *Analysis, 300.*

Lasserre, J. P., Fack, F., Revets, D., Planchon, S., Renaut, J., Hoffmann, L., Gutleb, A. C., Muller, C. P., & Bohn, T. (2009). Effects of the endocrine disruptors atrazine and PCB 153 on the protein expression of MCF-7 human cells. *Journal of Proteome Research, 8*(12), 5485–5496. 15353938. https://doi.org/10.1021/pr900480f. http://pubs.acs.org/doi/pdfplus/10.1021/pr900480f.

Li, J., & Heap, A. D. (2011). A review of comparative studies of spatial interpolation methods in environmental sciences: Performance and impact factors. *Ecological Informatics, 6*(3-4), 228–241, 15749541. Available from https://doi.org/10.1016/j.ecoinf.2010.12.003.

Li, X. F., Chen, Z. B., Chen, H. B., & Chen, Z. Q. (2011). Spatial distribution of soil nutrients and their response to land use in eroded area of South China. *Procedia Environmental Sciences, 10*, 14–19. 18780296. https://doi.org/10.1016/j.proenv.2011.09.004. http://www.sciencedirect.com/science/journal/18780296/10.

Liu, L., Wang, H., Dai, W., Lei, X., Yang, X., & Li, X. (2014). Spatial variability of soil organic carbon in the forestlands of northeast China. *Journal of Forestry Research, 25*(4), 867–876. 19930607. https://doi.org/10.1007/s11676-014-0533-3. http://link.springer.com/journal/11676.

Lopez-Granados, F. (2002). Spatial variability of agricultural soil parameters in southern Spain. *Plant and Soil, 246*(1), 97–105. Available from http://www.kluweronline.com/issn/0032-079X/contents.

Mamián, M., Torres, W., & Larmat, F. E. (2009). Electrochemical degradation of atrazine in aqueous solution at a platinum electrode. *Portugaliae Electrochimica Acta, 27*(3), 371–379, 08721904. Available from https://doi.org/10.4152/pea.200903371.

Marchetti, A., Piccini, C., Francaviglia, R., & Mabit, L. (2012). Spatial distribution of soil organic matter using geostatistics: A key indicator to assess soil degradation status in Central Italy. *Pedosphere, 22*(2), 230–242. 10001602. https://doi.org/10.1016/S1002-0160(12)60010-1. http://pedosphere.issas.ac.cn.

Mousavifard, S. M., Momtaz, H., Sepehr, E., Davatgar, N., & Sadaghiani, M. H. R. (2013). Determining and mapping some soil physico-chemical properties using geostatistical and GIS techniques in the Naqade region, Iran. *Archives of Agronomy and Soil Science, 59*(11), 1573–1589, 14763567. Available from https://doi.org/10.1080/03650340.2012.740556.

Mukhtar, M., Tahir, Z., Baloch, T. M., Mansoor, F., & Kamran, J. (2006). Entomological investigations of dengue vectors in epidemic-prone districts of Pakistan during. *Dengue Bull, 35*, 99–115.

Nixon, J. V. (2001). Introduction. *Religion and the Arts, 5*(1-2), 2–12. 16859252. https://doi.org/10.1163/156852901753498098. http://www.brill.com/religion-and-arts.

Panshin, S. Y., Carter, D. S., & Bayless, E. R. (2000). Analysis of atrazine and four degradation products in the pore water of the vadose zone, central Indiana. *Environmental Science and Technology, 34*(11), 2131–2137, 0013936X. Available from https://doi.org/10.1021/es990772z.

Parrón, T., Requena, M., Hernández, A. F., & Alarcón, R. (2011). Association between environmental exposure to pesticides and neurodegenerative diseases. *Toxicology and Applied Pharmacology, 256*(3), 379–385, 10960333. Available from https://doi.org/10.1016/j.taap.2011.05.006.

Satsuma, K. (2009). Complete biodegradation of atrazine by a microbial community isolated from a naturally derived river ecosystem (microcosm). *Chemosphere, 77*(4), 590–596. 00565354. https://doi.org/10.1016/j.chemosphere.2009.06.035. http://www.elsevier.com/locate/chemosphere.

Seyedmohammadi, J., Esmaeelnejad, L., & Shabanpour, M. (2016). Spatial variation modelling of groundwater electrical conductivity using geostatistics and GIS. *Modeling Earth Systems and Environment, 2*(4), 1–10. 36362121. https://doi.org/10.1007/s40808-016-0226-3. http://springer.com/journal/40808.

Shah, S., Yan, J., Ullah, I., Aslam, B., Tariq, A., Zhang, L., & Mumtaz, F. (2021). Classification of aquifer vulnerability by using the DRASTIC index and geo-electrical techniques. *Water, 13*(16), 2144. Available from https://doi.org/10.3390/w13162144.

Shahbeik, S., Afzal, P., Moarefvand, P., & Qumarsy, M. (2014). Comparison between ordinary kriging (OK) and inverse distance weighted (IDW) based on estimation error. case study: Dardevey iron ore deposit, NE Iran. *Arabian Journal of Geosciences, 7*(9), 3693–3704. 18665387. https://doi.org/10.1007/s12517-013-0978-2. http://www.springer.com/geosciences/journal/12517?cm_mmc = AD-_-enews-_-PSE1892-_-0.

Tariq, A., Riaz, I., Ahmad, Z., Yang, B., Amin, M., Kausar, R., Andleeb, S., Farooqi, M. A., & Rafiq, M. (2020). Land surface temperature relation with normalized satellite indices for the estimation of spatio-temporal trends in temperature among various land use land cover classes of an arid Potohar region using Landsat data. *Environmental Earth Sciences, 79*(1). 86662991. https://doi.org/10.1007/s12665-019-8766-2. https://link.springer.com/journal/12665.

Timm, L. C., Pires, L. F., Roveratti, R., Jacques Arthur, R. C., Reichardt, K., Martins De Oliveira, J. C., & Santos Bacchi, O. O. (2006). Field spatial and temporal patterns of soil water content and bulk density changes. *Scientia Agricola, 63*(1), 55–64. 678992X1. https://doi.org/10.1590/s0103-90162006000100009. http://www.scielo.br/pdf/sa/v63n1/27903.pdf.

Vonberg, D., Vanderborght, J., Cremer, N., Pütz, T., Herbst, M., & Vereecken, H. (2014). 20 years of long-term atrazine monitoring in a shallow aquifer in western Germany. *Water Research, 50*, 294–e306. 18792448. https://doi.org/10.1016/j.watres.2013.10.032. http://www.elsevier.com/locate/watres.

Wang, Y., Shao, M., Zhang, C., Liu, Z., Zou, J., & Xiao, J. (2015). Soil organic carbon in deep profiles under Chinese continental monsoon climate and its relations with land uses. *Ecological Engineering, 82*, 361–367. 09258574. https://doi.org/10.1016/j.ecoleng.2015.05.004. http://www.elsevier.com/inca/publications/store/5/2/2/7/5/1.

Wills, S. A. (2005). The spatial distribution of soil properties and prediction of soil organic carbon in Hayden Prairie and an adjacent agricultural field.

Xie, Z., Li, J., & Wu, W. (2008). Application of GIS and geostatistics to characterize spatial variation of soil fluoride on Hang-Jia-Hu Plain, China. *IFIP International Federation for Information Processing*, 15715736*258*, 253–266. Available from https://doi.org/10.1007/978-0-387-77251-6_28.

Yu, X. W., Wang, N., & Tang, T. T. (2009). Finite-element analysis of injectable calcium sulfate bone cement augmentation with dynamic hip screw system for the treatment of osteoporotic intertrochanteric fractures. *Shanghai Jiaotong Daxue Xuebao/Journal of Shanghai Jiaotong University*, 1006246*743*(11), 1813–1817.

Zaya, R. M., Amini, Z., Whitaker, A. S., Kohler, S. L., & Ide, C. F. (2011). Atrazine exposure affects growth, body condition and liver health in Xenopus laevis tadpoles. *Aquatic Toxicology, 104*(3-4), 243–253, 0166445X. Available from https://doi.org/10.1016/j.aquatox.2011.04.021.

Chapter 16

Comparison of "subjectivity" and "objectivity" in expert-based landslide susceptibility modeling

Christos Polykretis
Department of Natural Resources Management and Agricultural Engineering, Agricultural University of Athens, Athens, Greece

16.1 Introduction

According to Varnes (1984), natural hazard is defined a potentially damaging phenomenon which occurs in a given region and within a specified period of time. The particular environmental conditions due to the climate change, in combination with the not well-planned human interventions, have increased the frequency and severity of natural hazards worldwide over the recent decades (CRED, 2015). In this line, the scientific community and several stakeholders have shown a high interest in studying the natural hazards by assessing their spatial extents and intensities, and evaluating their current or potential effects.

Landslide is considered one of the major natural hazards, involving the movement of land mass (rocks, debris, earth, or a combination of them) from a given part of slope (depletion zone) to another (accumulation zone). Globally, the occurrence frequency of landslides seems to be on the rise; almost 5% of occurred natural hazards were related to landslides during 1995–2015 (CRED, 2015), while the same percentage was 11% during only the first semester of 2017 (CRED, 2017). Landslides seriously threaten both natural and man-made environments by mainly changes in terrain morphology and damages, or even destructions, of infrastructures (thus, economic losses). In some cases, they can also cause human injuries or life losses; during 1998–2017, landslides globally affected over 4.5 million people, with nearly 18,400 of them losing their life (CRED & UNISDR, 2018).

Due to the "without warning" occurrence, the determination of the time that a landslide is possibly to occur still constitutes a quite difficult task. Hence, most of the attention has been given to the more achievable task of

Geoinformatics for Geosciences. DOI: https://doi.org/10.1016/B978-0-323-98983-1.00017-X

determining the possible spatial location. Such spatial information can be obtained under the framework of landslide susceptibility (LS) assessment and mapping. LS refers to the possibility of landslide occurrence on the basis of particular terrain conditions (Brabb, 1984). The relevant assessment and mapping leads to the division of a given area of interest into homogeneous zones according to the degree of this possibility (Fell et al., 2008). The advancements in geoinformatic technologies, such as geographic information systems (GIS) and remote sensing (RS), as well as the wide availability of high-quality geospatial data, have significantly facilitated the LS assessment and mapping.

Nowadays, several modeling approaches are available for LS assessment at different spatial scales. Generally, they can be distinguished into two main groups: the qualitative and quantitative models. The qualitative models depend on knowledge and experience-based judgments of experts for weighting a given set of landslide-influencing, conditioning factors. They include the geomorphological analysis (Listo & Vieira, 2012), combination of index maps (Avtar et al., 2011) and logical models (Ahmed, 2015). On the other hand, the quantitative models depend on numerical expressions of the relationship between the factors and prior landslide occurrence. They include the bivariate or multivariate statistical analysis (Reichenbach et al., 2014; Mondal & Mandal, 2018; Pradhan & Kim, 2018; Wu & Song, 2018), geotechnical analysis (Vieira et al., 2018; Ciurleo et al., 2021), and machine learning models such as artificial neural networks, support vector machines, random forests, and decision trees (Su et al., 2015; Gorsevski et al., 2016; Hong et al., 2018; Taaleb et al., 2018).

Both qualitative and quantitative models incorporate the idea of weighting the conditioning factors considered for LS assessment. The basic difference between these two groups of models is the degree of objectivity. Due to their expert-based "nature," the qualitative models suffer from low objectivity (or high subjectivity) associated with the experts' judgments for weighting (Luo & Liu, 2018). The logical model namely analytical hierarchy process (AHP) is doubtless the most used of them (Chen et al., 2015; Sakkas et al., 2016; Mandal & Mandal, 2018). On the contrary, due to their landslide occurrence data-driven "nature," the quantitative models decrease bias in the weighting, resulting in high objectivity (Arabameri et al., 2019). The bivariate statistical analysis is one of the most preferable quantitative modeling approaches (Van Westen et al., 1997). It includes weighting in a "subfactor" level, that is, weights individually assigned to the classes of each factor, based on the observed densities of landslide data. Typical examples of bivariate statistical models are the frequency ratio (FR), information value (IV), and weight of evidence (WoE) (Kayastha, 2015; Youssef et al., 2016; Polykretis et al., 2021; Argyriou et al., 2022).

In some cases, the combination of the expert-based AHP with data-driven bivariate statistical models has been tested for LS assessment under the rationale of applying AHP for factor weighting and bivariate modeling for factor class weighting (Yang et al., 2015; Zhang et al., 2016; Zhou et al., 2016; Rehman et al., 2022). In this study, a different combination is proposed including the incorporation of three bivariate statistical models as individual objective "experts" in AHP modeling. Objective "judgments" derived from the data-driven FR, IV, and WoE models/"experts" were exploited for both factor and class weighting, resulting in objective AHP. The objective AHP was then compared with the typical AHP considering the judgments of actual (human) experts for factor and class weighting (subjective AHP). The two different AHP versions were tested in terms of LS assessment and mapping for a drainage basin of Greece. Furthermore, a validation procedure was conducted to comparatively evaluate the reliability of the resultant LS maps and thus the accuracy and prediction ability of the two applied modeling versions. Generally, as main purposes of the present study can be mentioned: (1) to develop a version of expert-based AHP modeling that improves the performance of the typical one, and (2) to highlight the impact of the subjectivity characterizing the (actual) experts' judgments on model's performance and LS outputs.

16.2 Materials and methods

16.2.1 Study area

The drainage basin of Selinous River—the largest Peloponnesian river, with a length of 49 km—was selected for investigation in this study. It is located in the central part of Greece and the northern part of geographical department of Peloponnese (Fig. 16.1), covering an extent of approximately 366 km^2. Administratively, it belongs to the Prefecture of Achaia, while the settlements situated within its boundaries offer residency to over 7000 residents according to the official 2011 census (Hellenic Statistical Authority, 2011). The basin's land is mostly covered by scrub vegetation, with a pronounced part being also used for agricultural purposes.

Sediments traversed by flow direction changes of the river in combination with the tectonic activity of transfer faults and the uplift rate (an average 2.2 mm/year) have significantly affected the topography of the basin. Its elevation rises steeply southwest-ward until 2145 m above sea level (a.s.l.) resulting in the existence of large and steep gorges. From a geological perspective, limestones and conglomerates constitute the dominant formations in the basin. Particularly, the upper reaches of the river flow across Pre-Neogene formations such as limestones (between 1000 and 1800 m a.s.l.), whereas its lower reaches mainly across Plio-Quaternary conglomeratic deposits (up to 1000 m a.s.l.) (Zelilidis, 2000).

Following the climatic regime of northern Peloponnese, the mean annual rainfall in the region ranges from 697 to 1178 mm, with the highest rainfall frequency and intensity being observed between November and January (Special

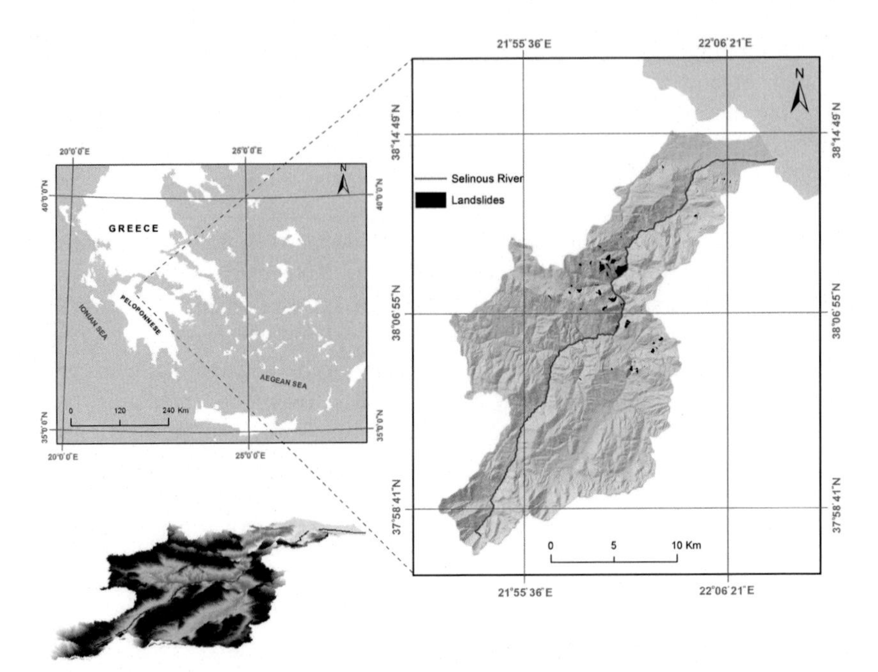

FIGURE 16.1 The landslide inventory map including a location map and a 3D presentation of the study area.

Secretariat for Water, 2012). During this wet period, the buildup of pore water pressures from the beating action of the rain loosens soil particles which are subsequently removed. Wide basins like this one drained by Selinous River generally contain perennial streams that rework the soil material from land movement events in their upper slopes (Karymbalis et al., 2016). An increased fluvial supply located around the active river mouths is the result of this material reworking.

16.2.2 Conditioning factors for overall modeling

In this study, seven conditioning factors that create suitable conditions for landslide activity by changing the state of a slope from stable to marginally stable (preparatory factors) were selected to be analyzed. These factors were the elevation, slope angle, curvature, distance to main roads, stream density, geology, and vegetation density. Literature suggestions (Pradhan and Kim, 2014; Su et al., 2015; Meng et al., 2016) for the specific spatial scale of analysis, the characteristics of the study area, and the data availability were taken into account for their selection.

In particular, presenting the height a.s.l., the elevation factor enables to recognize the local relief and locates points of maximum and minimum heights across a given terrain (Ayalew and Yamagishi, 2005). Hence, a digital representation of the terrain, such as the digital elevation model (DEM), and its derivatives of slope angle and curvature are of high usefulness. Regarding slope angle factor, its increase is typically correlated with an increased possibility of landsliding (Dai and Lee, 2002). Moreover, the curvature factor reflects the morphology of terrain, with its negative values representing the concave parts, zero values the flat parts, and positive values the convex parts (Saito et al., 2009). By retaining more rainfall water and erosion-induced sediment than convex, concave parts are correlated with a higher possibility of landsliding.

The factor of distance to roads constitutes an expression of the human impact on landslide activity, since road construction at the base of a slope tends to degrade its stability. Due to its effects on groundwater recharge, stream density is also an important factor for landslide activity. This factor determines the ratio of the total length of streams to the extent of a given region. A high stream density is correlated to low surface water infiltration and thus land movements with high velocity (Pradhan & Kim, 2014).

Geology can be considered one of the most crucial preparatory factors, since different geological formations have different slope instability performances in terms of strength and permeability. In addition, landslide activity is closely associated with vegetation density. Barren land is more prone to landslides as compared to land with higher vegetation density (Meng et al., 2016). The vegetation density is typically represented by the normalized difference vegetation index (NDVI). Ranging between -1 and 1 to indicate lack of vegetation and dense vegetation, respectively, this index is estimated using the near-infrared (NIR) and red (R) bands of satellite imagery data as follows (Karnieli et al., 2014):

$$\mathrm{NDVI} = \mathrm{NIR\text{-}R}/\mathrm{NIR} + \mathrm{R} \tag{16.1}$$

Various spatial datasets were collected and analyzed in GIS environment to represent the above conditioning factors. The main characteristics of them are shown in Table 16.1. Most of them were maintained in their primary raster format (grid), while others were converted from vector (point, line, or polygon features) to raster format with 20 m spatial resolution.

16.2.3 Landslide inventory for objective "experts"

Two different sources were firstly used for acquiring information that enable the spatial detection of the past landslides occurred in the study area. They consisted of: (1) a database with landslide records for northern Peloponnese during the period of 1906–2003, available by Tsagas (2011), and (2) a web platform maintained by the Laboratory of Engineering Geology at the Department of Geology at the University of Patras (accessed January 20, 2017), which provides information about landslide events occurred in

TABLE 16.1 Summary of the spatial datasets representing the conditioning factors.

Factors	Datasets	Data source	Primary format
Elevation	Vector layers of 5 m contours and elevation points	National Cadastre & Mapping Agency SA	Vector (lines and points)
Slope angle		DEM derived	Raster (grid)
Curvature		DEM derived	Raster (grid)
Distance to main roads	Main roads	OpenStreetMap	Vector (lines)
Stream density	Rivers and streams	General Use Map of Greece—1:50,000 (HMGS)	Vector (lines)
Geology	Geological formations	Geological Map of Greece—1:50,000 (IGME)	Vector (polygons)
Vegetation density (NDVI)	Sentinel-2 satellite images (10 m)	COPERNICUS	Raster (grid)

Prefectures of Achaia and Ilia (western Peloponnese) during the period of 1920–2015.

After the spatial detection, the delimitation of the detected landslides was conducted through the multitemporal interpretation of high-resolution satellites images provided by the Google Earth software package. Since it is not always possible to differentiate the boundaries of the landslide depletion and accumulation zones in an inventory map (Dagdelenler et al., 2016), these zones were assumed to be attributed to similar conditions and were mapped together generating a single polygon feature for each event. A total of 76 landslide events were eventually mapped as polygon features and included in the relevant inventory (Fig. 16.1). Following the classification of landslide types proposed by Varnes (1978), the mapped events in this case represent shallow rotational and translational slides as well as flows, varying in extent from some hundreds to several thousands of meters squared.

16.2.4 Analytical hierarchy process

The AHP originally developed by Saaty (1980) is a comprehensive decision-making approach which facilitates a complex multicriteria problem to be solved through a series of pairwise comparisons. Under each such comparison, two selected criteria are simultaneously evaluated by the significance (or preference) that is given to the one over the other from problem-related expert(s). The judgments of expert(s) are quantified using a value range of 1–9, as shown in Table 16.2; the lowest value (1) indicates an equal significance for the two selected criteria, and the highest value (9) indicates a extreme significance of the one over the other. The reciprocals of these values (between 1/2 and 1/9) indicate the inverse comparison results.

TABLE 16.2 Description of the value range used in the pairwise comparisons of AHP.

Value	Description
1	Equal significance
3	Moderate significance
5	Strong significance
7	Dominant significance
9	Extreme significance
2, 4, 6, 8	Intermediate significances
Reciprocals (1/2, 1/3, . . ., 1/9)	Inverse significances

AHP, analytical hierarchy process.

Based on the pairwise comparisons, a relevant matrix is generated to estimate weights for the involved criteria through its eigenvectors. To reduce the bias in decision-making, the calculation of a metric namely consistency ratio (*CR*) is required to indicate the probability that the matrix judgments were randomly generated (Zhou et al., 2016):

$$CR = CI/RI \tag{16.2}$$

where *RI* is a random index whose the values have been defined by Saaty (1980) depending on the number (*n*) of involved criteria (Table 16.3), and *CI* is a consistency index calculated as follows:

$$CI = \lambda_{max}\text{-}n/n\text{-}1 \tag{16.3}$$

where λ_{max} is the consistency vector, that is, the averaged value of each of calculated eigenvectors. A *CR* value greater than 0.10 reveals inconsistency, and thus the matrix judgments should be reconsidered.

16.2.5 Bivariate statistical models as objective "experts"

Three bivariate statistical models such as FR, IV, and WoE were considered as objective "experts" in this study. In terms of LS assessment, all the three models include class-level weight estimations based on the spatial association between the landslide inventory and each class of a given conditioning

TABLE 16.3 Random index (*RI*) values depending on the number (*n*) of involved criteria.

n	*RI*
1	0
2	0
3	0.52
4	0.89
5	1.11
6	1.25
7	1.35
8	1.4
9	1.45
10	1.49

factor. In particular, FR is defined by dividing the landslide ratio by the area ratio (Choi et al., 2012):

$$FR = \frac{Npix(Si)/\sum Npix(Si)}{Npix(Ni)/\sum Npix(Ni)} \tag{16.4}$$

where *Npix(Si)* is the number of landslide grid pixels in factor class *i*, and *Npix(Ni)* is the number of pixels in the same class. A *FR* value of 1 (average value) means that the landslide density in the class is proportional to the size of the class. If the value is greater than 1 then there is a high spatial association, whereas a value of less than 1 means a lower association.

The IV constitutes the result of the ratio of landslide density in a given factor class to the landslide density in entire factor (Van Westen, 1993):

$$IV = \ln\left(\frac{Npix(Si)/Npix(Ni)}{\sum Npix(Si)/\sum Npix(Ni)}\right) \tag{16.5}$$

The IV value can be either positive or negative; the higher (or lower) it is, the more (or less) significant the impact of the relevant factor class on landslide occurrence is.

The basic principles of WoE depend on Bayes' theorem (Bonham-Carter, 1994). Considering each factor class as an "evidence," two different weights are calculated according to the significance of its presence (W^+) or absence (W^-) on landslide occurrence (Guri et al., 2015):

$$W^+ = \ln\left(\frac{A1/A1 + A2}{A3/A3 + A4}\right) \tag{16.6}$$

$$W^- = \ln\left(\frac{A2/A1 + A2}{A4/A3 + A4}\right) \tag{16.7}$$

where *A1* is the number of landslide pixels present in a given factor class, *A2* is the number of landslide pixels not present in the class (or present in the rest of factor classes), *A3* is the number of pixels in the class in which no landslide pixels are present, and *A4* is the number of the pixels in which neither landslide nor the class is present. The difference between the two weights, known as contrast (*C*) determines quantitatively whether and how significant is the impact of each factor class on the occurrence of past landslides:

$$C = W^+ - W^- \tag{16.8}$$

A positive *C* value indicates a spatial association, whereas a negative value indicates that the association is lacking.

16.3 Landslide susceptibility assessment

16.3.1 Data (pre)processing

Two GIS-based data (pre)processing procedures took place under the general methodological framework. First, to apply the both AHP modeling approaches, the classification of the conditioning factors was required. The raster grids of the majority of factors being on a continuous numerical scale, such as elevation, slope angle, distance to main roads, and stream density, were divided into a number of discrete classes by the "natural breaks (jenks)" method (Fig. 16.2). In this classification method, class breaks identify the most similar within-group values and maximize the differences between classes according to the deviations about the median (Jenks, 1977). Exceptions to the specific classification option constituted the two factors of curvature and NDVI whose classification was executed in a manually way according to their numerical value meanings (Fig. 16.2). Moreover, the grid of the only factor being originally on a discrete classified scale, such as geology, was prepared by grouping its original classes based on more or less common characteristics (Fig. 16.2).

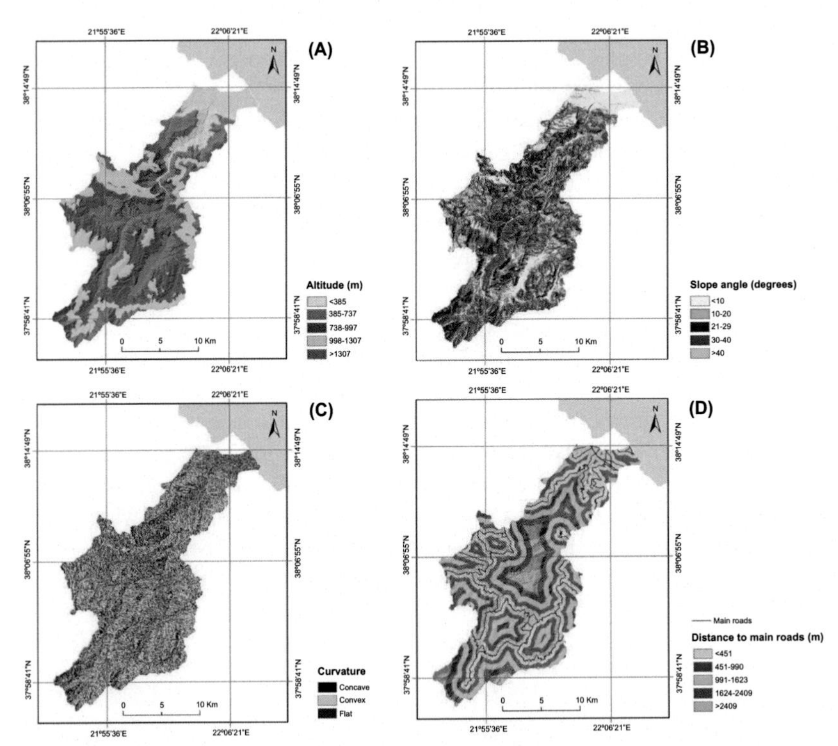

FIGURE 16.2 Conditioning factors (A) elevation, (B) slope angle, (C) curvature, (D) distance to main roads, (E) stream density, (F) vegetation density, and (G) geology.

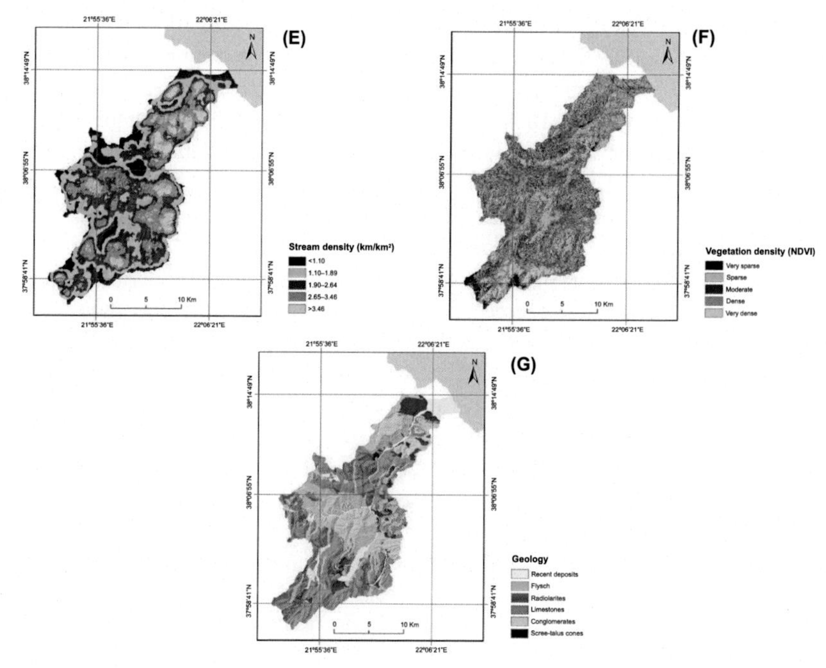

FIGURE 16.2 (Continued)

Then, for evaluating the LS outputs from the applied modeling approaches, a validation landslide dataset was necessary. In this line, the (pre)processing procedure of landslide data sampling was conducted. The sampling included the separation of the landslide inventory dataset into two subsets, the training and validation. In particular, among the total amount of 76 landslides contained in the inventory, 80% of them (61 in amount) were randomly selected for the training dataset. The remaining 20% (15 in amount) composed the validation dataset. Based on the sizes of mapped landslides and the spatial resolution of factor data, the entire study area was then tiled into grid pixels of 20×20 m as the basic analysis unit, resulting in 5140 training and 446 validation landslide pixels. It is worth mentioning that the training landslide dataset was beneficial only to the objective AHP modeling; specifically, for the implementation of the three data-driven statistical models considered as objective "experts."

16.3.2 Objective analytical hierarchy process modeling

After the data (pre)processing procedures, the implementation of the proposed objective AHP modeling approach included some distinct steps. First, by overlaying the landslide training dataset with each grid of conditioning factors, the appropriate bivariate statistics of the three different FR, IV, and WoE models were estimated for each factor class based on the relevant Eqs. (4, 5, and 8) (Table 16.4). These estimations enabled to quantify the impact of each factor class on landslide occurrence. The standard deviation of the class-level values was also evaluated to quantify the impact of each conditioning factor (Table 16.5). All the estimated values were normalized to be converted into a common scale.

A homogeneous group of "experts" represented by the three bivariate statistical models, with equal degree of importance for each one, was then considered. Accordingly, an overall "expert-based judgment" was the average of the normalized values of models. These quantitative (objective) "judgments" were subsequently expressed in the value range of 1–9 (and their reciprocals) as defined by AHP, to generate pairwise comparison matrices for the classes of each factor as well as the factors in themselves (Tables 16.6 and 16.7). Through the eigenvectors in matrices, both class-level and factor-level weights were derived (Tables 16.4 and 16.5). It is worth noting that the CR value calculated for each of the matrices was found to be lower than 0.10 revealing none inconsistency issue.

The extraction of the class-level and factor-level weights was followed by their aggregation in order to estimate the overall LS score. This step was achieved through a GIS-based weighted linear combination:

$$LS = \sum\nolimits_{i=1}^{n} fw_i \times cw_i^j \tag{16.9}$$

where fw_i is the weight of a given factor i, $cw^j{}_i$ is the weight for a given class j of factor i, and n is the number of factors. As last steps, the estimated score was divided into five discrete susceptibility classes ("very low," "low," "moderate," "high," and "very high" susceptibility) according to the "natural breaks (jenks)" option, and their spatial distribution was cartographically visualized by a LS map (Fig. 16.3).

16.3.3 Subjective analytical hierarchy process modeling

Under the typical subjective AHP modeling approach, three independent experts, with scientific background and experience in the fields of engineering geology and geosciences, were initially invited to express their opinion about the significance between the conditioning factors in relevant pairwise comparisons as well as about the level of association of their classes with landslide activity. For acquiring the experts' opinions, a web-based questionnaire was created (https://christos12.typeform.com/to/ihgk4B). In this case also, it was considered the experts to compose a homogeneous group with equal degree of importance for each one. Accordingly, the overall expert-based judgment was the average of the three (subjective) judgment opinions.

TABLE 16.4 Statistics from the FR, IV, and WoE models, and weights from the objective and subjective AHP modeling approaches for the classes of all conditioning factors.

Factor classes	FR	IV	C	Objective AHP weight	Subjective AHP weight
Elevation (m)					
[1] <385	0.82	−0.19	−0.22	0.122	0.044
[2] 385 − 737	1.44	0.37	0.50	0.459	0.076
[3] 738 − 997	1.02	0.02	0.03	0.194	0.144
[4] 998 − 1307	0.97	−0.03	−0.04	0.194	0.268
[5] >1307	0.03	−3.61	−3.70	0.031	0.468
Slope angle (degrees)					
[1] <10	0.05	−2.93	−3.12	0.041	0.042
[2] 10 − 20	0.14	−1.95	−2.23	0.062	0.062
[3] 21 − 29	0.34	−1.08	−1.32	0.099	0.132
[4] 30 − 40	2.09	0.74	1.12	0.251	0.260
[5] >40	7.22	1.98	2.45	0.548	0.504
Curvature					
[1] Concave	1.14	0.13	0.27	0.723	0.643
[2] Convex	0.88	−0.13	−0.24	0.216	0.283
[3] Flat	0.63	−0.46	−0.47	0.061	0.074
Distance to main roads (m)					
[1] <451	0.38	−0.97	−1.33	0.045	0.497
[2] 451 − 990	0.39	−0.94	−1.14	0.045	0.262
[3] 991 − 1623	1.01	0.01	0.01	0.107	0.136

(*Continued*)

TABLE 16.4 (Continued)

Factor classes	FR	IV	C	Objective AHP weight	Subjective AHP weight
[4] 1624 − 2409	4.39	1.48	2.05	0.565	0.069
[5] >2409	2.36	0.86	0.91	0.238	0.037
Stream density (km/km^2)					
[1] <1.10	0.53	− 0.64	− 0.72	0.044	0.059
[2] 1.10 − 1.89	1.48	0.39	0.57	0.280	0.097
[3] 1.90 − 2.64	0.53	− 0.63	− 0.82	0.044	0.159
[4] 2.65 − 3.46	0.87	− 0.14	− 0.18	0.101	0.259
[5] >3.46	2.29	0.83	0.98	0.532	0.426
Vegetation density (NDVI)					
[1] Very sparse	2.04	0.71	0.79	0.533	0.444
[2] Sparse	1.44	0.36	0.46	0.267	0.262
[3] Moderate	0.86	− 0.15	− 0.20	0.069	0.153
[4] Dense	0.68	− 0.39	− 0.52	0.037	0.089
[5] Very dense	0.96	− 0.04	− 0.05	0.094	0.053
Geology					
[1] Recent deposits	0.08	− 2.57	− 2.64	0.033	0.129
[2] Flysch	0.53	− 0.63	− 0.66	0.131	0.493
[3] Radiolarites	1.23	0.20	0.23	0.237	0.056
[4] Limestones	1.93	0.66	1.67	0.479	0.072
[5] Conglomerates	0.16	− 1.83	− 2.14	0.048	0.036
[6] Scree-talus cones	0.24	− 1.44	− 1.51	0.072	0.213

AHP, analytical hierarchy process; *FR*, frequency ratio; *IV*, information value; *WoE*, weight of evidence.

TABLE 16.5 Standard deviations of statistics from the FR, IV and WoE models, and weights from the objective and subjective AHP modeling approaches for all conditioning factors.

Factor classes	FR_{STDV}	IV_{STDV}	C_{STDV}	Objective AHP weight	Subjective AHP weight
Elevation	0.52	1.65	1.70	0.148	0.028
Slope angle	3.05	1.99	2.33	0.471	0.327
Curvature	0.25	0.29	0.38	0.029	0.064
Distance to main roads	1.70	1.08	1.42	0.148	0.116
Stream density	0.75	0.64	0.79	0.064	0.087
Vegetation density (NDVI)	0.55	0.44	0.53	0.042	0.049
Geology	0.74	1.24	1.60	0.097	0.327

AHP, analytical hierarchy process; *FR*, frequency ratio; *IV*, information value; *WoE*, weight of evidence.

TABLE 16.6 Indicative pairwise comparison matrices for the classes of elevation factor in objective and subjective AHP modeling approaches.

Elevation classes	[1]	[2]	[3]	[4]	[5]
Objective AHP					
[1] <385	1				
[2] 385 – 737	4	1			
[3] 738 – 997	2	1/3	1		
[4] 998 – 1307	2	1/3	1	1	
[5] >1307	1/6	1/9	1/7	1/7	1
Subjective AHP					
[1] <385	1				
[2] 385 – 737	2	1			
[3] 738 – 997	4	2	1		
[4] 998 – 1307	6	4	2	1	
[5] >1307	8	6	4	2	1

AHP, analytical hierarchy process.

TABLE 16.7 Pairwise comparison matrices for the conditioning factors in objective and subjective AHP modeling approaches.

Factors	[1]	[2]	[3]	[4]	[5]	[6]	[7]
Objective AHP							
[1] Elevation	1						
[2] Slope angle	5	1					
[3] Curvature	1/5	1/9	1				
[4] Distance to main roads	1	1/5	5	1			
[5] Stream density	1/3	1/7	3	1/3	1		
[6] Vegetation density (NDVI)	1/4	1/8	2	1/4	1/2	1	
[7] Geology	1/2	1/6	4	1/2	2	3	1
Subjective AHP							
[1] Elevation	1						
[2] Slope angle	8	1					
[3] Curvature	3	1/7	1				
[4] Distance to main roads	4	1/5	2	1			
[5] Stream density	4	1/5	2	1/2	1		
[6] Vegetation density (NDVI)	3	1/7	1/2	1/3	1/2	1	
[7] Geology	7	2	5	3	4	6	1

AHP, analytical hierarchy process.

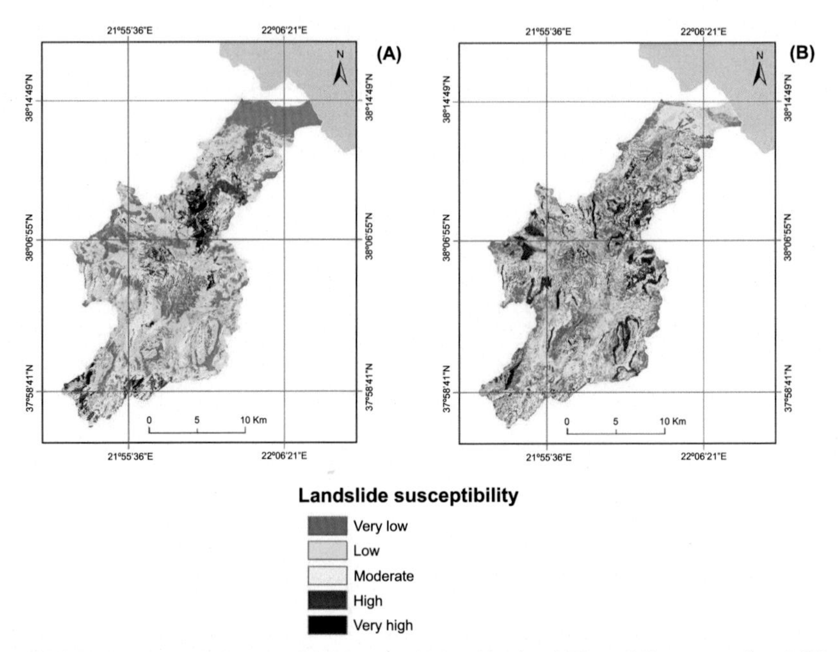

FIGURE 16.3 Landslide susceptibility maps (A) by objective AHP modeling approach and (B) by subjective AHP modeling approach. *AHP*, analytical hierarchy process.

By carrying out afterwards the same steps described previously for the objective approach, the pairwise comparison matrices were generated (Tables 16.6 and 16.7), the class-level and factor-level weights (Tables 16.4 and 16.5) were extracted, the overall LS score was estimated by Eq. (5), and the spatial distribution of relevant classes was cartographically visualized (Fig. 16.3). For one more time, the CR values calculated for the pairwise comparison matrices were found to be lower than 0.10 revealing consistency.

16.4 Results

The weights from the two different AHP approaches applied for this study are summarized in Tables 16.4 and 16.5. According to the proposed objective "expert"-based approach, among the conditioning factors, the highest weight was obtained from slope angle (0.471). Its class with the highest weight was the "greater than 40 degrees" (0.548). The slope angle was followed by distance to main roads and elevation (both with weight equal to 0.148). For these factors, the classes with the highest weights were the "1624 to 2409 m" (0.565) and "385 to 737 m a.s.l." (0.459), respectively. Curvature was the factor with the lowest weight (0.029). On the other hand, from the typical subjective expert-based approach, the highest factor-level weights were shown from slope angle (0.328) and geology (0.327). Their classes with the highest weights were the "greater than 40 degrees" (0.504) for the former, and the "flysch" (0.493) for the latter. Elevation was found to be the factor with the lowest weight (0.028).

The LS maps produced by the two modeling approaches are presented in Fig. 16.3. In the map from the objective AHP, the "high" and "very high" susceptibility seem to be mainly located in the central part of the drainage basin of Selinous River, with some large pockets of "high" susceptibility in the southern part. These two susceptibility classes cover 7% and 5%, respectively, of the study area (Fig. 16.4). In the map from the subjective AHP, large pockets of "high" and "very high" susceptibility are detected almost throughout the basin. The coverage percentages for these susceptibility classes are 10% and 6%, respectively (Fig. 16.4).

Moreover, the cross-comparison of the two resultant maps in terms of coverage differences between their classes (Table 16.8) indicated that the highest difference percentages (10% and 9%, respectively) were shown between the ''very low'' and ''low'' susceptibility classes of the two modeling approaches. It is also worth noting that no coverage difference was observed between the ''high'' and ''very high'' classes of the objective approach, and the ''very low'' class of the subjective approach. On the contrary, 4% and 2% of the ''high'' and ''very high'', classes of the subjective approach were characterized with ''very low'' susceptibility in the map from the objective approach. Concerning the coverage similarities, it is derived that 1% of the study area was characterized by similar susceptibility for each of two, "high" and "very high", classes. The corresponding percentages were 10%, 18%, and 12% for the other three classes of "moderate", "low," and "very low" susceptibility, respectively.

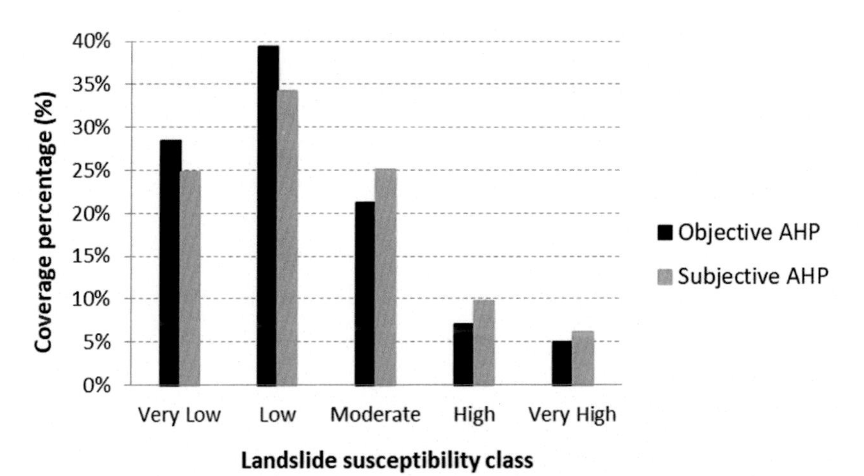

FIGURE 16.4 Coverage percentages of landslide susceptibility classes.

TABLE 16.8 Coverage (%) cross-comparison for the landslide susceptibility classes between the objective and subjective AHP modeling approaches.

Objective AHP			Subjective AHP		
	Very low	Low	Moderate	High	Very high
Very low	12	10	5	4	2
Low	9	18	7	1	1
Moderate	3	4	10	3	1
High	–	2	3	1	1
Very high	–	–	–	1	1

AHP, analytical hierarchy process.

16.4.1 Validation

To evaluate the performance of LS modeling, a validation step is required. Since it can provide information about the accuracy and prediction ability of modeling, and thus the reliability of its LS output, this step is considered crucial. A standard validation procedure widely used in LS assessment studies is the receiver operating characteristics (ROC) analysis (Cervi et al., 2010; Pradhan & Kim, 2018; Zhou et al., 2018). More details about ROC analysis can be found in Fawcett (2006).

In terms of LS modeling, ROC includes the development of a curve graph as a result of linking a validation dataset (representing both landslide and nonlandslide locations) to the LS output of modeling. In this graph, the term of sensitivity is plotted against the term of 1−specificity. The sensitivity is defined as (Althuwaynee et al., 2014):

$$Sensitivity = n(TP)/[n(TP) + n(FN)] \qquad (16.10)$$

where *n(TP)* is the number of true positives, and *n(FN)* is the number of false negatives; true positive is the linking of a landslide location to one of "high" and "very high" susceptibility classes, whereas false negative is the linking of a nonlandslide location to the same classes. On the contrary, the specificity is defined as:

$$Specificity = n(TN)/[n(TN) + n(FP)] \qquad (16.11)$$

where *n(TN)* is the number of true negatives, and *n(FP)* is the number of false positives; true negative is the linking of a nonlandslide location to one of "very low", "low" and "moderate" susceptibility classes, whereas false positive is the linking of a landslide location to the same classes. Therefore a high sensitivity indicates a high number of correctly linked cases, and a low 1−specificity (related to high specificity) indicates a low number of incorrectly linked cases. A metric known as area under curve (AUC) is also calculated to reveal the prediction ability of modeling. Ranging from 0.5 to 1.0, the higher this value is, the better its prediction ability is.

In this study, the ROC analysis was implemented by using the "independent" (not used for the training of bivariate statistical models) landslide validation dataset to evaluate how well the two applied modeling approaches can predict the distribution of future landslides. Since ROC requires additionally nonlandslide locations, an equal amount (446) of pixels from the not landslide-affected part of study area were randomly selected. The target values of 0 and 1 were then assigned to the nonlandslide and landslide pixels, respectively, composing the final validation dataset (totally 892 pixels). The ROC curves and the relevant AUC values are presented in Fig. 16.5. The proposed objective modeling approach was found to have an AUC value equal to 0.84, followed by the typical subjective approach with a value of 0.72.

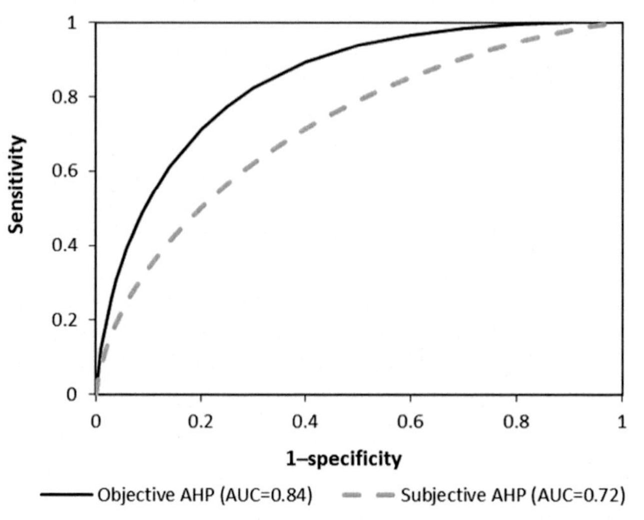

FIGURE 16.5 ROC curves of the applied modeling approaches. *ROC*, receiver operating characteristics.

16.5 Discussion

Due to the observed upward tendency in landslide occurrence, the knowledge about the spatial probability of potential events as well as the conditioning factors that may affect or even cause them is an essential need. The acquisition of the specific knowledge is achieved by the LS assessment and mapping. Several LS assessment models have been developed and applied, with their own advantages and disadvantages (Du et al., 2017). Over the recent years, the combination of these individual models is increasingly tested to enhance their benefits and overcome their weaknesses.

In the present study, a such combination is proposed including the incorporation of three data-driven quantitative models in an expert-based modeling approach. In particular, the bivariate statistical models of FR, IV, and WoE were incorporated as individual objective "experts" in AHP modeling. The three bivariate statistical models depend on observed landslide densities to expose the spatial association between the landslide occurrence and the classes of several conditioning factors. The main advantage of this type of modeling can be detected to the fact that missing or undersampled landslide data does not significantly affect its outputs (Regmi et al., 2010). On the other hand, the AHP modeling is typically subjective depending on expert opinions. Experimental evidence suggests that opinions from a group of experts appear capable to develop more reliable estimates than an individual opinion (Lee & Jones, 2004). However, group opinions may be different for every individual member and thus may be subjected to cognitive limitations with high uncertainty. In this line, the applied combinatorial modeling approach of objective AHP was expected to reduce the uncertainty of the typical subjective AHP—which was also comparatively applied—and improve the reliability of LS assessment for the Greek drainage basin of Selinous River.

By applying the two different AHP modeling approaches, the impacts of seven conditioning factors (elevation, slope angle, curvature, distance to main roads, stream density, geology, and vegetation density) and their classes on landslide activity were quantitatively determined. Focusing on the factors, both similarities and differences were identified. Particularly, the factor of slope angle was found to have the greatest impact on landslide activity according to both modeling approaches. This finding seems to be in agreement with findings from other relevant studies for Greece which also highlighted slope angle as one of the most influential factors (Sakkas et al., 2016; Polykretis & Chalkias, 2018). In contrast to the similarity of the two approaches, a notable difference was also observed including the ranking of elevation factor among the highly influential factors in objective AHP and as the least influential factor in subjective AHP. Between these opposite findings, this one from the objective approach seems to be in line with the majority of relevant studies which have indicated elevation among the most significant factors for LS assessment (Ayalew & Yamagishi, 2005). From a subfactor (class-level) perspective, the steep parts of study area, and especially those being at low elevation and very close to roads, can be recognized as more prone to landsliding, based on the objective approach. The corresponding parts are steep and geologically covered by flysch, based on the subjective approach.

The output maps of the two modeling approaches visualized the spatial distribution of the estimated LS in the study area. Both maps revealed a significant cluster of "high" and "very high" susceptibilities in the central part of the study area. However, the total coverage of these two susceptibility classes in the map of subjective approach seems to be much higher than in the map of objective approach. Considering that and also the relevant coverage cross-comparison findings between the two maps, it could be mentioned that the typical subjective AHP may be implicated in a possibility of overestimation of the LS levels.

With regard to the validation procedure through ROC analysis and the resultant AUC values, the proposed objective AHP approach presented a far more than satisfactory prediction ability, considering the scale of analysis. Moreover, its specific ability was found to be significantly better than the typical subjective AHP approach. It could be noted that, among others, this finding confirms the expected superiority of combinatorial against the individual modeling, and is in agreement with previous studies (Chen et al., 2019; Li & Wang, 2019).

16.6 Conclusions

In terms of LS assessment for the study area, the proposed objective "expert"-based modeling was observed to perform significantly better compared to the typical subjective expert-based modeling. Due to this confirmed superiority, it proves that the assessment provided by the objective statistical "experts" efficiently incorporates the knowledge of the study area and the conditioning factors involved in the analysis. Therefore the main issue in subjective expert-based modeling approaches should not be to invite as many experts as possible with experience and knowledge of the landslide phenomenon and its generalized associations with the various conditioning factors, but to invite experts with detailed knowledge of the landslide activity in the area under investigation in order to ensure an accurate evaluation about the importance of the factors involved.

Although a LS map presents only the predicted spatial and not also the temporal probability of landsliding, the more reliable map produced by the objective modeling approach could constitute an essential base in the primary stage of space-focused landslide risk management and mitigation for the study area. In particular, this map could aid the local authorities and decision makers to spatially determine potential landslide instances and mitigate, or even prevent, the damage and losses that they may cause to infrastructures and properties. Knowledge about the potential landsliding in a region is also valuable for planners and developers, as it allows them to select safe locations under land use planning and implementation of construction projects.

References

Ahmed, B. (2015). Landslide susceptibility modelling applying user-defined weighting and data-driven statistical techniques in Cox's Bazar Municipality, Bangladesh. *Natural Hazards, 79*, 1707–1737.

Althuwaynee, O. F., Pradhan, B., Park, H. J., & Lee, J. H. (2014). A novel ensemble decision tree-based CHi-squared Automatic Interaction Detection (CHAID) and multivariate logistic regression models in landslide susceptibility mapping. *Landslides, 11*, 1063–1078.

Arabameri, A., Pradhan, B., Rezaei, K., & Lee, C.-W. (2019). Assessment of landslide susceptibility using statistical- and artificial intelligence based FR–RF integrated model and multiresolution DEMs. *Remote Sensing, 11*, 999.

Argyriou, A. V., Polykretis, C., Teeuw, R. M., & Papadopoulos, N. (2022). Geoinformatic analysis of rainfall-triggered landslides in Crete (Greece) based on spatial detection and hazard mapping. *Sustainability, 14*, 3956.

Avtar, R., Singh, C. K., Singh, G., Verma, R. L., Mukherjee, S., & Sawada, H. (2011). Landslide susceptibility zonation study using remote sensing and GIS technology in the Ken-Betwa River Link area, India. *Bulletin of Engineering Geology and the Environment, 70*, 595–606.

Ayalew, L., & Yamagishi, H. (2005). The application of GIS-based logistic regression for landslide susceptibility mapping in the Kakuda-Yahiko Mountains, Central Japan. *Geomorphology, 65*, 15–31.

Bonham-Carter, G. F. (1994). *Geographic information systems for geoscientists: Modeling with GIS*. Ottawa, ON, Canada: Pergamon Press.

Brabb, E. E. (1984). Innovative approaches to landslide hazard mapping. In *Proceedings of the 4th International Symposium on Landslides* (pp. 307–324). Torornto, ON, Canada: Canadian Geotechnical Society.

Cervi, F., Berti, M., Borgatti, L., Ronchetti, F., Manenti, F., & Corsini, A. (2010). Comparing predictive capability of statistical and deterministic methods for landslide susceptibility mapping: a case study in the northern Apennines (Reggio Emilia Province, Italy). *Landslides*, *7*, 433–444.

Chen, W., Sun, Z., & Han, J. (2019). Landslide susceptibility modeling using integrated ensemble weights of evidence with logistic regression and random forest models. *Applied Sciences*, *9*, 171.

Chen, X., Chen, H., You, Y., & Liu, J. (2015). Susceptibility assessment of debris flows using the analytic hierarchy process method—A case study in Subao river valley, China. *Journal of Rock Mechanics and Geotechnical Engineering*, *7*, 404–410.

Choi, J., Oh, H.-J., Lee, H.-J., Lee, C., & Lee, S. (2012). Combining landslide susceptibility maps obtained from frequency ratio, logistic regression, and artificial neural network models using ASTER images and GIS. *Engineering Geology*, *124*, 12–23.

Ciurleo, M., Mandaglio, M. C., & Moraci, N. (2021). A quantitative approach for debris flow inception and propagation analysis in the lead up to risk management. *Landslides*, *18*, 2073–2093.

CRED. (2017). *CRED CRUNCH 48—Disaster data: A balanced perspective*. (p. 2) Brussels, Belgium: Université Catholique de Louvain.

CRED. (2018). *UNISDR—United Nations International Strategy for Disaster Reduction. Economic Losses, Poverty & Disasters (1998–2017)* (p. 33) Belgium: Université Catholique de Louvain.

CRED (Centre for Research on the Epidemiology of Disasters). (2015). *The human cost of weather related disasters 1995-2015* (p. 30) Brussels, Belgium: Université Catholique de Louvain.

Dagdelenler, G., Nefeslioglu, H. A., & Gokceoglu, C. (2016). Modification of seed cell sampling strategy for landslide susceptibility mapping: An application from the Eastern part of the Gallipoli Peninsula (Canakkale, Turkey). *Bulletin of Engineering Geology and the Environment*, *75*, 575–590.

Dai, F. C., & Lee, C. F. (2002). Landslide characteristics and slope instability modeling using GIS, Lantau Island, Hong Kong. *Geomorphology*, *42*, 213–228.

Du, G.-L., Zhang, Y.-S., Iqbal, J., Yang, Z.-H., & Yao, X. (2017). Landslide susceptibility mapping using an integrated model of information value method and logistic regression in the Bailongjiang watershed, Gansu Province, China. *Journal of Mountain Science*, *14*, 249–268.

Fawcett, T. (2006). An introduction to ROC analysis. *Pattern Recognition Letters*, *27*, 861–874.

Fell, R., Corominas, J., Bonnard, C., Cascini, L., Leroi, E., & Savage, W. Z. (2008). Guidelines for landslide susceptibility, hazard and risk zoning for land use planning. *Engineering Geology*, *102*, 85–98.

Gorsevski, P. V., Brown, M. K., Panter, K., Onasch, C. M., Simic, A., & Snyder, J. (2016). Landslide detection and susceptibility mapping using LiDAR and an artificial neural network approach: A case study in the Cuyahoga Valley National Park, Ohio. *Landslides*, *13*, 467–484.

Guri, P. K., Champati ray, P. K., & Patel, R. C. (2015). Spatial prediction of landslide susceptibility in parts of Garhwal Himalaya, India, using the weight of evidence modelling. *Environmental Monitoring and Assessment*, *187*, 324.

Hellenic Statistical Authority. (2011). *Population and Housing Census: Resident Population.* Available online: https://www.statistics.gr/el/statistics/pop. Accessed on 30 June 2022.

Hong, H., Liu, J., D., Tien Bui, Pradhan, B., Acharya, T. D., Pham, B. T., Zhu, A-X., Chen, W., & Ahmad, B. B (2018). Landslide susceptibility mapping using J48 Decision Tree with AdaBoost, Bagging and Rotation Forest ensembles in the Guangchang area (China). *Catena, 163*, 399–413.

Jenks, G. F. (1977). *Optimal Data Classification for Choropleth Maps.* Lawrence, KS, USA: University of Kansas.

Karnieli, A., Qin, Z., Wu, B., Panov, N., & Yan, F. (2014). Spatio-temporal dynamics of land-use and land-cover inthe Mu Us Sandy Land, China, using the change vector analysis technique. *Remote Sensing*, *6*, 9316–9339.

Karymbalis, E., Ferentinou, M., & Giles, P. T. (2016). *Use of morphometric variables and self-organizing maps to identify clusters of alluvial fans and catchments in the north Peloponnese. Geology and geomorphology of alluvial and fluvial fans: Terrestrial and planetary perspectives.* (p. 440) Greece: Geological Society of London, Special Publications.

Kayastha, P. (2015). Landslide susceptibility mapping and factor effect analysis using frequency ratio in a catchment scale: A case study from Garuwa sub-basin, East Nepal. *Arabian Journal of Geosciences*, *8*, 8601–8613.

Laboratory of Engineering Geology, Department of Geology, University of Patras. Landslide Inventory Database. Available online: http://landslide.engeolab.gr/. Accessed on 20 January 2017.

Lee, E. M., & Jones, D. K. C. (2004). *Landslide Risk Assessment* (p. 161) London, UK: Thomas Telford.

Li, R., & Wang, N. (2019). Landslide susceptibility mapping for the Muchuan county (China): A comparison between bivariate statistical models (WoE, EBF, and IoE) and their ensembles with logistic regression. *Symmetry*, *11*, 762.

Listo, F. D. L. R., & Vieira, B. C. (2012). Mapping of risk and susceptibility of shallow-landslide in the city of Sao Paulo, Brazil. *Geomorphology*, 169–170, 30-44.

Luo, W., & Liu, C.-C. (2018). Innovative landslide susceptibility mapping supported by geomorphon and geographical detector methods. *Landslides*, *15*(201), 465–474.

Mandal, B., & Mandal, S. (2018). Analytical hierarchy process (AHP) based landslide susceptibility mapping of Lish river basin of eastern Darjeeling Himalaya, India. *Advances in Space Research*, *62*, 3114–3132.

Meng, Q., Miao, F., Zhen, J., Wang, X., Wang, A., Peng, Y., & Fan, Q. (2016). GIS-based landslide susceptibility mapping with logistic regression, analytical hierarchy process, and combined fuzzy and support vector machine methods: a case study from Wolong Giant Panda Natural Reserve, China. *Bulletin of Engineering Geology and the Environment*, *75*, 923–944.

Mondal, S., & Mandal, S. (2018). RS & GIS-based landslide susceptibility mapping of the Balason River basin, Darjeeling Himalaya, using logistic regression (LR) model. *Georisk*, *12*, 29–44.

Polykretis, C., & Chalkias, C. (2018). Comparison and evaluation of landslide susceptibility maps obtained from weight of evidence, logistic regression, and artificial neural network models. *Natural Hazards*, *93*, 249–274.

Polykretis, C., Grillakis, M. G., Argyriou, A. V., Papadopoulos, N., & Alexakis, D. D. (2021). Integrating multivariate (GeoDetector) and bivariate (IV) statistics for hybrid landslide susceptibility modeling: A case of the Vicinity of Pinios Artificial Lake, Ilia, Greece. *Land*, *10*, 973.

Pradhan, A. M. S., & Kim, Y.-T. (2014). Relative effect method of landslide susceptibility zonation in weathered granite soil: A case study in Deokjeok-ri Creek, South Korea. *Natural Hazards*, *72*, 1189–1217.

Pradhan, A. M. S., & Kim, Y.-T. (2018). GIS-based landslide susceptibility model considering effective contributing area for drainage time. *Geocarto International*, *33*, 810–829.

Special Secretariat for Water. (2012). *Management plan for the river catchments of drainage district of Northern Peloponnese.* (p. 456) Athens, Greece: Energy and Climate Change.

Regmi, N. R., Giardino, J. R., & Vitek, J. D. (2010). Modeling susceptibility to landslides using the weight of evidence approach: Western Colorado, USA. *Geomorphology*, *115*, 172–187.

Rehman, A., Song, J., Haq, F., Mahmood, S., Ahamad, M. I., Basharat, M., Sajid, M., & Mehmood, M. S. (2022). Multi-hazard susceptibility assessment using the analytical hierarchy process and frequency ratio techniques in the Northwest Himalayas, Pakistan. *Remote Sensing*, *14*, 554.

Reichenbach, P., Busca, C., Mondini, A. C., & Rossi, M. (2014). The influence of land use change on landslide susceptibility zonation: The Briga Catchment Test Site (Messina, Italy). *Environmental Management*, *54*, 1372–1384.

Saaty, T. L. (1980). *The Analytic Hierarchy Process.* New York, USA: McGraw-Hill.

Saito, H., Nakayama, D., & Matsuyama, H. (2009). Comparison of landslide susceptibility based on a decision-tree model and actual landslide occurrence: The Akaishi Mountains, Japan. *Geomorphology*, *109*, 108–121.

Sakkas, G., Misailidis, I., Sakellariou, N., Kouskouna, V., & Kaviris, G. (2016). Modeling landslide susceptibility in Greece: A weighted linear combination approach using analytic hierarchical process, validated with spatial and statistical analysis. *Natural Hazards*, *84*, 1873–1904.

Su, C., Wang, L., Wang, X., Huang, Z., & Zhang, X. (2015). Mapping of rainfall-induced landslide susceptibility in Wencheng, China, using support vector machine. *Natural Hazards*, *76*, 1759–1779.

Taaleb, K., Cheng, T., & Zhang, Y. (2018). Mapping landslide susceptibility and types using random forest. *Big Earth Data*, *2*, 159–178.

Tsagas, D. (2011). *Geomorphological observations, and gravity movements in northern Peloponnesus. PhD dissertation.* Athens, Greece: National and Kapodistrian University of Athens.

Van Westen, C. J. (1993). *Application of geographical information system to landslide hazard zonation. International Institute for Geo-Information Science and Earth Observation* (p. 245) Enschede, The Netherlands: International Institute for Geo-Information Science and Earth Observation, ITC-Publication No. 15.

Van Westen, C. J., Rengers, N., Terlien, M. T. J., & Soeters, R. (1997). Prediction of the occurrence of slope instability phenomena through GIS-based hazard zonation. *Geologische Rundschau*, *86*, 404–414.

Varnes, D. J. (1978). *Slope movement types and processes. Landslides: Analysis and control.* (pp. 11–33). Washington, USA: Transportation Research Board, Special Report 176.

Varnes, D. J. (1984). *Landslide hazard zonation: A review of principles and practice. United Nations Educational Scientific and Cultural Organization* (p. 63) Paris, France: UNESCO.

Vieira, B. C., Fernandes, N. F., Filho, O. A., Martins, T. D., & Montgomery, D. R. (2018). Assessing shallow landslide hazards using the TRIGRS and SHALSTAB models, Serra do Mar, Brazil. *Environmental Earth Sciences*, *77*, 260.

Wu, H., & Song, T. (2018). An evaluation of landslide susceptibility using probability statistic modeling and GIS's spatial clustering analysis. *Human and Ecological Risk Assessment: An International Journal, 24*, 1952–1968.

Yang, Z-h, Lan, H-x, Gao, X., Li, L-p, Meng, Y-s, & Wu, Y-m (2015). Urgent landslide susceptibility assessment in the 2013 Lushan earthquake-impacted area, Sichuan Province, China. *Natural Hazards, 75*, 2467–2487.

Youssef, A. M., Pourghasemi, H. R., El-Haddad, B. A., & Dhahry, B. K. (2016). Landslide susceptibility maps using different probabilistic and bivariate statistical models and comparison of their performance at Wadi Itwad Basin, Asir Region, Saudi Arabia. *Bulletin of Engineering Geology and the Environment, 75*, 63–87.

Zelilidis, A. (2000). Drainage evolution in a rifted basin, Corinth graben, Greece. *Geomorphology, 35*, 69–85.

Zhang, G., Cai, Y., Zheng, Z., Zhen, J., Liu, Y., & Huang, K. (2016). Integration of the statistical index method and the analytic hierarchy process technique for the assessment of landslide susceptibility in Huizhou, China. *Catena, 142*, 233–244.

Zhou, C., Yin, K., Cao, Y., Ahmed, B., Li, Y., Catani, F., & Pourghasemi, H. R. (2018). Landslide susceptibility modeling applying machine learning methods: A case study from Longju in the Three Gorges Reservoir area, China. *Computers & Geosciences, 112*, 23–37.

Zhou, S., Chen, G., Fang, L., & Nie, Y. (2016). GIS-based integration of subjective and objective weighting methods for regional landslides susceptibility mapping. *Sustainability, 8*, 334.

Chapter 17

Remote sensing and geographic information system for soil analysis—vulnerability mapping and assessment

Mohamed A.E. AbdelRahman
Division of Environmental Studies and Land Use, National Authority for Remote Sensing and Space Sciences (NARSS), Cairo, Egypt

17.1 Introduction

Remote sensing has a wide range of applications in many different fields such as:

- Coastal applications: monitoring shoreline changes, tracking sediment transport, and mapping coastal features. The data can be used for coastal mapping and erosion prevention (Abascal Zorrilla et al., 2018; Dabuleviciene et al., 2018; De Sanjosé Blasco et al., 2018; Jiang et al., 2016; Pradhan et al., 2018; Tahsin et al., 2018; Ventura et al., 2018).
- Ocean applications: monitor ocean circulation and current systems, measure ocean temperature and wave height, and track sea ice. The data can be used to better understand the oceans and how best to manage their resources (Bouzaiene et al., 2021; Casella et al., 2020; Chen et al., 2021; Ciani et al., 2021; Cotroneo et al., 2021; Farach-Espinoza et al., 2021; Fifani et al., 2021; Isern-Fontanet et al., 2021; Kim et al., 2021; Menna et al., 2022; Poulain et al., 2021; Timmermans & Marshall, 2020; Toyota, 2009).
- Risk assessment: track hurricanes, earthquakes, erosion, and floods. The data can be used to assess the effects of a natural disaster and create preparedness strategies to use before and after a hazardous event (Joyce et al., 2009; van Westen, 2000).
- Natural resource management: land use monitoring, wetland mapping, and wildlife habitat mapping. The data can be used to reduce the damage urban growth does to the environment and help determine how best to protect natural resources.

Geoinformatics for Geosciences. DOI: https://doi.org/10.1016/B978-0-323-98983-1.00018-1

17.2 Orbits

Satellites can be placed in several types of orbits around the Earth. The three common classes of orbits are low Earth orbit (about 160–2000 km above Earth), medium Earth orbit (about 2000–35,500 km above Earth), and high Earth orbit (above 35,500 km above Earth). Satellites orbiting at an altitude of 35,786 km are located at an altitude whose orbital speed coincides with the rotation of the planet and are in the so-called geosynchronous orbit. In addition, a satellite in geostationary orbit directly above the equator will have a geostationary orbit. The geostationary orbit allows the satellite to maintain its position directly above the same place on the Earth's surface (Gupta et al., 2022).

LEO is a commonly used orbit because satellites can follow many orbital paths around the planet. For example, satellites orbiting around the poles by about 90 degrees tilt to the equatorial plane and travel from pole to pole as the Earth rotates. This allows sensors on the satellite to quickly obtain data about the entire world, including the Polar Regions. Many of the satellites that orbit around the poles are considered sun-synchronous, which means that the satellite passes through the same location at the same solar time each cycle. One example of a polar sun-synchronous satellite is NASA's Aqua satellite, which orbits 705 km above the Earth's surface (Glover, 1997; Kidder, 2015).

On the other hand, nonpolar LEO satellites do not provide global coverage but instead cover only a partial range of latitudes. The NASA/Japan Aerospace Exploration Agency's Joint Core Observatory for Global Precipitation Measurement (GPM) is an example of an asynchronous low Earth satellite. Its orbital path acquires data between latitudes 65 degrees north and south at an altitude of 407 km above the planet (Furukawa et al., 2011).

The medium Earth orbit satellite takes about 12 hours to complete its orbit. Within 24 hours, the satellite crosses the same two points on the equator every day. This orbit is fixed and highly predictable. As a result, this is an orbit used by many telecommunications and GPS satellites. One example of a constellation of medium Earth orbit satellites is the European Space Agency's Galileo Global Navigation Satellite System (GNSS), which orbits 23,222 km above Earth (Camacho-Lara, 2013).

17.3 Observing with the electromagnetic spectrum

Electromagnetic energy, caused by the vibration of charged particles, is transmitted in the form of waves through the atmosphere and the vacuum of space. These waves have different wavelengths (the distance from the crest of the wave to the crest of the wave) and their frequencies; a shorter wavelength means a higher frequency. Some, such as radio waves, microwaves, and infrared, have a longer wavelength, whereas others, such as ultraviolet, X-rays, and gamma rays, have a much shorter wavelength. Visible light is in the middle of this range from long to short wave radiation. This tiny bit of energy is all that the human eye can detect. Some waves are absorbed or reflected by components of the atmosphere, such as water vapor and carbon dioxide, while some wavelengths allow unimpeded movement through the atmosphere. Visible light has wavelengths that can travel through the atmosphere. Microwave energy has wavelengths that can pass through clouds, a feature used by many weather and communications satellites (Zumdahl & Zumdahl, 2003) (Fig. 17.1).

The main source of energy observed by satellites is the sun. The amount of the sun's reflected energy depends on the surface's roughness and whiteness, which is the way the surface, reflects light rather than absorbing it. Snow, for example, has a high degree of albedo and reflects up to 90% of incoming solar radiation. On the other hand, the ocean reflects only about 6% of the incoming solar radiation and absorbs the rest. Often, when energy is absorbed, it is re-emitted, usually at longer wavelengths. For example, the energy absorbed by the ocean is re-emitted as infrared radiation (Purcell & Morin, 2013).

All things on Earth reflect, absorb, or transmit energy, the amount of which varies with wavelength. Just as your fingerprint is unique to you, everything on Earth has a unique spectral fingerprint. Researchers can use this information to identify different Earth features as well as different types of rocks and minerals. The number of spectral bands detected by a particular instrument, and their spectral resolution determines the amount of differentiation a researcher can determine between materials (Browne, 2013) (Fig. 17.2).

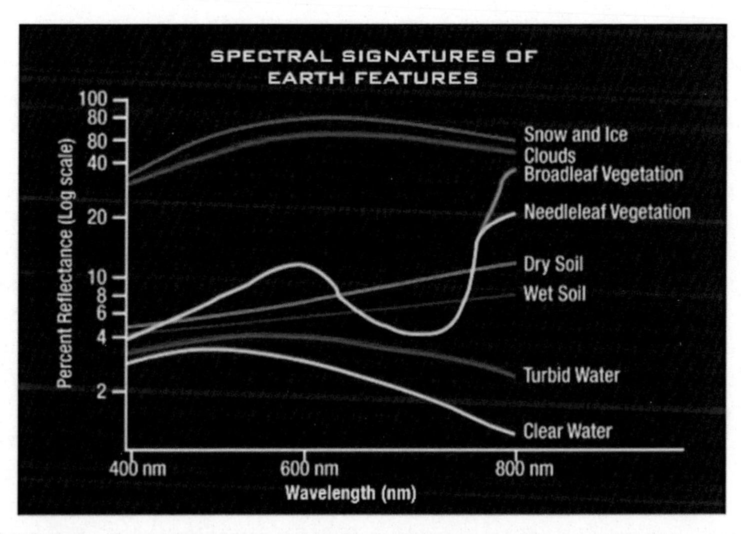

FIGURE 17.1 Spectral signatures of different Earth features within the visible light spectrum. *Jeannie Allen.*

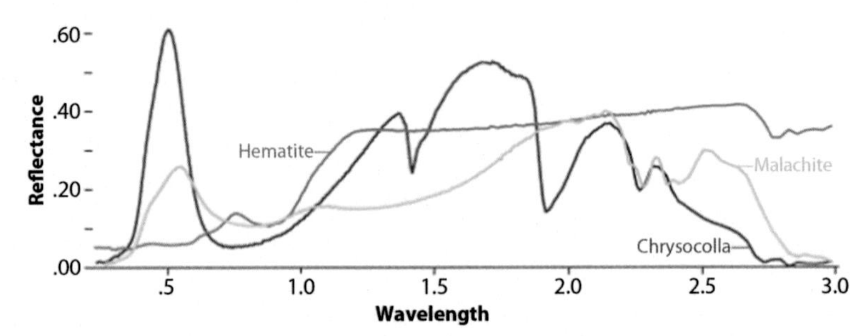

FIGURE 17.2 Just as iron and copper look different in visible light, iron- and copper-rich minerals reflect varying amounts of light in the infrared spectrum. This graph compares the reflectance of hematite (an iron ore) with malachite and chrysocolla (copper-rich minerals) from 200 to 3000 nm. *NASA image by Robert Simmon, using data from the USGS Spectroscopy Lab.*

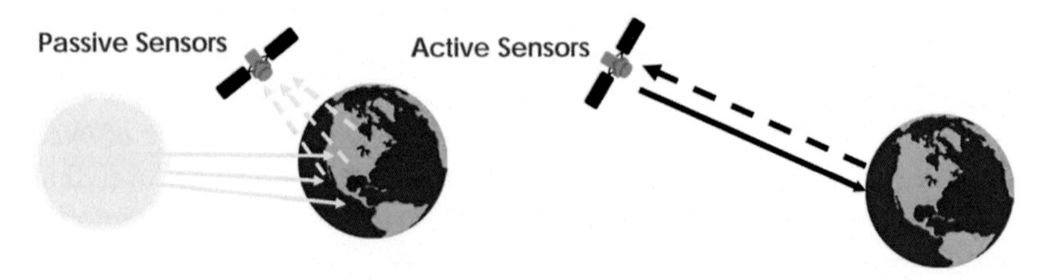

FIGURE 17.3 Diagram of a passive sensor versus an active sensor. *NASA Applied Sciences Remote Sensing Training Program.*

17.4 Sensors

Sensors or instruments aboard satellites and planes use the sun as their light source or provide their own light source and measure the energy that is reflected back. Sensors that use natural energy from the sun are called passive sensors. Those that provide their own power source are called active sensors (Zhu et al., 2017) (Fig. 17.3).

Passive sensors include various types of radiometers (instruments that quantitatively measure the intensity of electromagnetic radiation in selected ranges) and spectrometers (devices designed to detect, measure, and analyze the spectral content of reflected electromagnetic radiation). Most passive systems used by remote sensing applications operate in the visible, infrared, thermal infrared, and microwave portions of the electromagnetic spectrum. These sensors measure land and sea surface temperatures, vegetation characteristics, cloud characteristics, aerosols, and other physical features. Most passive sensors cannot penetrate dense cloud cover and therefore have limitations in monitoring areas such as the tropics where dense cloud cover is frequent (Kasai, 2008).

Active sensors include various types of radio detection and range sensors, altimeters, and scatters. The majority of active sensors operate in the microwave range of the electromagnetic spectrum, giving them the ability to penetrate the atmosphere in most conditions. These types of sensors are useful for measuring vertical profiles of aerosols, forest structure, precipitation and winds, sea surface topography, and ice, among others (Blais, 2004; Kasai, 2008; Melesse et al., 2007; Toth & Jóźków, 2016)

17.5 Resolution

The decision plays a role in how the data from the sensor is used. Accuracy can vary depending on the satellite's orbit and sensor design. There are four types of accuracy that must be considered in any data set—radiometric, spatial, spectroscopic, and temporal (Blais, 2004; Kasai, 2008; Melesse et al., 2007; Toth & Jóźków, 2016)

Radiometric resolution is the amount of information in each pixel, that is, the number of bits that represent the recorded energy. Each bit registers an exponent of power 2. For example, an 8-bit precision is 28, which indicates that the sensor has 256 possible numeric values (0–255) to store information. Thus the higher the accuracy of the radiation measurement, the more values are available for storing information, which provides better discrimination even among the slightest differences in energy. For example, when evaluating water quality, radiative accuracy is necessary to distinguish between subtle differences in ocean color (Blais, 2004; Kasai, 2008; Melesse et al., 2007; Toth & Jóźków, 2016).

Spatial resolution is determined by the size of each pixel within a digital image and the area on the Earth's surface that that pixel represents. For example, the majority of the bands observed by the Medium Resolution Imaging Spectroradiometer (MODIS) have a spatial resolution of 1 km; each pixel represents a 1 km × 1 km area on Earth. MODIS also includes ranges with a spatial resolution of 250 m or 500 m. The finer the resolution (the lower the number), the more detail you can see (Blais, 2004; Kasai, 2008; Melesse et al., 2007; Toth & Jóźków, 2016).

Spectral resolution is the ability of the sensor to distinguish finer wavelengths, that is, the presence of more and narrower bands. Many sensors are considered multispectral, which means they have 3–10 bands. Some sensors have hundreds or even thousands of bands and are considered ultra-spectrum. The narrower the range of wavelengths for a given band, the finer the spectral resolution. For example, the Airborne Visible/Infrared Imaging Spectrometer (AVIRIS) captures information in 224 spectral channels. The cube on the right represents the details within the data. At this level of detail, the types of rocks, minerals, types of vegetation, and other features can be distinguished. In the cube, the small, highly responsive region in the right corner of the image is in the red portion of the visible spectrum (~700 nm), due to the presence of 1 cm (half an inch) red brine shrimp in the evaporation pond. (Blais, 2004; Kasai, 2008; Melesse et al., 2007; Toth & Jóźków, 2016).

Temporal resolution is the time it takes a satellite to complete its orbit and revisit the same observation area. This accuracy depends on the orbit, sensor characteristics, and bandwidth. Because geostationary satellites match the rate of Earth's rotation, the temporal resolution is much finer. Satellites in a polar orbit have a time resolution that can vary from 1 to 16 days. For example, the MODIS sensor aboard NASA's Terra and Aqua satellite has a time resolution of 1–2 days, allowing the sensor to visualize the Earth as it changes day by day. On the other hand, the Operational Earth Imager (OLI) aboard, the joint NASA/USGS Landsat 8 satellite, has a narrower width of space and a time resolution of 16 days; it does not show daily changes but rather bi-monthly changes (Ashraf et al., 2011).

17.6 Data processing, interpretation, and analysis

Remote sensing data acquired from instruments aboard satellites require processing before the data are usable by most researchers and applied science users. Most raw NASA Earth observation satellite data are processed at NASA's Science Investigator-led Processing Systems (SIPS) facilities. All data are processed to at least a Level 1, but most have associated Level 2 (derived geophysical variables) and Level 3 (variables mapped on uniform space-time grid scales) products. Many even have Level 4 products.

Once the data is processed, it can be used in a variety of applications, from agriculture to water resources to health and air quality. A single sensor will not address all of the search questions in a given application. Users often need to make use of various sensors and data products to address their questions, given the data limitations offered by different spectral, spatio-temporal and resolution.

17.7 Creating satellite imagery

Many sensors obtain data at different spectral wavelengths. For example, Band 1 from OLI aboard Landsat 8 gets data at 0.433–0.453 μm, whereas MODIS Band 1 gets data at 0.620–0.670 μm. OLI has a total of 9 bands, whereas MODIS has 36 bands, all of which measure different regions of the electromagnetic spectrum. Scopes can be combined to produce images of data to reveal different features in a scene. Data images are often used to distinguish the characteristics of the area being studied or to identify the area of study.

True-color images show the Earth as it appears to the human eye. To obtain a true-color image from Landsat 8 OLI (red, green, blue [RGB]), the four (red), three (green), and two (blue) sensor bands are combined. Other spectral band combinations can be used for specific scientific applications, such as flood monitoring, urban planning, and vegetation mapping. For example, the creation of a False Color Infrared Imaging Radiometer Array (VIIRS, aboard the Suomi National Polar-orbiting Partnership [Suomi NPP]) using bands M11, I2, and I1 is useful for distinguishing burn scars from low vegetation or Bare the soil as well to expose the flooded areas. To see more band combinations from Landsat sensors, check out NASA Scientific Visualization Studio's Landsat Band Remix video or NASA's Many Hues of London Earth Observatory article. For other common band combinations, see How to Interpret Common False Color Imagery at NASA's Earth Observatory, which provides common band groupings along with insight into image interpretation.

17.8 Image interpretation

Once the data has been processed and transformed into images with groups of different ranges, these images can aid in resource management and disaster assessment decisions. This requires a correct interpretation of the images. There are some strategies to get started (adapted from the NASA Earth Observatory article How to Interpret a Satellite Image: Five Tips and Strategies):

1. Know the scale—There are different scales that depend on the spatial resolution of the image and each scale provides different features of interest. For example, when tracking a flood, a detailed high-resolution view will show homes and businesses surrounded by water. The wider landscape view shows which parts of a province or urban area were flooded and possibly where the water came from. A wider view may show the entire area—a submerged river system or mountain ranges and valleys that control the flow. A hemispherical view will show the movement of weather systems associated with floods.
2. Look for patterns, shapes, and textures—Many features are easy to identify based on their pattern or shape. For example, agricultural areas are generally geometric in shape, usually circles or rectangles. Usually straight lines are man-made structures, such as roads or canals.
3. Defining colors—When using color to distinguish features, it is important to know the range of range used to create the image. Real or natural color images are created using strip combinations that simulate what we see with our eyes if we look down from space. Water absorbs light so it usually appears black or blue in true-color photos; reflection of sunlight from the surface of the water may cause it to appear gray or silver. Sediment can make water appear browner, while algae can make water appear greener. The color of the vegetation varies depending on the season: in spring and summer, it is usually bright green; fall may be orange, yellow and tan; winter may have more brown. Bare land is usually a shade of brown, although this depends on the mineral composition of the sediment. Urban areas are usually gray from the heavy use of concrete. Ice and snow are white in true-to-color images, as are clouds. When

using color to identify objects or features, it is also important to use surrounding features to put objects in context.

4. Consider what you know—Knowing the area you are monitoring helps identify these features. For example, knowing that an area was recently burned by a wildfire can help determine why vegetation appears differently in a remote sensing image.

17.9 Quantitative analysis

Different types of land cover can be distinguished more easily using image classification algorithms. Image classification uses the spectral information of the individual pixels of an image. Software that uses image classification algorithms can automatically group pixels into what is called unsupervised classification. The user can also select areas of a known land cover type to "train" the program to cluster as pixels; this is called a supervised classification. Maps or images can also be integrated into a geographic information system (GIS) and then each pixel can be compared with other GIS data, such as census data (Ashraf et al., 2011).

Satellites often also carry a variety of sensors that measure biophysical parameters, such as sea surface temperature, nitrogen dioxide, or other atmospheric pollutants, winds, aerosols, and biomass. These parameters can be evaluated by statistical and spectral analysis techniques.

17.10 Soil science

The comprehensive study of soil, including soil formation, categorization, and mapping, as well as soil physical, chemical, biological, and fertility qualities, is known as soil science in geography. All for the benefit of better managing and using soils (Johnson et al., 1982).

17.11 Importance of studying soil science

Healthy soil is a key component of healthy biodiversity, which we must maintain and preserve. But how does having good soil benefit our economy and society? More crops will flourish on healthy, well-balanced soil, providing more food for the world's expanding population (Adamchuk et al., 2004; Chang et al., 2021; Xiuwei et al., 2016).

To acquire the best yield from a crop, soil scientists improve soil health, which in turn increases food output. They create techniques to stop erosion to maintain sturdy river banks, which enhance the quality of the water. Numerous industries depend on soil, a nonrenewable resource (Adamchuk et al., 2004; Chang et al., 2021; Xiuwei et al., 2016).

They categorize soil types so that land can be zoned appropriately for managing floods, or they aid in our understanding of how toxins flow through the soil. Soil science is essential to the agriculture sector, the forestry sector, the environment, and even city planners. What essential role do GIS and remote sensing play in soil science, then?

17.12 Geographic information system technologies uses in soil science

First off, GIS is also the result of data collection and analysis, not only the creation of maps. To further soil research, GIS is frequently utilized in conjunction with GPS and remote sensing technology. Monitoring or spotting physical changes in the characteristics of the Earth from a distance is called remote sensing. The utilization of ground penetrating radar (GPR), satellite images, aerial imagery, or LiDAR data is frequently required for this. Depending on the project budget and the level of information required, these various technologies are capable of detecting various aspects at various resolutions. While aerial imagery can provide extremely high-resolution imagery of a particular location, satellite imagery typically delivers large-scale imagery at a lesser resolution. Light detection and ranging, or LiDAR, is a technique that efficiently produces 3D images of the earth's surface. To examine subterranean features or the soil substrate, GPR shoots radar pulses into the ground. GPR is a noninvasive and frequently highly economical method of analyzing the characteristics of soil. The ability to integrate together the mapping capabilities with the pictures or data collected on a GPS itself is where GPS technology comes into play. GPS can be hand-held, installed on machinery, or attached to vehicles to collect data-like borders or distinctive characteristics. GIS software may be used to map and analyze all of this remote sensing data collecting. Land managers can then use these computerized models to assist in making these wise choices (Arunkumar et al., 2020).

17.13 Geographic information system and soil mapping

The process of building geographically referenced soil databases using spatially explicit environmental variables and field surveys is known as digital soil mapping (DSM), and it uses GIS to map soils. Remote sensing methods are used to gather the data used to create these DSM. Traditional soil surveys, which were conducted solely in the field before the use of remote sensing, were labor-intensive, time-consuming, and expensive. In a conventional soil survey, soil samples would be taken all around the property at regular intervals. But because to improvements in GIS software, many tasks may now be completed without the necessity of field surveys. This does not imply that conventional soil surveys are wholly obsolete, as most DSM surveys will involve some fieldwork and data collecting. The fieldwork is far less, although. Nowadays, the majority of soil scientists will take a rugged tablet or handheld computer into the field with digital imagery already loaded on it. The majority of the work will already have been completed by the various remote sensing technologies, so the scientist only needs to collect soil samples or take pictures (Melesse & Wang, 2007; Nageswara Rao, 2005; Patel et al., 2002; Seelan et al., 2003; Steininger, 1996).

17.14 Importance of soil maps

The foundation for many aspects of land use planning and management decision making is having accurate soil maps. For instance, based on soil types and other variables, farm managers might use GIS to pinpoint the best locations for water storage systems (such as dams). Understanding a region's soil substrate can also aid in the management and conservation of groundwater resources in the future. There are numerous uses for soil maps, including enhancing grasslands, estimating crop yields, managing erosion, and planning flood mitigation (Arunkumar et al., 2002, 2004; Gopal Krishan et al., 2009; Karale, 1992; Kudrat et al., 1992; Melesse & Wang, 2007; Nageswara Rao, 2005; Philipson & Lindell, 2003; Seelan et al., 2003; Steininger, 1996; Zhang, 1994; Lal, 2001; Morgan et al., 1984; Patel et al., 2002; Saha et al., 1991; Spanner et al., 1983; Williams, 1975).

17.15 Geographic information system to monitor soil loss and erosion

Estimating the danger of soil erosion and the level of land degradation, which is significant globally in agriculture, the environment, and urban areas, can be done using GIS and remote sensing. Information like rainfall data, soil type, land use, vegetation cover, and the digital elevation model can be integrated to determine the erosion risk of an area in regions of the world where soil loss is hurting people, nature, and food production. Sedimentation and soil loss on land are two factors that are threatening the Great Barrier Reef in Australia. Land managers and the local government have been evaluating gully erosion and focusing land repair operations using a combination of airborne and land-based LiDAR measurements. They expect to prevent soil erosion, increase the quality of the water, and improve the general health of the grassland by doing this. It frequently takes many years of data to compare when monitoring the risk of soil loss and erosion. Many people find high-resolution photography to be prohibitively expensive, but because to ongoing technological advancements, free GIS software, and improved data collection techniques, these costs are declining.

17.16 Geographic information system and soil contamination

Urban regions, waste sites, and mine sites are all areas where soil contamination and cleanup are becoming hot topics. If the contamination from metals and other compounds cannot be found and removed, the soil may become poisonous for years to come. The amount of field effort necessary to assess land pollution as equivalent to soil assessments presents a challenge. Powerful GIS software allows for more accurate contamination spread measurement, decreased field work, and more effective treatment method targeting.

17.17 Geographic information system and soil moisture

Scientists must examine the soil in addition to the water when analyzing and mitigating flood risk. Understanding flood risk requires an understanding of the soil substrate and soil moisture content. Land managers and city planners must make full use of the technology at their disposal to better understand the catchments they work in, since climate change is placing increased pressure on water resources worldwide. Understanding soil moisture content is crucial for a number of reasons since different soil types or substrates will hold water in different ways. Growing crops is clearly advantageous to farmers, but it also aids in our understanding of groundwater recharge, which helps us manage our water resources more effectively. Data on soil moisture can be gathered on a big or small scale, depending on the needs of individual farms or regions. NASA published soil moisture maps at the world scale in 2014 using satellite-based technologies. This information could be utilized for a variety of purposes, including weather forecasting and drought monitoring.

17.18 Geographic information system, soil science, and agriculture

GIS is just one example of how the agriculture sector is quickly embracing technology and benefiting from it. Farmers may now boost output and yields, cut expenses, and manage land resources more effectively thanks to GIS. In most of the developed world, precision agriculture is the norm, with tractors and other farm equipment linked with GPS and sensors to continuously transmit data back to farm management. Nowadays, farms can sometimes be managed almost entirely from the comfort of a couch (Patel et al., 2002; Saha et al., 1991; Spanner et al., 1983).

17.19 Geographic information system and crop production

Soil maps, along with information gathered from distant sensors and GPS technologies, can significantly increase agricultural production. Data on crop health, soil moisture, temperature, nutrition, or the presence of pests and invasive species can all be gathered using remote sensing. This data can be integrated with GIS software to create different fertilization rates throughout a field, ensuring that more fertilizer is given to the areas of the field that require it the most. Because of this, farmers may use less fertilizer, which saves them money and lessens the negative impacts of excess fertilizer runoff. The same methodology can be used for applying pesticides and herbicides as well as watering rates.

17.20 Geographic information system and grassland management

The grass quality is crucial while producing animals in large-scale production systems. The production values will be impacted by a variety of factors, but food quality is a top concern. Land managers must comprehend the characteristics of the soil that the grass grows to manage grasslands sustainably. Managers can use geospatial analysis to assess the appropriateness of various land types for grazing, the ideal stocking levels, and how the grassland will react to various environmental factors (Arunkumar et al., 2002, 2004; Melesse & Wang, 2007; Nageswara Rao, 2005; Philipson & Lindell, 2003; Seelan et al., 2003; Steininger, 1996).

17.21 The future of geographic information system in soil science

Numerous sectors are centered on soil science, and it is obvious that managing soil sustainability offers many advantages to both our way of life and the earth. An emphasis on raising crop yields and the sustainable use of agricultural land and natural resources has been spurred by predictions of an increase in the world's population and the ensuing pressure on food supply. Although much has already been accomplished in the field of soil science GIS, there is still more to be done. Future developments are likely to include more robust modeling programs that are more accurate and detailed (Lal, 2001; Morgan et al., 1984; Nageswara Rao, 2005; Patel et al., 2002; Saha et al., 1991; Spanner et al., 1983; Williams, 1975).

References

Abascal Zorrilla, N., Vantrepotte, V., Gensac, E., Huybrechts, N., & Gardel, A. (2018). The Advantages of Landsat 8-OLI-derived suspended particulate matter maps for monitoring the subtidal extension of Amazonian Coastal Mud Banks (French Guiana). *Remote Sensing*, *10*, 1733. Available from https://doi.org/10.3390/rs10111733.

Adamchuk, V. I., Hummel, J. W., Morgan, M. T., & Upadhyaya, S. K. (2004). On-the-go soil sensors for precision agriculture. *Computers and Electronics in Agriculture*, *44*(1), 71–91. Available from https://doi.org/10.1016/j.compag.2004.03.002.

Arunkumar, V., Natarajan, S., & Sivasamy, R. (2002). Characterisation and classification of soils of lower Palar Manimuthar Watershed. *Agropedology*, *12*, 97–103.

Arunkumar, V., Natarajan, S., & Sivasamy, R. (2004). Soil resource mapping of Vellamadai village, Coimbatore district using fused (IRS 1C LISS III and PAN merged) space borne multispectral data. *Madras Agricultural Journal*, *91*, 399–405.

Arunkumar, V., Pandiyan, M., & Yuvaraj, M. (2020). A review on remote sensing and GIS applications in soil resource management. *International Journal of Current Microbiology and Applied Sciences*, *9*(5), 1063–1075. Available from https://doi.org/10.20546/ijcmas.2020.905.117.

Ashraf, M. A., Maah, M. J., & Yusoff, I. (2011). Introduction to remote sensing of biomass. In I. Atazadeh (Ed.), *Biomass and remote sensing of biomass [Internet]*. London: IntechOpen. Available from https://www.intechopen.com/chapters/19222, https://doi.org/10.5772/16462.

Blais, F. (2004). Review of 20 years of range sensor development. *Journal of Electronic Imaging*, *13*(1), 231–240.

Bouzaiene, M., Menna, M., Elhmaidi, D., Dilmahamod, A. F., & Poulain, P. M. (2021). Spreading of Lagrangian particles in the Black Sea: A comparison between drifters and a high-resolution ocean model. *Remote Sensing*, *13*, 2603. Available from https://doi.org/10.3390/rs13132603.

Browne, M. (2013). *Physics for engineering and science* (2nd ed.). New York: McGraw Hill/Schaum, ISBN 978-0-07-161399-6.

Camacho-Lara, S. (2013). Current and future GNSS and their augmentation systems. In J. N. Pelton, S. Madry, & S. Camacho-Lara (Eds.), *Handbook of satellite applications*. New York, NY: Springer. Available from https://doi.org/10.1007/978-1-4419-7671-0_25.

Casella, D., Meloni, M., Petrenko, A. A., Doglioli, A. M., & Bouffard, J. (2020). Coastal current intrusions from satellite altimetry. *Remote Sensing*, *12*, 3686. Available from https://doi.org/10.3390/rs12223686.

Chang, L., Gang, L., Hairu, L., Xiaokang, W., Hong, C., Chenxi, D., Enshuai, S., & Chengbo, S. (2021). Using ground-penetrating radar to investigate the thickness of mollic epipedons developed from loessial parent material. *Soil and Tillage Research*, *212*, 105047. Available from https://doi.org/10.1016/j.still.2021.105047.

Chen, Y. R., Paduan, J. D., Cook, M. S., Chuang, L. Z.-H., & Chung, Y. J. (2021). Observations of surface currents and tidal variability off of Northeastern Taiwan from shore-based high frequency radar. *Remote Sensing*, *13*, 3438. Available from https://doi.org/10.3390/rs13173438.

Ciani, D., Charles, E., Buongiorno Nardelli, B., Rio, M. H., & Santoleri, R. (2021). Ocean currents reconstruction from a combination of altimeter and ocean colour data: A feasibility study. *Remote Sensing*, *13*, 2389. Available from https://doi.org/10.3390/rs13122389.

Cotroneo, Y., Celentano, P., Aulicino, G., Perilli, A., Olita, A., Falco, P., Sorgente, R., Ribotti, A., Budillon, G., Fusco, G., & Pessini, F. (2021). Connectivity analysis applied to mesoscale eddies in the Western Mediterranean Basin. *Remote Sensing*, *13*, 4228. Available from https://doi.org/10.3390/rs13214228.

Dabuleviciene, T., Kozlov, I. E., Vaiciute, D., & Dailidiene, I. (2018). Remote sensing of coastal upwelling in the South-Eastern Baltic sea: Statistical properties and implications for the coastal environment. *Remote Sensing*, *10*, 1752. Available from https://doi.org/10.3390/rs10111752.

De Sanjosé Blasco, J. J., Gómez-Lende, M., Sánchez-Fernández, M., & Serrano-Cañadas, E. (2018). Monitoring retreat of coastal sandy systems using geomatics techniques: Somo beach (Cantabrian Coast, Spain, 1875–2017). *Remote Sensing*, *10*, 1500. Available from https://doi.org/10.3390/rs10091500.

Farach-Espinoza, E. B., López-Martínez, J., García-Morales, R., Nevárez-Martínez, M. O., Lluch-Cota, D. B., & Ortega-García, S. (2021). Temporal variability of oceanic mesoscale events in the Gulf of California. *Remote Sensing*, *13*, 1774. Available from https://doi.org/10.3390/rs13091774.

Fifani, G., Baudena, A., Fakhri, M., Baaklini, G., Faugère, Y., Morrow, R., Mortier, L., & d'Ovidio, F. (2021). Drifting speed of Lagrangian fronts and oil spill dispersal at the ocean surface. *Remote Sensing*, *13*, 4499. Available from https://doi.org/10.3390/rs13224499.

Furukawa, K. et al. (2011). Proto-flight test of the Dual-frequency Precipitation Radar for the Global Precipitation Measurement, 2011 IEEE International Geoscience and Remote Sensing Symposium, pp. 1279−1282, https://doi.org/10.1109/IGARSS.2011.6049433.

Glover, D. R. (1997). NASA experimental communications satellites, 1958−1995. In Andrew J. Butrica (Ed.), *Beyond the ionosphere: Fifty years of satellite communication*. NASA. Bibcode.

Gopal Krishan, S., Kushwaha, P. S., & Velmurugan, A. (2009). Land degradation mapping in the upper catchment of River Tons. *Journal of the Indian Society of Remote Sensing*, *37*, 49−59.

Gupta, R. K., Jain, A., Wang, J., Singh, V. P., & Bharti, S. (2022). *Artificial intelligence of things for weather forecasting and climatic behvioral analyses*. IGI GLOBAL publisher of timely knowledge. Available from https://doi.org/10.4018/978-1-6684-3981-4.

Isern-Fontanet, J., García-Ladona, E., González-Haro, C., Turiel, A., Rosell-Fieschi, M., Company, J. B., & Padial, A. (2021). High-resolution ocean currents from sea surface temperature observations: The Catalan Sea (Western Mediterranean). *Remote Sensing*, *13*, 3635. Available from https://doi.org/10.3390/rs13183635.

Jiang, D., Hao, M., & Fu, J. (2016). Monitoring the coastal environment using remote sensing and GIS techniques. In Maged Marghany (Ed.), *Applied studies of coastal and marine environments*. London: IntechOpen. Available from https://doi.org/10.5772/62242.

Johnson, R. W., Glasscum, R., & Wojtasinski, R. (1982). Application of ground penetrating radar to soil survey. *Soil Survey Horizons*, *23*(3), 7−25. Available from https://doi.org/10.2136/sh1982.3.0017.

Joyce, K. E., Wright, K. C., Samsonov, S. V., & Ambrosia, V. G. (2009). *Remote sensing and the disaster management cycle. Advances in geoscience and remote sensing*. London, United Kingdom: IntechOpen. Available from https://www.intechopen.com/chapters/9556, https://doi.org/10.5772/8341.

Karale, R. L. (1992). Remote sensing with IRS 1A in soil studies: Developments, status and prospects. In R. L. Karale (Ed.), *Natural resourcecs management—A new perspective* (pp. 128−143). Bangalore: National Natural Resources Management System (NNRMS), Department of Space, Govt. of India.

Kasai, Y. (2008). Introduction to terahertz-wave remote sensing. *Journal of the National Institute of Information and Communications Technology*, *55*(1), 65−67.

Kidder, S. Q. (2015). Satellites and satellite remote senssing: [vague] → Orbits. In North Gerald, Pyla John, & Zhang Fuqing (Eds.), *Encyclopedia of atmospheric sciences* (2 ed., pp. 95−106). Elsiver. Available from https://doi.org/10.1016/B978-0-12-382225-3.00362-5, ISBN 978-0-12-382225-3.

Kim, T., Jo, H. J., & Moon, J. H. (2021). Occurrence and evolution of mesoscale thermodynamic phenomena in the northern part of the East Sea (Japan Sea) derived from satellite altimeter data. *Remote Sensing*, *13*, 1071. Available from https://doi.org/10.3390/rs13061071.

Kudrat, M., Tiwari, A. K., Saha, S. K., & Bhan, S. K. (1992). Soil resource mapping using IRS 1A LISS II digital data—A case study of Khandi area adjacent to Chandigarh, India. *International Journal of Remote Sensing*, *13*, 3287−3302.

Lal, R. (2001). Soil degradation by erosion. *Land Degradation & Development*, *12*, 519−539.

Melesse, A., & Wang, X. (2007). *Impervious surface area dynamics and storm runoff response. Remote sensing of impervious surfaces* (19, pp. 369−384). CRC Press/Taylor and Francis.

Melesse, A. M., Weng, Q., Thenkabail, P. S., & Senay, G. B. (2007). Remote sensing sensors and applications in environmental resources mapping and modelling. *Sensors*, *7*(12), 3209–3241.

Menna, M., Gačić, M., Martellucci, R., Notarstefano, G., Fedele, G., Mauri, E., Gerin, R., & Poulain, P. M. (2022). Climatic, decadal, and interannual variability in the upper layer of the Mediterranean Sea using remotely sensed and in-situ data. *Remote Sensing*, *14*, 1322. Available from https://doi.org/10.3390/rs14061322.

Morgan, R. P. C., Morgan, D. D. V., & Finney, H. J. (1984). A predictive model for the assessment of soil erosion risk. *Journal of Agricultural Engineering Research*, *30*, 245–253.

Nageswara Rao, P. P. (2005). Applications of Remote Sensing and Geographical Indication System in Land Resources Management North Eastern Space Applications Centre, Umiam-793103, Meghalaya.

Patel, N. R., Suresh, K., Prasad, J., & Pande, L. M. (2002). Soil erosion risk assessment and land use adjustment for soil conservation planning using remote sensing and GIS. *Asian Journal of Geoinformatics*, *1*(2), 47–55.

Philipson, P., & Lindell, T. (2003). Can coral reefs be monitored from space? *Ambio*, *32*, 586–593.

Poulain, P. M., Centurioni, L., Özgökmen, T., Tarry, D., Pascual, A., Ruiz, S., Mauri, E., Menna, M., & Notarstefano, G. (2021). On the structure and kinematics of an Algerian eddy in the Southwestern Mediterranean Sea. *Remote Sensing*, *13*, 3039. Available from https://doi.org/10.3390/rs13153039.

Pradhan, B., Rizeei, H. M., & Abdulle, A. (2018). Quantitative assessment for detection and monitoring of coastline dynamics with temporal RADARSAT images. *Remote Sensing*, *10*, 1705. Available from https://doi.org/10.3390/rs10111705.

Purcell., & Morin. (2013). Harvard University *Electricity and magnetism* (3rd ed., p. 820), New York: Cambridge University Press ISBN 978-1-107-01402-2.

Saha, S. K., Kudrat, M., & Bhan, S. K. (1991). *Erosional soil loss prediction using digital satellite data and USLE. Applications of remote sensing in Asia and Oceania- environmental change monitoring* (pp. 369–372). Asian Association of Remote Sensing.

Seelan, S. K., Laguette, S., Casady, G. M., & Seielstad, G. A. (2003). Remote sensing applications for precision agriculture: A learning community approach. *Remote Sensing of Environment*, *88*, 157–169.

Spanner, M. A., Strahler, A. H., Estes, J. E. (1983). Soil loss prediction in a Geographic Information System Format. In: Papers Selected for Presentation at the Seventeenth International Symposium on Remote Sensing of Environment, pp. 89–102. 2–9 June 1982. Buenos Aires, Argentina. Ann Arbor, Mich.

Steininger, M. K. (1996). Tropical secondary forest regrowth in the Amazon: age, area and change estimation with Thematic Mapper data. *International Journal of Remote Sensing*, *17*, 9–27.

Tahsin, S., Medeiros, S. C., & Singh, A. (2018). Assessing the resilience of coastal wetlands to extreme hydrologic events using vegetation indices: A review. *Remote Sensing*, *10*, 1390. Available from https://doi.org/10.3390/rs10091390.

Timmermans, M., & Marshall, J. (2020). Understanding Arctic Ocean circulation: A review of ocean dynamics in a changing climate. *Journal of Geophysical Research: Oceans*, *125*, 1–35.

Toth, C., & Jóźków, G. (2016). Remote sensing platforms and sensors: A survey. *ISPRS Journal of Photogrammetry and Remote Sensing*, *115*, 22–36.

Toyota, T. (2009). Application of remote sensing to the estimation of sea ice thickness distribution. In Gary Jedlovec (Ed.), *Advances in geoscience and remote sensing*. London: IntechOpen. Available from https://doi.org/10.5772/8332.

van Westen, C. J. (2000). Remote sensing for natural disaster management. *International Archives of Photogrammetry and Remote Sensing*, *XXXIII*, 1609–1617, Part B7. Amsterdam.

Ventura, D., Bonifazi, A., Gravina, M. F., Belluscio, A., & Ardizzone, G. (2018). Mapping and classification of ecologically sensitive marine habitats using unmanned aerial vehicle (UAV) imagery and object-based image analysis (OBIA). *Remote Sensing, 10*, 1331. Available from https://doi.org/10.3390/rs10091331.

Williams, J. R. (1975). *Sediment yield prediction with Universal Equation using runoff energy factor. Present and prospective technology for predicting sediment yields and sources* (pp. 244–252). Agricultural Research Service, US Department of Agriculture.

Zhang, L. (1994). A comparison of the efficiency of the three models to estimate water yield changes after forced catchment conversion, M.Sc. (Forest) Thesis, University of Melbourne, Australia.

Zhu, L., Suomalainen, J., Liu, J., Hyyppä, J., Harri Kaartinen, H., & Haggren, H. (2017). A review: Remote sensing sensors. In R. B. Rustamov, S. Hasanova, & M. H. Zeynalova (Eds.), *Multi-purposeful application of geospatial data [Internet]*. London: IntechOpen, [cited 2022 Jul 25]. Available from https://www.intechopen.com/chapters/57384, https://doi.org/10.5772/intechopen.71049.

Zumdahl, S. S., & Zumdahl, S. A. (2003). *Atomic structure and periodicity. Chemistry* (6th ed., pp. 290–294). Boston, MA: Houghton Mifflin Company.

Chapter 18

Multiparameter analysis of the flood of November 15, 2017 in west Attica using satellite remote sensing

Alexia Tsouni[1], Sylvia Antoniadi[1], Emmanouela Ieronimidi[1], Katerina Karagiannopoulou[1], Nikos Mamasis[2], Demetris Koutsoyiannis[2] and Charalampos Kontoes[1]

[1]Operational Unit "BEYOND Centre for Earth Observation Research and Satellite Remote Sensing", Institute for Astronomy, Astrophysics, Space Applications and Remote Sensing, National Observatory of Athens, Athens, Greece, [2]School of Civil engineering, Department of Water Resources and Environmental Engineering, Laboratory of Hydrology and Water Resources, National Technical University of Athens, Athens, Greece

18.1 Introduction

In Greece, in the period of 2000–2020, more than 380 flooding episodes with negative social and economic impacts were recorded, with 132 human casualties, but also material damage and destruction of infrastructure, transport problems, etc. Attica was affected in 30% of these cases. Floods in Attica are usually the sudden result of heavy rainfall, to which the urbanized environment of Attica is particularly vulnerable. Most episodes (73 of 112, 65%) were recorded in the highly urbanized areas, that is, Athens, Piraeus, and the suburbs. Floods in the period of 2000–2020 claimed the lives of 38 people in the Attica region. The autumn months, starting with November, have been the most catastrophic, which is not surprising as heavy rains and storms are particularly high in this season. The flooding in Mandra in November 2017 with 24 deaths was the deadliest episode (Papagiannaki et al., 2021).

18.2 Multiparameter flood analysis

The FloodHub team of the Operational Unit BEYOND/IAASARS/NOA was immediately activated following the Mandra flood, analyzed the flood event,

Geoinformatics for Geosciences. DOI: https://doi.org/10.1016/B978-0-323-98983-1.00019-3

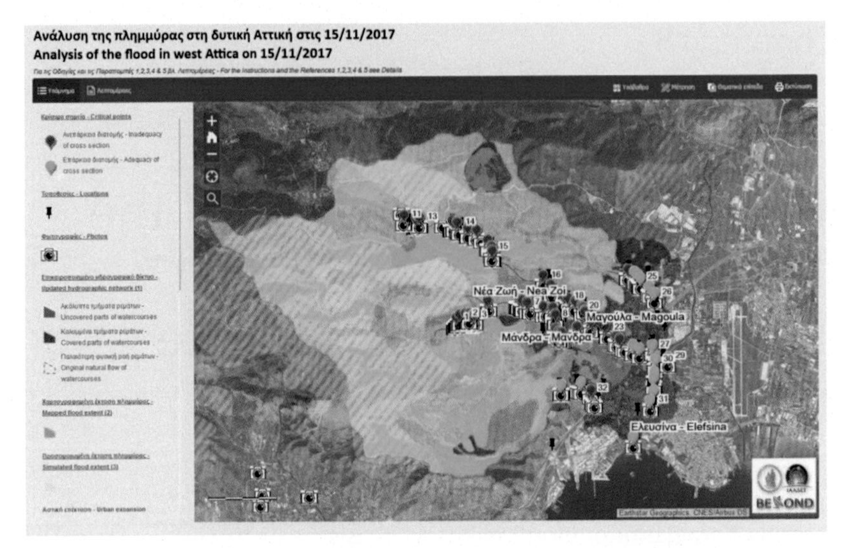

FIGURE 18.1 Interactive web application for the analysis of the flooding in west Attica on November 15, 2017.

performed simulations, and studied the area, both by using satellite remote sensing and photo-interpretation and by visiting the area for data collection and more detailed analysis.

A dedicated interactive web application (BEYOND/IAASARS/NOA, n.d.) was created (Fig. 18.1), bilingual (in Greek and English) and easy to use (friendly also for nonspecialized users), where all the data of the multiparameter flood analysis were collected.

More specifically, the following features were produced and are uploaded on the web application:

- Critical points

The critical point is colored red in cases where the stream cross-section was found to be inadequate for the particular flood and green when the cross-section was found to be adequate. At each critical point, a window containing a description of the situation opens, and, for each case of inadequate cross-section, appropriate mitigation measures are proposed to prevent future failures and disasters (Fig. 18.2). A total of 66 critical points were identified and examined during the on-site inspection, of which two characteristic ones are presented in Chapter 8.

- Locations

Of note, 52 characteristic locations known to the general public are highlighted (Fig. 18.3), to facilitate its geographical orientation in the interactive web application.

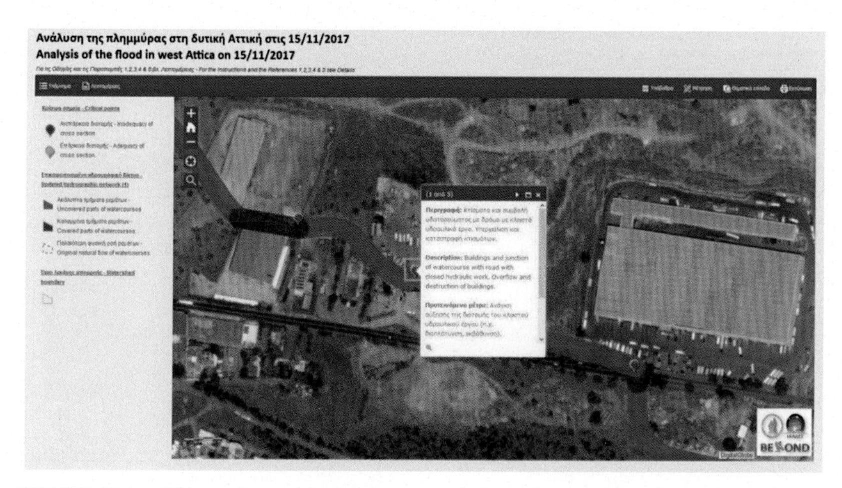

FIGURE 18.2 Critical point, accompanied by a description of the situation and proposed measures.

FIGURE 18.3 Characteristic location, known to the general public.

- Photographs

Around each critical point and other points of interest, photographs with coordinates and zoom function (Fig. 18.4), taken during the on-site inspection, are available to document and facilitate the analysis of the situation and the formulation of proposed measures. A total of 307 photographs were taken during the on-site inspection, of which two characteristic ones are presented in Chapter 8.

- River basin boundary (magenda)

The river basin boundary is depicted in magenda, and it was determined using the EU-DEM digital terrain model (25 m resolution) in a GIS environment.

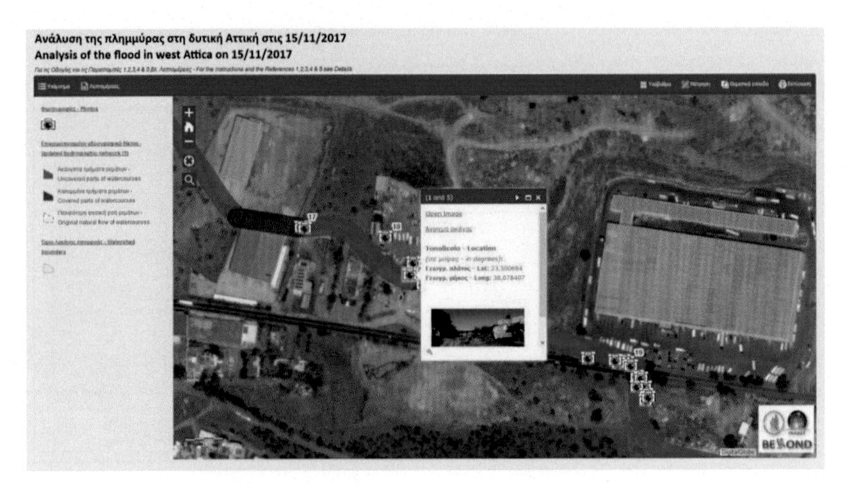

FIGURE 18.4 Photograph taken during the on-site inspection.

- Updated hydrographic network

The identification of the uncovered (blue) and covered (red) parts of the streams was done by field research during the on-site inspection in the area (November 21–23, 2017), while historical satellite images of Google Earth Pro (2002–2017), historical aerial photographs (1945–2007), as well as the hydrographic network as depicted in the conclusion of the Public Administration Inspectors and HYDROSCOPE were also taken into account for the assessment of the former natural riverbank of the streams (dotted blue). A more detailed description of the methodology for creating this information layer is given in Chapter 2.

- Mapped flood area

The mapping of the maximum extent of the flooding (magenda) was done using satellite remote sensing (processing of the WorldView-4 very high-resolution 0.31 m image of November 21, 2017), photo interpretation, and utilization of the data collected during the on-site inspection in the area (November 21–23, 2017) as well as additional data that were published. A more detailed description of the methodology for creating this information layer is given in Chapter 6.

- Simulated flood area

The simulation of the maximum flood extent (blue) was performed using the HEC-RAS software (version 5.0.1), simulating two-dimensional flow in the EU-DEM digital terrain model (25 m resolution), assuming a 6-hour rainfall with a return period of 1000 years, and taking into account land use (from CORINE 2012), urban expansion (from aerial photographs and WorldView-4 satellite imagery), and diachronic burnt areas (from the FireHub service of BEYOND/IAASARS/NOA) within the catchment. A more detailed description of the methodology for creating this information layer is given in Chapter 7.

- Urban expansion in the last 20 years (yellow)

Sources: Two sets of six aerial photographs, each with a spatial resolution of 1 m for the years 1997 and 2007, and a multispectral WorldView-4 satellite image with a very high resolution of 0.31 m in the panchromatic and 2 m in the multispectral data for the year 2017. A more detailed description of the methodology for creating this information layer is given in Chapter 3.

- Diachronic burnt areas in the last 30 years

Source: Diachronic Burnt Scar Mapping from the FireHub service of BEYOND/IAASARS/NOA. A more detailed description of the methodology for creating this information layer is given in Chapter 4.

- Updated land cover (Corine land cover)

A more detailed description of the methodology for creating this information layer is given in Chapter 5.

For the creation of this interactive web application, the "Map Tools" of the ArcGIS Online platform was chosen as a template. ArcGIS Online is a web-based GIS environment that enables the use, creation and sharing of maps, scenes, applications, spatial information layers, etc. It includes interactive maps that allow the general public to explore and understand the geographic data provided. Subsequently, the user can change the original appearance of the template using HyperText Markup Language 5 (HTML5), Cascading Style Sheet (CSS), and JavaScript markup languages. ArcGIS Online is an integral part of the ArcGIS system and thus it is possible to extend the capabilities of ArcGIS Desktop, ArcGIS Enterprise, ArcGIS Web APIs, and ArcGIS Runtime SDKs (Hall and Morgan, 2022).

18.3 Update of the hydrographic network

The web application depicts the updated hydrographic network in the areas of Mandra—Magoula—Eleusis (Fig. 18.5) and Nea Peramos (Fig. 18.6) as mapped today, with the uncovered and covered parts of the streams as they have been shaped after human interventions, and with a dotted line the former natural riverbank of the streams that has been disturbed and lost among the built-up areas.

The covered sections of the streams include all the closed sections of the streams (bridges and culverts), as well as the diversion work to Sarandapotamos after the confluence of the Soures and Mikro Katerini streams (downstream of Mandra and Magoula and upstream of Eleusis).

The identification of the uncovered and covered parts of the streams was done by field research during the on-site inspection in the area (November 21–23, 2017), whereas for the estimation of the former natural riverbank of

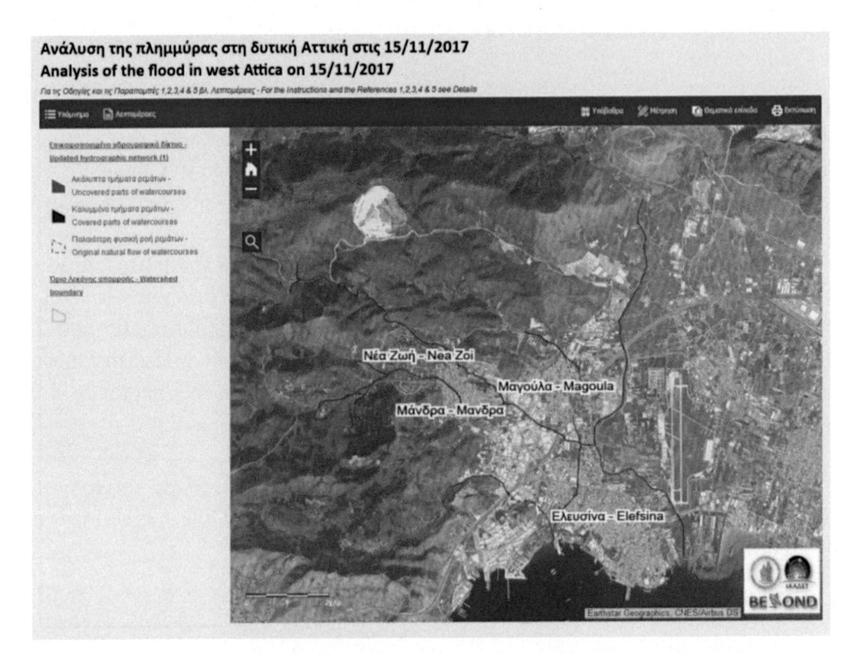

FIGURE 18.5 Updated hydrographic network in the areas of Mandra—Magoula—Eleusis.

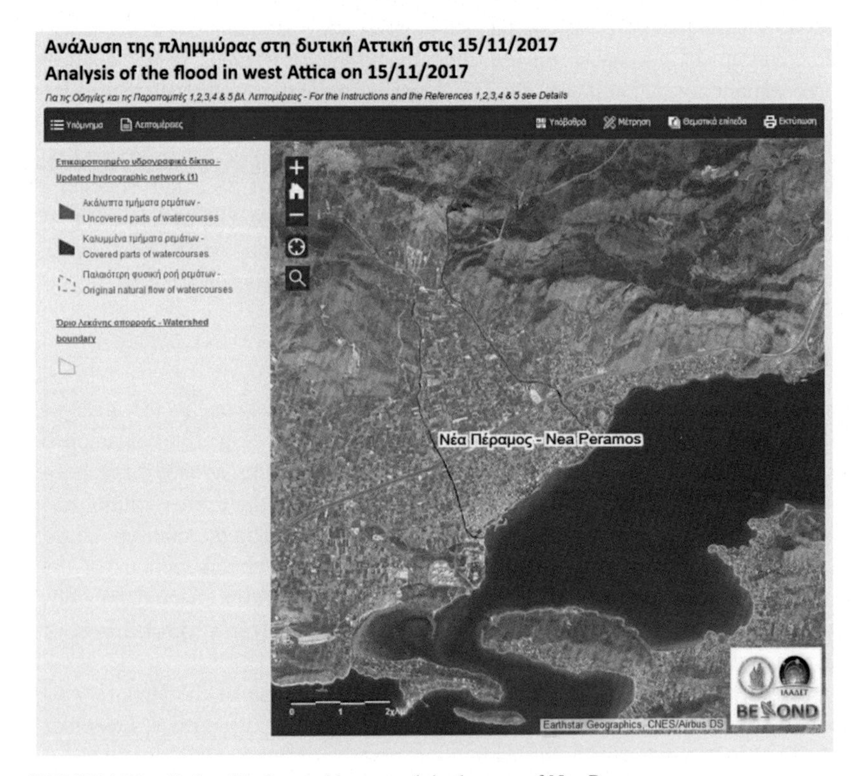

FIGURE 18.6 Updated hydrographic network in the area of Nea Peramos.

FIGURE 18.7 A series of historical aerial photographs of the southern part of the urban area of Mandra.

the streams, historical satellite images of Google Earth Pro (2002–2017), historical aerial photographs (1945–2007), as well as the hydrographic network as depicted in the Conclusion of the Inspectors of Public Administration (2017) and HYDROSCOPE (n.d.) were also taken into account.

FIGURE 18.8 A series of historical aerial photographs of the industrial area north of the urban area of Mandra around the current construction site of the Municipality of Mandra.

The series of historical aerial photographs used for the Agia Aikaterini stream covers the southern part of the urban area of Mandra and covers the years 1970, 1982, 1988, and 1995 (Fig. 18.7). It clearly shows the former natural riverbank that was parallel to Koropouli Street (in detail the junction with Philippou Street where there was a bridge) and the whole evolution until the current full coverage of the stream.

The series of historical aerial photographs used for the Soures stream covers the industrial area north of the urban area of Mandra around the current construction site of the Municipality of Mandra and covers the years 1945, 1970, 1982, 1988, 1988, 1995, and 2007 (Fig. 18.8). The former natural riverbank is clearly visible, and the whole evolution is up to the current situation where the building of municipal and private industrial facilities obstructs the flow of the stream.

Overall, the current situation is as follows:

- On the one hand, there is a series of arbitrary human interventions within the riverbank, inadequacy of existing technical works (either due to construction or due to lack of cleaning/maintenance) or complete absence of flood protection measures and road drainage in some areas, factors that have exacerbated the disaster. Indicatively, the following typical examples are highlighted: (1) the urban area of Mandra is built in the natural riverbank of the Agia Aikaterini/Katsimidi stream without any riverbank regulation or any flood protection works in this area (e.g., diversion) and with the underground drain of Koropouli Street blocked, (2) the natural riverbank of the Soures stream is obstructed in various places by private (e.g., D. Vakontios SA) and municipal facilities (construction site of the Municipality of Mandra), (3) there are paved roads in the west and north of the urban area of Mandra that cross the streams without any riverbank regulation or technical work (e.g., culvert, bridge), and (4) the natural Loutsa stream continues as a road under the junction of Attiki Street and Olympia Street, south of the industrial area of Mandra.

- On the other hand, there is a series of technical works that functioned adequately and averted further destruction. The following typical examples are highlighted: (1) the settlement of the Soures stream with a twin open channel of rectangular concrete cross-section, east of the Eleusis—Thebes National Road, in the industrial area, was generally adequate; (2) the technical works in the riverbank of the Mikro Aikaterini stream north and east of the urban area of Magoula were adequate, as well as its enclosure in a closed rectangular concrete culvert, upstream of the Attiki Odos; (3) the confluence of the regulated streams Soures and Mikro Aikaterini was sufficient (but marginal); and (4) the diversion of the regulated streams Soures and Mikro Aikaterini, after their confluence, through a closed concrete rectangular cross-section technical work and their discharge into Sarandapotamos was sufficient and absorbed the flooding of Eleusis.

18.4 Urban expansion

The urban expansions that occurred in the areas of Mandra—Magoula in the last 20 years (1997–2017) were identified using two sets of six aerial photographs, each with a spatial resolution of 1 m for the years 1997 and 2007, while for the year 2017 a multispectral WorldView-4 satellite image with a spatial resolution of 0.31 m in the panchromatic and 2 m in the multispectral data was used. The acquisition date of the WorldView-4 image is November 21, 2017, and the spectral information of three channels of the visible segment (R: 673 nm, G: 545 nm, B: 480 nm) and one of the near infrared (NIR: 850 nm) is provided (WorldView-4 satellite image, 2017).

Data preprocessing included the creation of two mosaics, each consisting of the aerial photographs corresponding to the years 1997 and 2007, and the definition of a common projection system for all three data types (WGS 84 UTM 34N). Subsequently, the mapping of the urban expansions was performed by photo-interpretation, and the generated result is presented in the web application, as shown in Fig. 18.9.

It is evident that urban expansion in the last 20 years has further obstructed the flow in the case of the Soures stream, north of the urban area

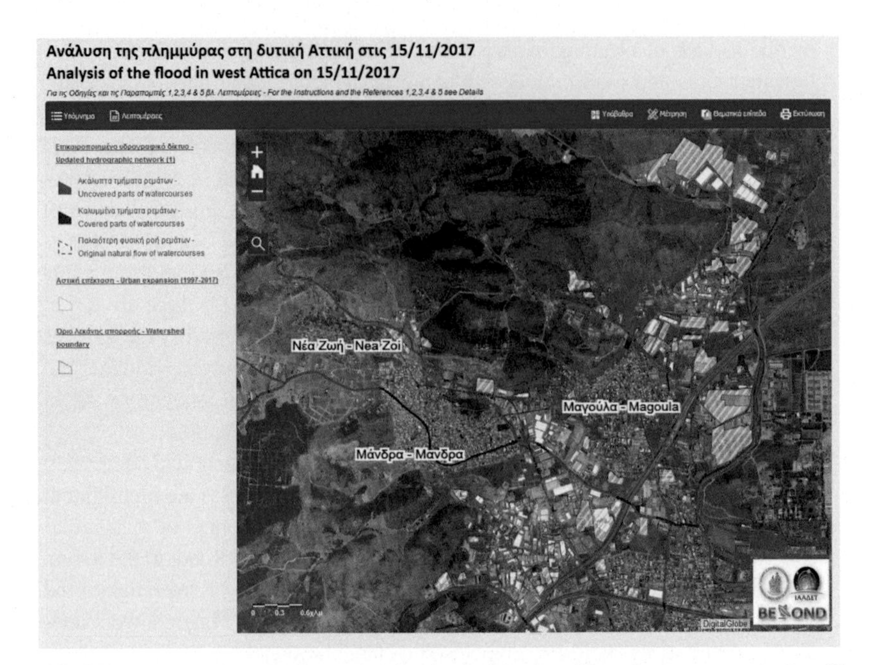

FIGURE 18.9 Mapping of urban expansion in the Mandra—Magoula areas over the last 20 years (marked with yellow polygons).

of Mandra (as confirmed by the historical aerial photographs in Fig. 18.8). However, the existing construction 20 years ago already obstructed the flow of the Agia Aikaterini/Katsimidi stream within the urban area of Mandra (as confirmed by the historical aerial photographs in Fig. 18.7).

18.5 Diachronic burnt areas

The diachronic burnt areas over the last 30 years (1986–2017) for the area of interest were obtained from the Diachronic Burnt Scar Mapping from the FireHub service of BEYOND/IAASARS/NOA (n.d.) and are depicted in the web application with striping and coloring for historicity, as shown in Fig. 18.10.

For each fire, as shown in Fig. 18.11, the following information is provided: location, year, before and after status, and area. To assess the condition, historical Google Earth Pro satellite imagery from 2002 to the present was photo-interpreted with very high spatial resolution (up to 20 cm).

The data of the major fires within the catchment and the percentage of the area of each fire in relation to the total area of the catchment (75.511368 km^2) are given in Table 18.1. The total percentage of burnt areas in relation to the catchment area is 34.56%. However, this is essentially due to the earlier fire on Mount Patera in 1986, with a percentage of 29.04%, an area which has now been reforested. Similarly, the next largest burnt area from

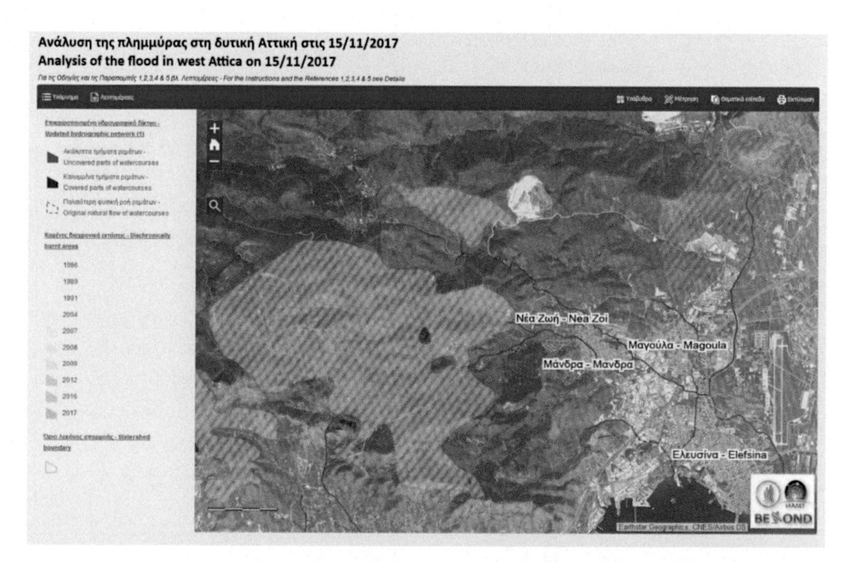

FIGURE 18.10 Mapping of the diachronic burnt areas of the last 30 years in the wider area.

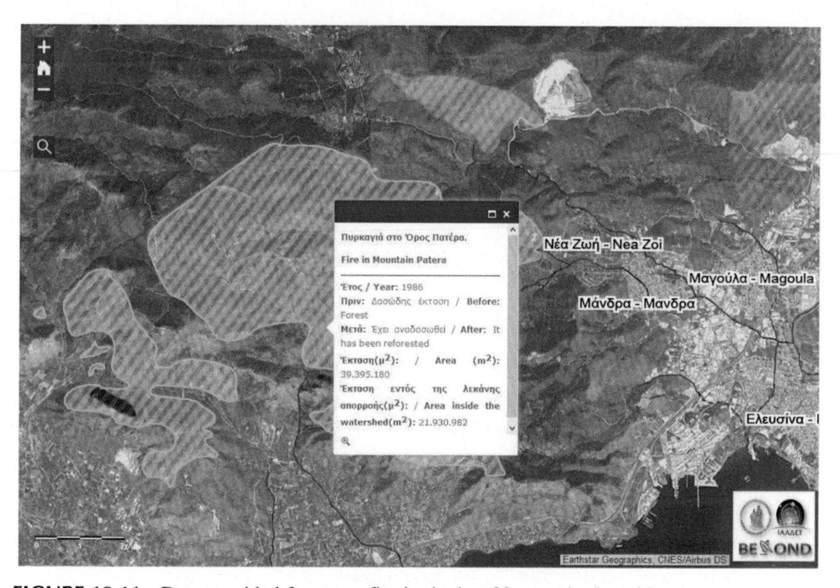

FIGURE 18.11 Data provided for every fire in the last 30 years in the wider area.

the fire east of the Paleokountura community has been reforested with a percentage of 3.50% in the year 1989. All other fires combined, from 1989 to the present, have burned only 2.02% of the watershed. Therefore the influence of the burned areas over time on flooding is considered to be small.

TABLE 18.1 Details of the major fires within the catchment area.

Fire location	Year	Status before	Status after	Area inside the catchment (m^2)	Percentage in relation to the catchment area (%)
Fire 3.30 km northeast of the community of Vlychada	2017	High altitude forests	It remains burnt land	14,922	0.02
Fire on Mount Pateras	2012	High altitude forests	It has become a bushy area	117,761	0.16
Fire near the Papastratos tobacco factory	2009	Shrubby vegetation	It has remained bare ground	616,677	0.82
Fire 1.83 km north of the Mandra tolls	2008	Low altitude forests	It has become a bushy area	634,409	0.84
Fire 1.80 km north of the Mandra tolls	2004	Low altitude forests	It has become a bushy area	126,323	0.17
Fire 1.88 km north of the Mandra tolls	1991	Low altitude forests	It has become a bushy area	13,510	0.02
Fire 2.00 km east of the community of Paleokoundoura	1989	Low altitude forests	It has been reforested	2,639,695	3.50
Fire on Mount Pateras	1986	Forest area	It has been reforested	21,930,982	29.04
Total				26,094,279	34.56

18.6 Land cover update

Taking into account the urban expansion and the current situation of the diachronic burnt areas, as discussed above, the Corine Land Cover database (Corine Land Cover 2012, version 18.5.1), provided by the European Environment Agency (n.d.), was updated. The updated land cover map of the wider Mandra—Magoula area is presented in Fig. 18.12.

18.7 Mapping of the maximum flood extent of November 15, 2017

The mapping of the maximum extent of the flooding (Fig. 18.13) was done using satellite remote sensing (processing of the WorldView-4 very high-resolution 0.31 m image of November 21, 2017), photo interpretation, and utilization of the data collected during the on-site inspection in the area (November 21–23, 2017), as well as additional data that were published.

First of all, it turns out that the largest part of the urban area of Mandra was flooded, basically the zone on either side of the former natural riverbank of the Agia Aikaterini/Katsimidis stream, primarily alongside Koropouli Street. A very large part of the industrial area on both sides of the Eleusis—Thebes National Road, from upstream of the Municipality of Mandra's construction site to downstream of Attiki Road, was also flooded. Finally, an area west of the urban area

FIGURE 18.12 Updated land cover map in the catchment area of the wider Mandra—Magoula area.

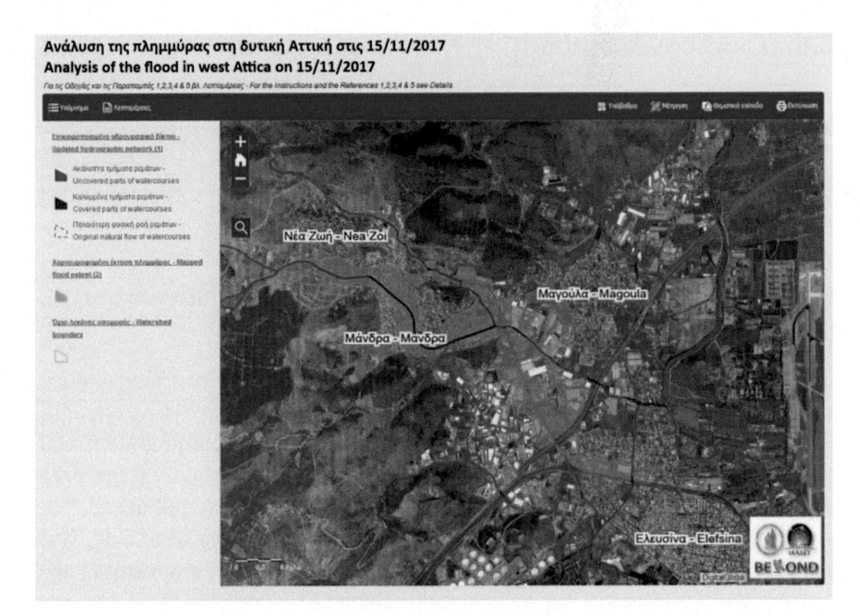

FIGURE 18.13 Mapping of the maximum extent of the flood in west Attica on November 15, 2017 using satellite remote sensing.

of Magoula was also flooded. In contrast, neither Magoula nor Eleusis was generally flooded. These results are to be expected given the updated situation of the hydrographic network, as analyzed above, and confirm that: (1) on the one hand, the unprecedented disaster was favored by arbitrary interventions in the riverbank, the inadequacy or even nonexistence of technical works in some areas; (2) on the other hand, the disaster would have been even greater if it had not been for a series of technical works that proved to be sufficient and did not fail.

The mapping of the maximum extent of the flooding was first conducted using satellite remote sensing. An indicative comparison of Planet satellite images in the area of interest before (November 09, 2017) and after (November 15, 2017) the disaster with the flooded areas visible is shown in Fig. 18.14.

The mapping of the flooded areas was mainly based on a multispectral image from the optical receiver of the WorldView-4 satellite, which was acquired on November 21, 2017 (Fig. 18.15). This is a high spatial resolution image with a resolution of 31 cm.

The depiction of flooded areas was based on a combined use of unsupervised image classification and photo-interpretation methods. These procedures were performed on appropriate combinations of spectral channels of the satellite image to achieve an optimal and more detailed mapping of the flooded areas.

For these purposes, a pseudochromatic image (Fig. 18.16) was used with a combination of spectral channels: near infrared (R)—red (G)—green (B). Areas of increased soil moisture are shown in olive green (detailed in Fig. 18.17).

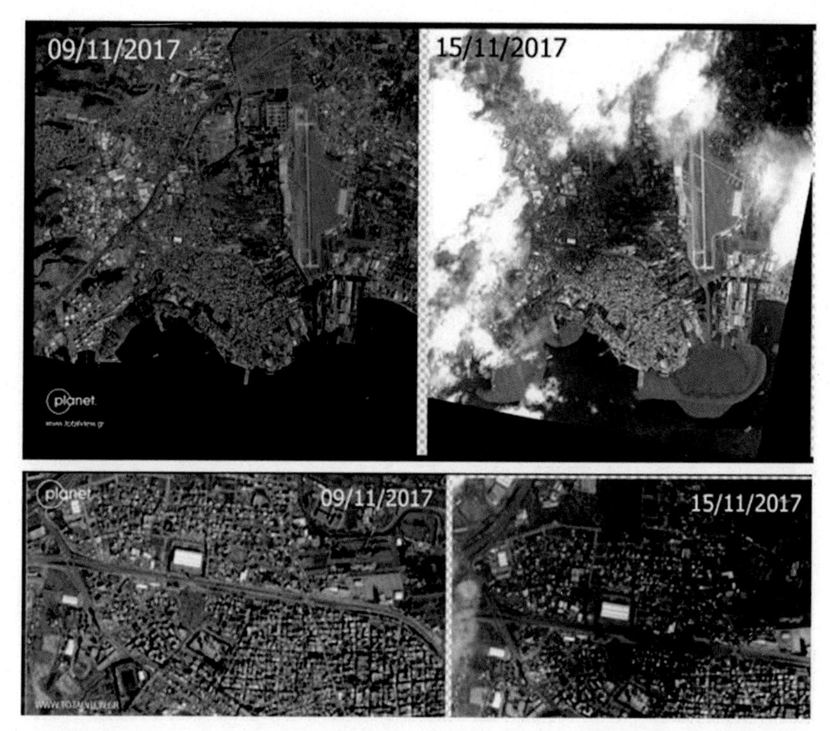

FIGURE 18.14 Indicative comparison of Planet satellite images in two areas of the wider area of interest before (November 09, 2017) and after (November 15, 2017) the flood.

FIGURE 18.15 The WorldView-4 satellite image used to map the extent of the flooded areas (date of acquisition November 21, 2017, resolution 31 cm).

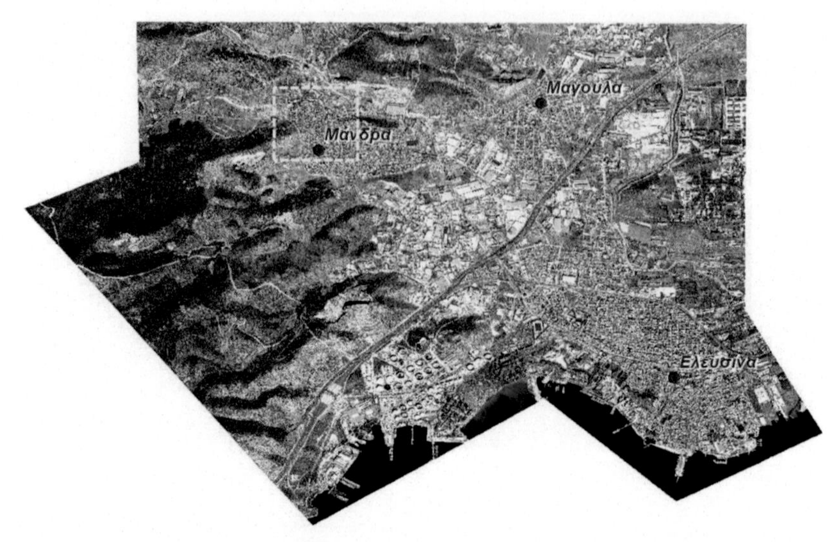

FIGURE 18.16 Pseudocolour satellite image (R:NIR, G:R, B:G) of the wider area of interest. The area in the yellow box is shown enlarged in the next figure.

FIGURE 18.17 Detail of the pseudochromatic satellite image (R:NIR, G:R, B:G) in the center of Mandra. Areas of increased soil moisture are shown in olive green.

The difference between the absolute brightness values of the red and green spectral channels was also exploited to highlight areas of water-saturated soils (Fig. 18.18). Areas of increased soil moisture are shown in very dark shades of gray to black (detailed in Fig. 18.19).

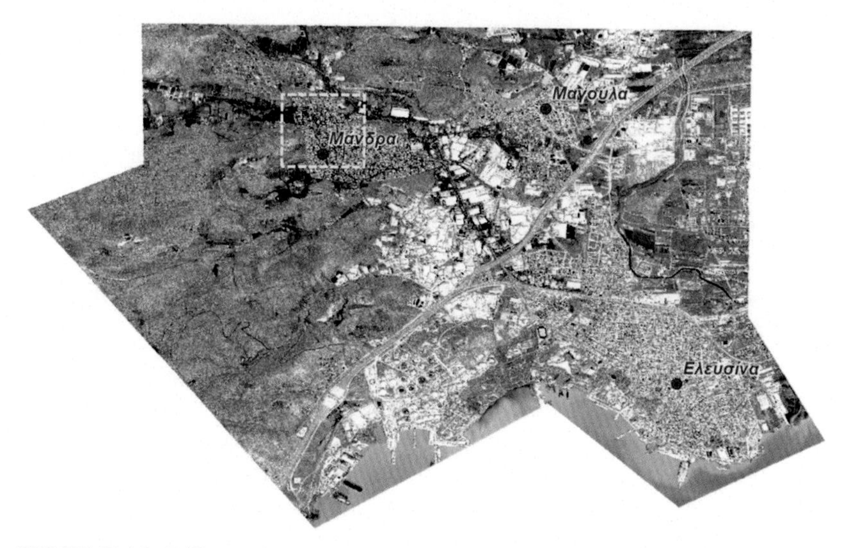

FIGURE 18.18 Difference in absolute brightness values of the red and green spectral channels of the satellite image. The area in the yellow box is magnified in the next figure.

FIGURE 18.19 Detail of the difference between the absolute brightness values of the red and green spectral channels of the satellite image in the center of Mandra. The areas of increased soil moisture are shown in very dark shades of gray to black.

The web application presents the final mapping of the maximum flood extent using satellite remote sensing (Fig. 18.13) as a hybrid result of satellite image classification with automatic unsupervised classification (with an algorithm that identifies water-saturated soils), corrections by photo-interpretation (in cases of confusion of the automatic process), and the use of data collected during the on-site inspection (November 21–23, 2017) and additional published data.

18.8 Simulation of the maximum flood extent of November 15, 2017

The simulation of the maximum flood extent (Fig. 18.20) was performed using the HEC-RAS software (version 5.0.1), simulating two-dimensional flow in the EU-DEM digital terrain model (25 m resolution), assuming a 6-hour rainfall with a return period of 1000 years, and taking into account land use, by updating the CORINE 2012 Database according to landscape changes, both due to urban expansion over the last 20 years (from historical aerial photographs and very high-resolution satellite data—WorldView-4 image), and due to the burnt areas over the last 30 years as provided by the FireHub Diachronic Inventory of Forest Fires of the Operational Unit BEYOND/IAASARS/NOA.

The result of this model seems to approximate well the result of the mapping using satellite remote sensing, and in addition it depicts flooded zones in

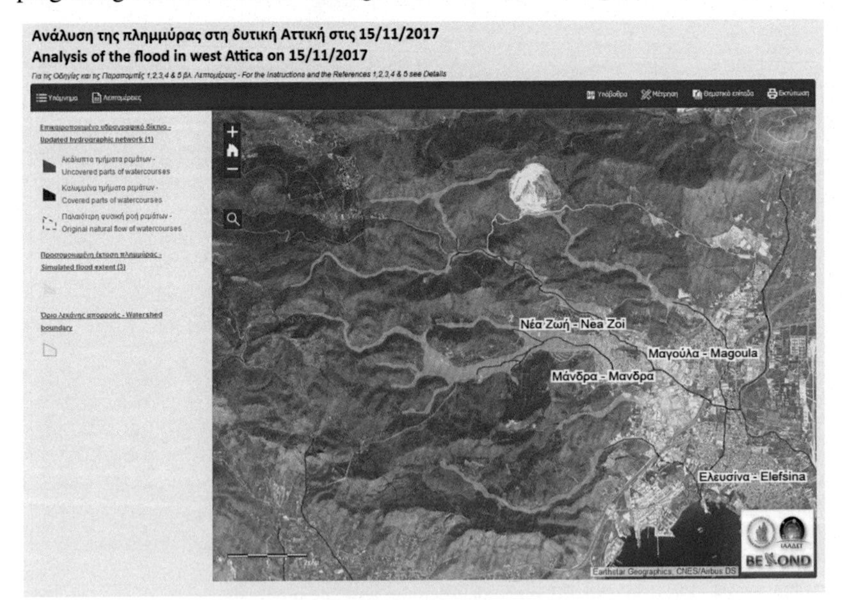

FIGURE 18.20 Simulation of the maximum flood extent in west Attica on November 15, 2017 using HEC-RAS software (up to Attiki Odos).

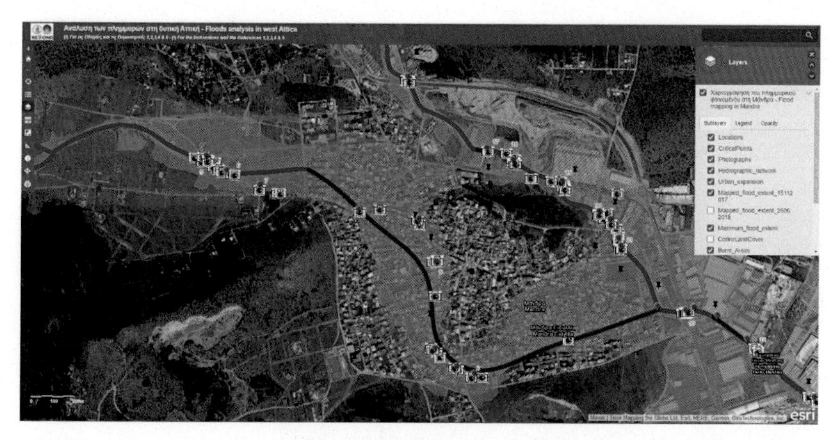

FIGURE 18.21 Detail in Mandra city: mapping of the flood on November 15, 2017 with high-resolution Sentinel-2 satellite data (magenda) and simulation of the maximum flood extent (blue).

the upstream riverbanks. The simulation is obviously influenced by the level of accuracy of the input data and subject to a number of assumptions; however, it provides a first picture of the maximum flood inundation that reasonably approximates reality. A detail in Mandra city is presented in Fig. 18.21.

18.8.1 Digital terrain model and geomorphology of the catchment area

To investigate geomorphologically the wider area of Mandra, which was affected by the flood on November 15, 2017, the necessary data was the Digital Elevation Model, which was selected from a set of available EU-DEM (n.d.) data covering Europe and produced within the Copernicus Programme with funding from the European Union. The Digital Elevation Model (DEM) used is of 25 m resolution and is a hybrid product based on SRTM and ASTER GDEM data.

The parameters that reflect the geomorphological profile of the area, and at the same time influence the processes during runoff, were calculated directly from the DEM. The steps of the geomorphometric analysis of the DEM were carried out in a GIS environment and include the correction of the model, through the identification and removal of topographic depressions (Fill Sinks), and the definition of the drainage network, using flow routing algorithms (Flow Direction and Flow Accumulation).

The catchment under study belongs to the Attica Water District (GR06), where the average annual rainfall is 411 mm and the average temperature ranges from 16°C to 18°C, according to data published by the Special Secretariat of Waters (SSW) (Attica Water District, n.d.). The basin is mostly developed in a lowland topography, while in the west the topography becomes

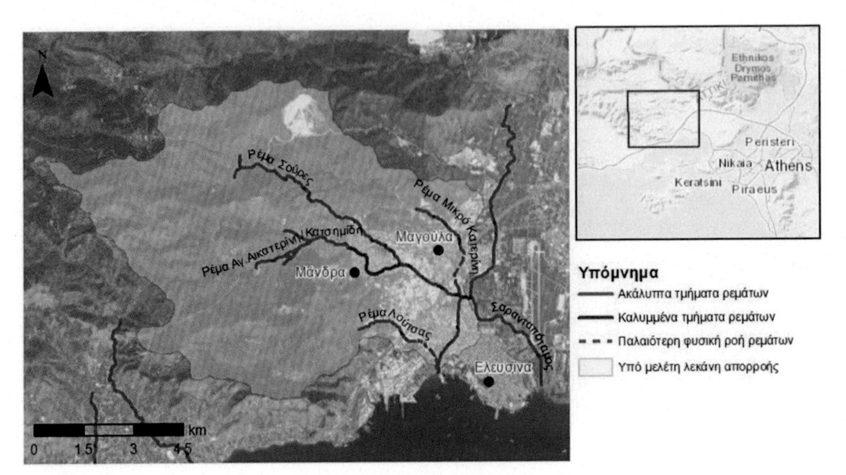

FIGURE 18.22 The catchment area (light blue) with the main streams (the uncovered parts in blue and the covered parts in red).

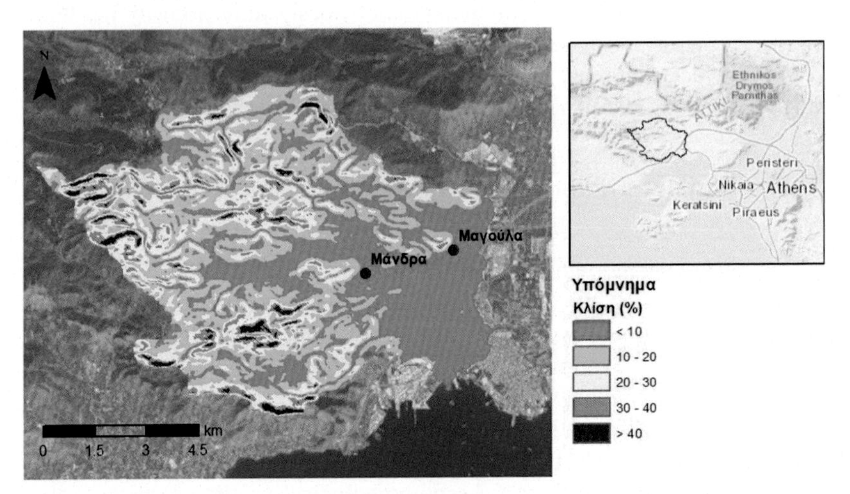

FIGURE 18.23 The slopes in the catchment area.

semimountainous–mountainous due to Mount Pateras, with a maximum elevation of 1132 m. Fig. 18.22 shows the catchment area, which constitutes the area of interest, with the main streams, which are the Ag. Aikaterini/Katsimidi stream, the Soures stream, the Mikro Katerini stream, and the Loutsa stream.

The delimitation of the catchment from its watershed, as derived from the EU-DEM, results in a basin area of 75 km^2, with an average and maximum elevation of 246 m and 813 m, respectively.

Another parameter that affects rainfall-runoff processes is the gradient. Most of the catchment is characterized by gentle slopes (Fig. 18.23), with an average slope of 15%.

A key temporal parameter associated with peak flow is the time of concentration, which is a typical input to flood models. Concentration time is defined as the time required for surface-flowing water to reach from the hydraulically furthest point in the basin to the outlet cross-section (Dingman, 1994, p. 397). From this definition, it follows that this time quantity is empirically related to the peak flow path of water within the basin. The maximum flow path in this basin covers a length of approximately 18 km and is depicted in a horizontal topography in Fig. 18.24 and in a cross-section in Fig. 18.25.

Due to the lack of data and the complexity of determining the concentration time, simple empirical relationships have been proposed to estimate it as a function of basic basin parameters (slope, area, elevation, etc.). The most

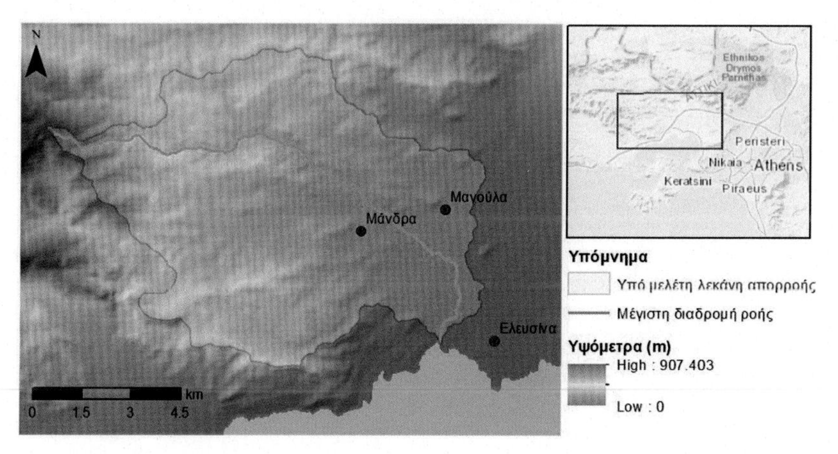

FIGURE 18.24 Horizontal topography of the maximum flow path (blue) in the catchment.

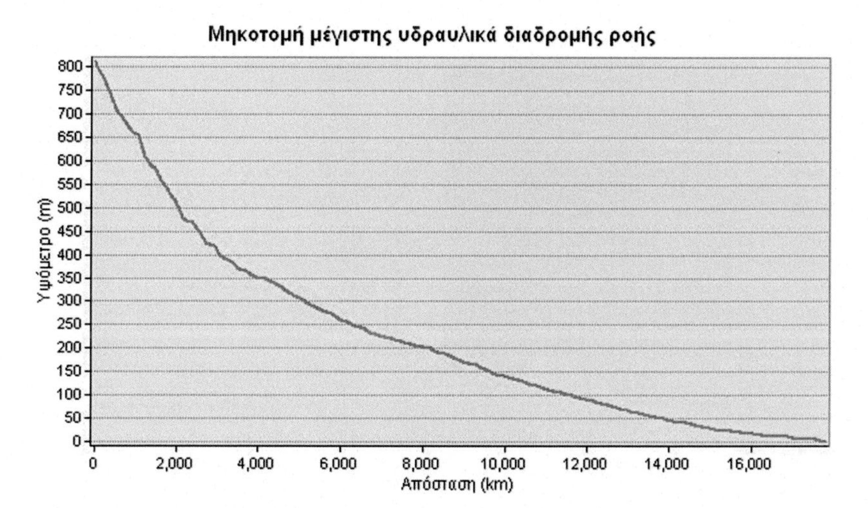

FIGURE 18.25 Cross-section of the maximum flow path in the catchment.

widespread relationship in Greek practice, which is recommended by the specifications of hydraulic works (Presidential Decree 696/1974), is the relationship of Giandotti (1934):

$$t_c = \frac{4\sqrt{A} + 1.5L}{0.8\sqrt{\Delta H}}.$$

where t_c is the concentration time (h), A is the basin area (km^2), L is the length of the main gorge (m), and ΔH is the difference between the average elevation of the basin and its outlet elevation (m).

Applying this relationship to the basin under study, the concentration time was estimated to be about 5 hours. However, this quantity, as an expression of the basin's temporal response to the rainfall event, decreases in higher intensity rainfall episodes. Within the DEUCALION project (n.d.) and following analyses of observed flood flows, it has been shown that the time decreases with increasing flow and thus the return period. This dependence is integrated with the Giandotti time reduction, according to the following empirical relationship (Efstratiadis et al., 2014):

$$t_c(T) = t_c\sqrt{i(5)/i(T)}.$$

where $i(5)$ is the critical rainfall intensity corresponding to a return period $T = 5$ years, and $i(\mathrm{T})$ is the rainfall intensity corresponding to the return period of the study.

Thus, for a return period $T = 1000$ years, the concentration time based on the above relationship is reduced to about 3 hours for the basin under study.

18.8.2 Soil, hydrolithology, and runoff in the catchment area

As can be seen from the SSW soil map for the Attica Water District (Fig. 18.26), the soils of the area are primarily of soil type A (sandy, clay-sandy, sandy-loamy,

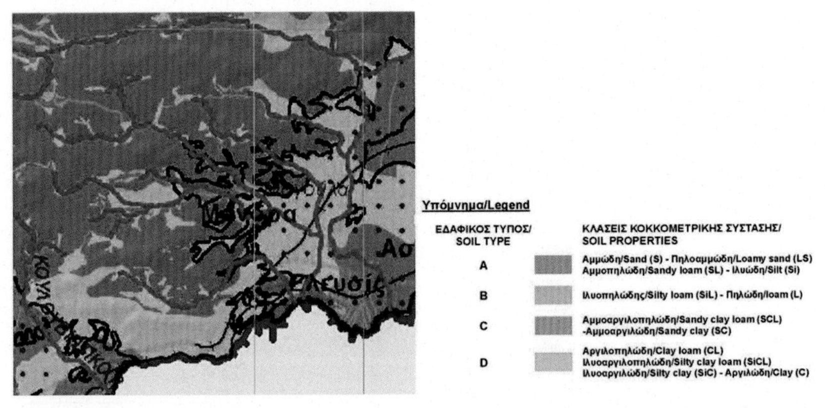

FIGURE 18.26 Soil map of the area of interest, from the corresponding flood risk management plan on behalf of SSW.

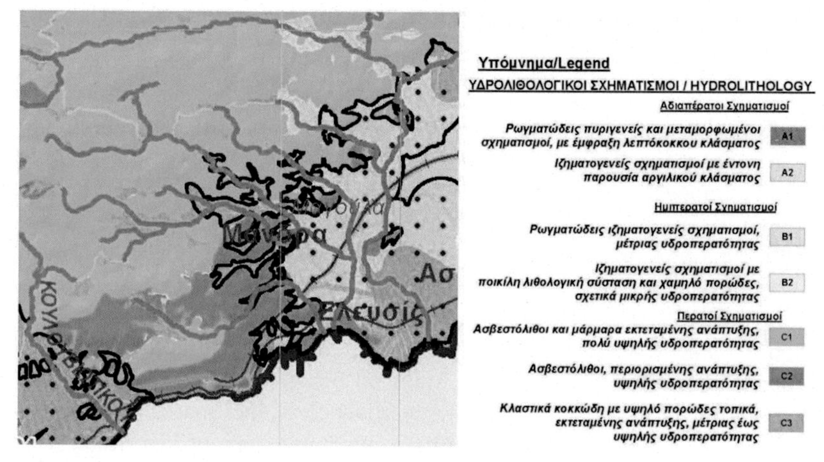

FIGURE 18.27 Hydro-lithological map of the area of interest, from the respective flood risk management plan on behalf of SSW.

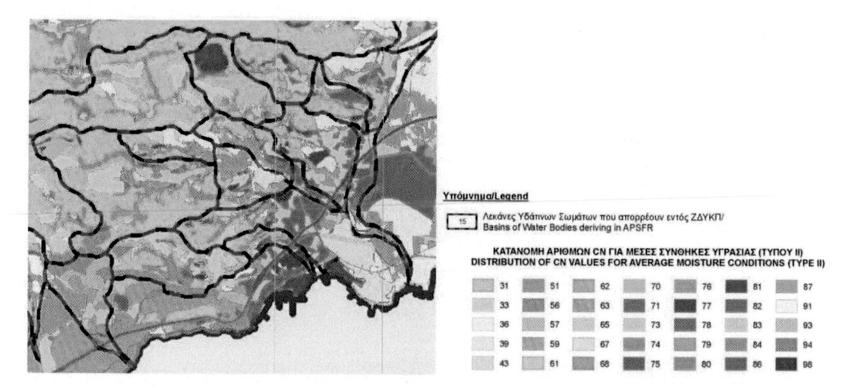

FIGURE 18.28 Map of distribution of CN numbers for average humidity conditions (Type II) in the area of interest, from the respective flood risk management plan on behalf of SSW.

and silty), secondarily B (silty-loamy), while areas of soil type C (sandy-loamy, sandy-clayey) also appear locally (Soil map for the Attica Water District, n.d.).

As can be seen from the hydro-lithological map of SSW for the Attica Water District (Fig. 18.27), the soils of the area consist mainly of permeable formations (limestone, marble, and classic granular), while semipermeable formations (sedimentary) also appear locally (Hydro-lithological map for the Attica Water District, n.d.). These soils have a medium (B1) to very high (C1) water permeability.

The distribution of the *CN* runoff curve numbers for average humidity conditions (Type II) in the area is given by the corresponding map of SSW for the Attica Water District (n.d.) (Fig. 18.28).

18.8.3 Available rainfall and flood data of November 15, 2017

Due to the lack of other precipitation data, the analysis of this particular rain event was based on the data of the press release of the National Observatory of Athens (NOA) on November 20, 2017 (Press release, 2017). According to the analysis of the XPOL meteorological radar data available at the Institute for Environmental Research and Sustainable Development (IERSD), which recorded the event with a spatial resolution of 150 m and a temporal resolution of 2 minutes, the accumulated rainfall at the core of the event exceeded 200 mm in a period of 6 hours (Fig. 18.29).

The area was also affected by floods in the past, with the most significant one of recent years on January 01, 1996, as published in the relevant Technical Report on Flood Hazard Maps on behalf of SSW (n.d.). As confirmed by the flood hazard maps, the confluence of the Soures stream with the Katsimidis stream in Mandra forms a vulnerable zone of high flood risk. SSW in the framework of the Flood Risk Management Plan of the River Basins of the Attica Water District has presented, among others, a Flood

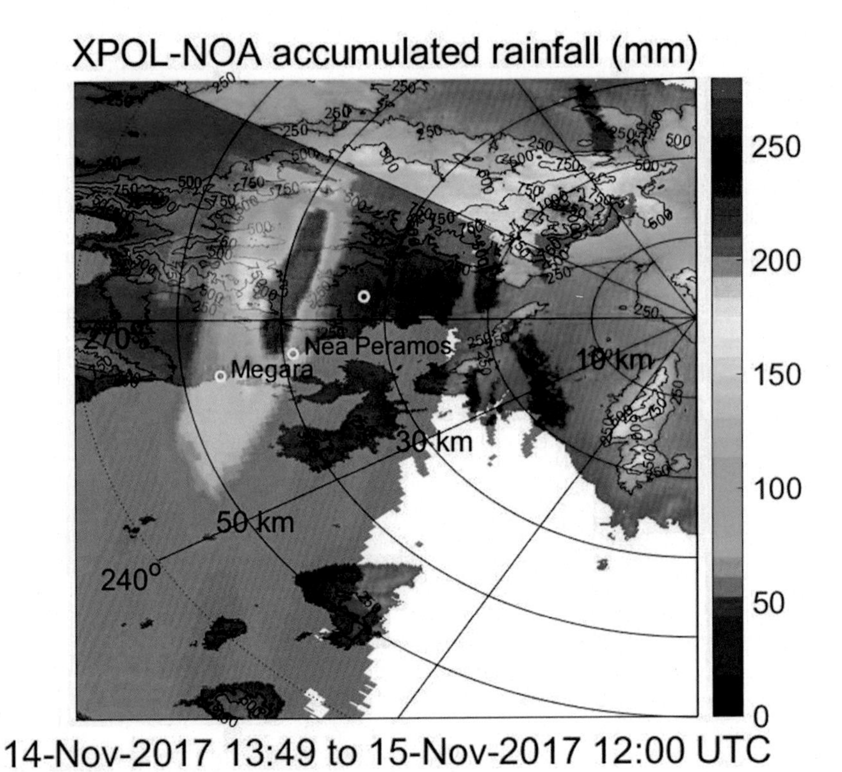

FIGURE 18.29 High-resolution spatial imaging of accumulated rainfall from the XPOL meteorological radar of IERSD-NOA.

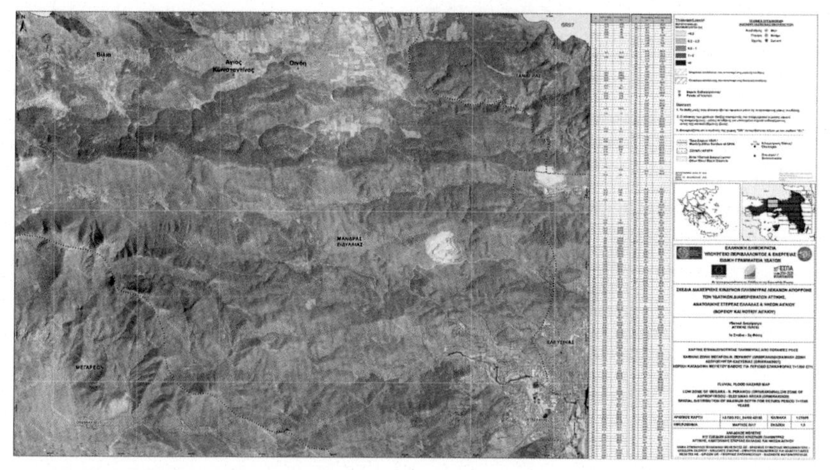

FIGURE 18.30 Flood hazard map from river flows in the area of interest with spatial distribution of maximum depth for a return period $T = 1000$ years, from the respective flood risk management plan on behalf of SSW.

Hazard Map for the adverse scenario of the return period $T = 1000$ years (Flood Hazard Map for the Attica Water District, n.d.), where the percentage of flooding in the city of Mandra reaches almost 80% (Fig. 18.30). This scenario is largely confirmed by the flood of November 15, 2017.

18.8.4 Hydrological and hydraulic simulation of the event of November 15, 2017

Hydrological and hydraulic simulation is a fairly complex process, which requires appropriate input data that will more realistically approximate reality. In this study, an attempt is firstly made to estimate the total amount of rainfall precipitated in the basin, assuming that this particular flood is a low probability exceedance event, with a return period of 1000 years and a rainfall duration of 6 hours (assumptions based on the data collected and presented in Sections 18.8.1 and 18.8.3). The general ombrian curve equation (Koutsoyiannis et al., 2010), which has the following form, is applied:

$$i(d,T) = \frac{\lambda'(T^{\kappa} - \psi')}{(1 + d/\theta)^{\eta}}.$$

where d is the duration of the rainfall, T is the return period, κ is the shape parameter, λ' is the scale parameter, ψ' is the position parameter of the distribution function, and θ, η are the parameters of the duration function.

The values of the parameters (κ, λ', ψ', θ, η) were taken from the Technical Report on the methodology of construction of ombrian curves at country level (n.d.), posted on the relevant website of SSW, and refer to the rain gauge station of Mandra. This station was chosen as it is the closest station to the basin of interest. The point rainfall intensity calculated for a return period of 1000 years and a rain duration of 6 hours, resulted in 32.65 mm/h corresponding to 196 mm accumulated rainfall, which is quite close to the rainfall magnitude estimate for this event.

Using the reduction factor φ, the point rainfall is reduced to a surface rainfall, taking into account the basin area and the duration of the rainfall. The surface reduction factor φ was determined from the following analytical expression (Koutsoyiannis and Xanthopoulos, 1999):

$$\varphi = max\left(1 - \frac{0.048A^{0.36-0.01\ln A}}{d^{0.35}}, 0.25\right).$$

where φ is the surface reduction factor (dimensionless), A is the area in km^2, and d is the duration of the rain in h.

After reduction by the reduction factor, the accumulated rainfall is 176 mm.

The Soil Conservation Service (SCS, 1972) method was used to calculate the hydrological deficits and thus determine the active (net) precipitation. The method uses two main parameters, the maximum potential retention S (mm) and the initial deficit h_{ao} (mm), which is expressed as a percentage of S ($h_{a0} = a \times S$). In the literature, it is recommended to apply a loss rate (retention and evaporation) of 20% ($a = 0.2$), which has been derived from observational data.

The net rainfall is estimated from the following empirical relationship:

$$h_e = \begin{cases} 0 & h \leq h_{a0} \\ \dfrac{(h - h_{a0})^2}{(h - h_{a0} + S)} & h > h_{a0} \end{cases}.$$

The application of the method requires the calculation of the runoff curve number (CN), which incorporates the physiographic characteristics of the basin into a single value and is related to the maximum potential retention S, by the following relationship:

$$S = 254(100/CN - 1).$$

The CN parameter takes values from 0 to 100 and is influenced by soil and land use conditions in the catchment, as well as antecedent soil moisture conditions.

Initially, SCS classifies soils into four groups, depending on their permeability:.

Group A: Soils with high infiltration rates, for example, sandy and gravelly soils with a very low percentage of silt and clay.

Group B: Soils with medium infiltration rates, for example, sandy loam.

Group C: Soils with low infiltration rates, for example, clay loam soils, soils with a significant percentage of clay, soils poor in organic matter.

Group D: Soils with very low infiltration rates, for example, soils that swell significantly when wetted, plastic clays, shallow soils with almost impermeable horizons near the surface.

It then defines three types of antecedent soil moisture conditions:.

Type I: Dry conditions, corresponding to less than 13 mm of rainfall in the previous 5 days (or 35 mm for a vegetated area under growing conditions).

Type II: Average conditions, corresponding to rainfall over the previous five days of between 13 and 38 mm (or 35–53 mm for an area with vegetation cover in growing conditions).

Type III: Wet conditions, corresponding to rainfall over the previous five days of more than 38 mm (or 53 mm for an area with vegetation cover in growing conditions).

In the analysis for the area of interest, initial type II soil moisture conditions were assumed.

The values of the *CN* runoff curve number by SCS for initial soil moisture conditions of type II are given by hydrological soil type and land use category (Koutsoyiannis, 2011, p. 126).

Land use information in the area of interest was retrieved from the updated land cover map (Corine Land Cover 2012), with an allowance for urban expansion and the current status of diachronic burnt areas, as discussed above in Chapter 5.

The hydrologic soil type information in the area of interest was retrieved from the soil and hydro-lithological map of SSW, as presented in Section 18.8.2.

The *CN* runoff curve number values in the area of interest were assigned based on the relevant SCS range and the value distribution map of SSW (Fig. 18.28).

To estimate a weighted average runoff curve number for the whole basin, a surface integration was performed with a weighting factor of the area of each sub-area of different land use categories, and the weighted value of *CN* was obtained equal to 50. The process of determining this value is influenced by the accuracy of the land use data and is a rough estimate, as it refers to the entire area of the basin under study, which is heterogeneous in its physiographic characteristics.

The temporal distribution of the net rainfall was chosen to be done according to the worst profile method (Koutsoyiannis, 1994), as recommended in the technical specifications (paragraph 3.4.2.) of the study "Flood Risk Management Plan for the River Basins of the Attica Water District," which was conducted for SSW, for floods of low probability of exceedance, that is, with return periods of 1000 and 10,000 years. This method is also

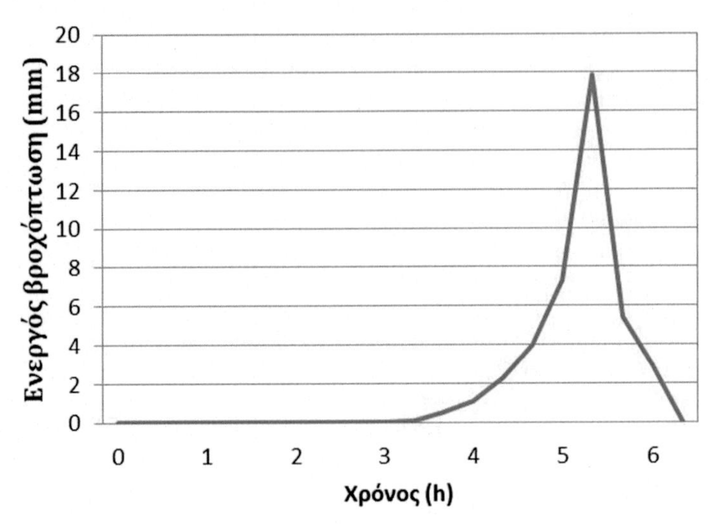

FIGURE 18.31 Hydrograph of active precipitation with a duration of 6 hours and a return period of 1000 years for the study catchment.

referred to in a relevant hydrological study (Hydrological Study of the Xiria Basin of Magnesia, n.d.).

The design hydrograph with a 6 hours rainfall duration and a return period of 1000 years is shown in the following Fig. 18.31 (active rainfall distribution versus time—the cumulative rainfall is 196 mm).

With the above rain scenario and the above assumptions ($d = 6$ hours, $T = 1000$ years, and $CN = 50$), using the EU-DEM digital elevation model (as mentioned in Section 18.8.1) and with Manning roughness coefficient values for the updated land cover (as described in Chapter 5), a simulation of two-dimensional flow on the soil was performed using the HEC-RAS software (version 5.0.1) (n.d.), under the assumption of nonpermanent flow and uniform precipitation in the catchment, in the absence of more accurate data.

The output of this model is obviously influenced by the level of accuracy of the input data (rainfall, digital elevation model, land cover, etc.) and is subject to a number of assumptions that need further investigation. However, it provides a first picture of the maximum flood inundation (Figs. 18.32 and 18.33), which seems to be a good approximation of the mapped flood extent as derived using satellite remote sensing, and in addition depicts flooded zones in the upstream riverbanks (Fig. 18.13).

A more accurate simulation of the flooding event would require the development of a hydraulic model based on the geometric data of the study area (riverbank layout, side embankments, cross-sections at characteristic points, determination of the Manning roughness coefficient along the riverbanks, bridges and other structures, failures, obstacles, debris, etc.) and

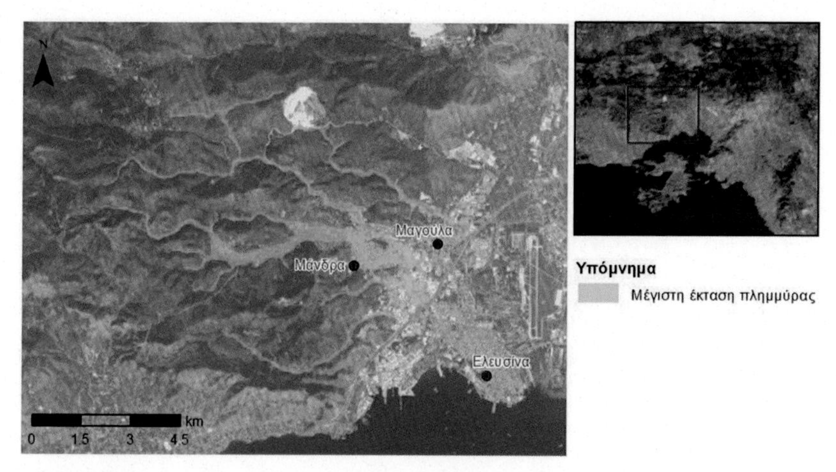

FIGURE 18.32 Maximum flood extent for the simulated scenario in the wider area of interest up to Attiki Odos, with satellite imagery as background.

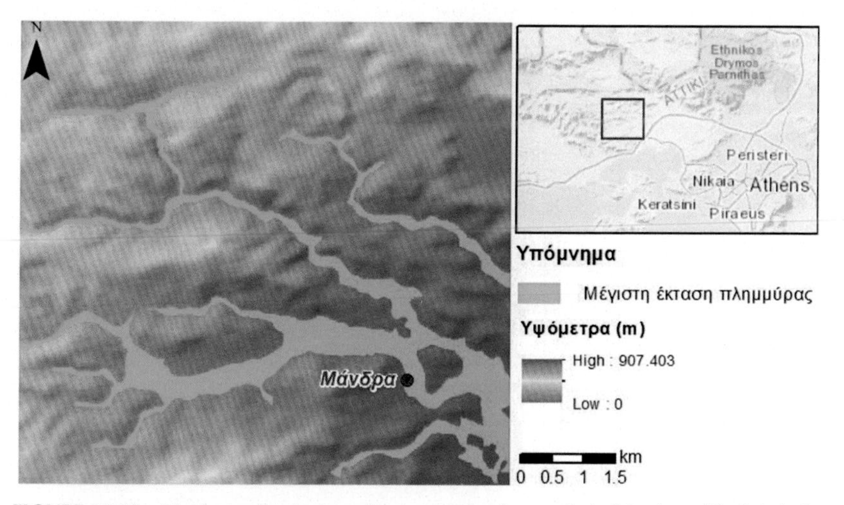

FIGURE 18.33 Maximum flood extent for the simulated scenario in Mandra, with digital elevation model as background.

hydrological data in order to determine the flood flows at subbasin level and to channel the flows to the drainage network.

18.9 Critical points and proposed measures

A number of critical points were identified during the on-site inspection in the area of interest, concerning both cases of inadequacy and cases of adequacy of the cross-section of the streams and engineering works, where they existed.

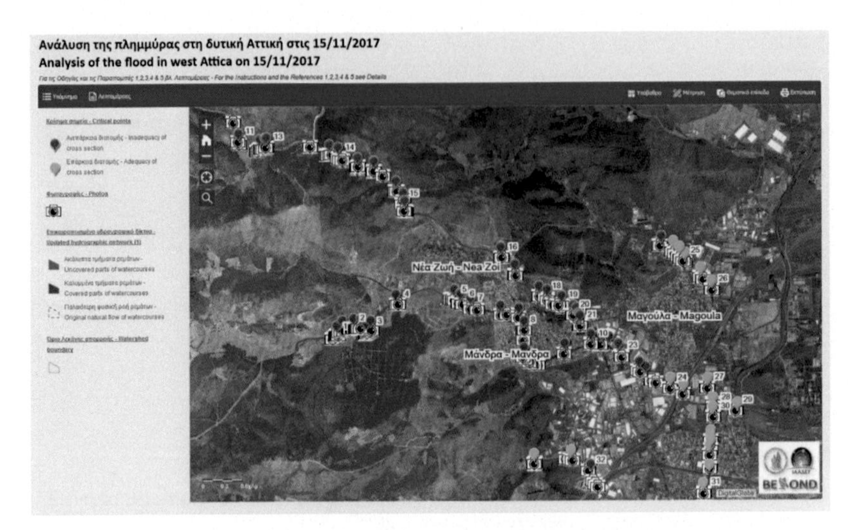

FIGURE 18.34 Critical points, accompanied by photographs and mitigation measures.

1  **Inadequacy**

Status description:
Confluence of a non-regulated watercourse with a road. Overflow and destruction of road and fencing.

Proposed measure:
Construction of an enclosed hydraulic structure (e.g. culvert) to drain the watercourse under the road.

Upstream of Mandra to the west, extension of Agia Aikaterini Street
Lat: 23,464959 / Long: 38,073667

24 **Adequacy**

Status description:
Confluence of a regulated watercourse with a road. Adequacy of cross-section.

Proposed measure:
Study on the adequacy of the cross-section of the hydraulic work, especially if the Agia Aikaterini stream is diverted into it.

Twin open channel upstream of Attiki Odos
Lat: 23,521632 / Long: 38,066755

FIGURE 18.35 Example of two critical points.

In cases of inadequacy, appropriate mitigation measures are proposed to prevent future failures and disasters.

The set of 66 critical points examined, 307 photographs taken, and the proposed measures formulated are available in the interactive web application (Fig. 18.34).

In this chapter, two typical critical points are presented as an example (Fig. 18.35). For each critical point, the situation is described, a mitigation measure is proposed (if a failure was found), and a photograph is given (with the location coordinates).

18.10 Discussion

With this multiparameter analysis of the FloodHub service/BEYOND/IAASARS/NOA, the areas affected by the flood are accurately derived from the detailed mapping of the maximum flood extent using satellite remote sensing (WorldView-4 very high-resolution 0.31 m image processing of November 21, 2017), photo interpretation, and utilization of data collected during the on-site inspection in the area (November 21–23, 2017), as well as additional published data. They are also validated by the simulation of the maximum flood extent performed using the HEC-RAS software (version 5.0.1), simulating two-dimensional flow in the EU-DEM digital elevation model (25 m resolution), assuming a 6-hour rainfall duration and a 1000-year return period, taking into account the land use, with an update of the CORINE 2012 Database according to landscape changes, both due to urban expansion over the last 20 years (from historical aerial photographs and very high-resolution satellite data—WorldView-4 image), and due to burnt areas over the last 30 years (from the diachronic burnt scar mapping of the FireHub service/BEYOND/IAASARS/NOA). The result of this model is a good approximation to the result of the mapping using satellite remote sensing.

Moreover, this analysis also highlights the critical factors that contributed to the huge disaster, which are: arbitrary human interventions within the riverbanks, the inadequacy of existing technical works (either due to construction or due to lack of cleaning/maintenance) or the complete absence of flood protection measures and road drainage in some areas, and partly landscape changes due to some small burnt areas upstream, and mainly due to urban expansion where building obstructs the flow of the streams. On the other hand, it is confirmed that the construction of a series of technical works, that have worked adequately, has mitigated further damage.

Furthermore, the analysis identifies a number of critical points in the area of interest, concerning both cases of inadequacy and cases of adequacy of the cross-section of the streams and engineering works, where they exist. In cases of inadequacy, appropriate mitigation measures are proposed to prevent future failures and disasters.

18.11 Conclusion

This FloodHub service of BEYOND/IAASARS/NOA can be activated:

- POST-disaster: for any flood event, in any river basin, so that:
- on the one hand, estimate the maximum flood extent, both by mapping using satellite remote sensing (very high resolution image processing and photo interpretation) and by simulation (using software, processing of available meteorological data, and some assumptions),
- and to carry out a more detailed study of the area of interest in order to identify the critical factors influencing flooding (e.g., human interventions, engineering works, landscape changes due to burnt areas and urban expansions), and to propose appropriate measures for mitigation and prevention of future failures and disasters.
- PRE-disaster: proactively, for different rainfall scenarios, in any given river basin, with updated data and a set of assumptions, to be used by the competent authorities as a tool for planning and preparing for flood risk management, especially in the areas most at risk.

Overall, this multidisciplinary approach, with the combined use of satellite remote sensing and specialised data analysis and event simulation models, is an extremely useful tool for the civil protection authorities and decision makers in support of their actions towards disaster resilience for the benefit of society as a whole.

18.12 Acknowledgments

Thanks to totalView (www.totalview.gr) for kindly providing us with the very high-resolution 31 cm WorldView-4 satellite image of the flood of West Attica on November 21, 2017.

References

Attica Water District, Special Secretariat for Water. Available at: http://floods.ypeka.gr/index.php/23-ydatika-diamerismata/gr06/240-gr06.

BEYOND/IAASARS/NOA. (n.d.). FireHub service. Available at: http://ocean.space.noa.gr/diachronic_bsm/.

BEYOND/IAASARS/NOA. (n.d.). FloodHub interactive web platform. Available at: https://www.arcgis.com/apps/View/index.html?appid = 35da5148b81a41eb84140314c854ad98.

Conclusion of the Inspectors of Public Administration. (2017). Available from https://www.efsyn.gr/arthro/porisma-fotia-gia-tis-fonikes-plimmyres-sti-mandra.

DEUCALION project. (n.d.). Available at: http://deucalionproject.itia.ntua.gr/.

Dingman, S. L. (1994). *Physical hydrology*. Englewood Cliffs, New Jersey: Prentice Hall.

Distribution of CN runoff curve numbers for average humidity conditions (Type II) for the Attica Water District, SSW. Available at: http://thyamis.itia.ntua.gr/egyfloods/gr06/gr06_maps_jpg_p04/GR06_P04_S1_CN.jpg.

Efstratiadis, I., Koukouvinos, A., Michaelidi, E., Galiouna, E., Tzouka, K., Kousis, A. D., Mamasis, N., & Koutsoyiannis, D. (2014). *Technical report describing regional relationships for the estimation of characteristic hydrological parameters.DEUCALION—Flood flow estimation in Greece under hydroclimatic variability: Development of a physically established conceptual-probabilistic framework and computational tools, September 2014* (p. 146). Contractors: ETME: Peppas & Co., Machairas Office, Water Resources and Environment Sector - National Technical University of Athens, National Observatory of Athens. Available at: https://www.itia.ntua.gr/el/getfile/1495/1/documents/Report_3_3.pdf.

EU-DEM. (n.d.). Available at: https://www.eea.europa.eu/data-and-maps/data/eu-dem#tab-metadata.

European Environment Agency. (n.d.). Corine Land Cover 2012, version 18.5.1. Available at: https://www.eea.europa.eu/data-and-maps/data/clc-2012-raster.

Flood Hazard Map for the Attica Water District for the return period T = 1000 years on behalf of SSW. Available at: http://thyamis.itia.ntua.gr/egyFloods/gr06/gr06_maps_jpg_p05/max-depth/GR06_P05_S3_MD_T1000_04400-42150.jpg.

Giandotti, M. (1934). *Previsione delle piene e delle magre dei corsi d'acqua, Vol. 8* (pp. 107—117). Istituto Poligrafico dello Stato.

Hall, K., & Morgan, D. (2022). *Using the map tools in ArcGIS Online*. Esri UK. Available at: https://storymaps.arcgis.com/stories/e7f016d60b304877a3a84eeacf253f1a. Accessed May 05, 2023.

HEC-RAS Software (version 5.0.1), U.S. Army Corps of Engineers. Available at: http://www.hec.usace.army.mil/software/hec-ras/downloads.aspx.

Hydro-lithological map for the Attica Water District, SSW. Available at: http://thyamis.itia.ntua.gr/egyfloods/gr06/gr06_maps_jpg_p01/GR06_P01_S5_hydrolithology.jpg.

Hydrological Study of the Xiria Basin of Magnesia. Available at: https://www.itia.ntua.gr/el/getfile/966/1/documents/Xhrias_flood_final5.pdf.

HYDROSCOPE. (n.d.). Available at: https://www.hydroscope.gr/.

Koutsoyiannis, D. (1994). *Worst profile method, U.S. Department of the Interior, 1977* (p. 817).

Koutsoyiannis, D. National Technical University of Athens. (2011). *Design of urban sewerage networks, Vol. 4* (p. 180).

Koutsoyiannis, D., & Xanthopoulos, Th.National Technical University of Athens. (1999). *Technical hydrology, Vol. 3* (p. 418).

Koutsoyiannis, D., Markonis, Y., Koukouvinos, A., Papalexiou, S. M., Mamassis, N., & Dimitriadis, P. (2010). *Hydrological study of severe rainfall in the Kephisos basin, Greece*. Available at: http://www.itia.ntua.gr/el/docinfo/970/.

Papagiannaki, K., Lagouvardos, K., & Kotroni, V. NOA-Penteli. (2021). *Flood incidents in Attica in the period 2000—2020*. Available at: https://www.meteo.gr/article_view.cfm?entryID = 1971. Accessed October 20, 2021.

Press release. (2017). NOA, 20/11/2017. Available at: http://www.noa.gr/index.php?option = com_content&view = article&id = 1074:deltio-typou-ethnikou-asteroskopeiou-athinon&catid = 86:news-eaa-greek&lang = el&Itemid = 428.

Soil Conservation Service (SCS). (1972). *National Engineering Handbook, Section 4, Hydrology*. Washington: U.S. Department of Agriculture.

Soil map for the Attica Water District, SSW. Available at: http://thyamis.itia.ntua.gr/egyfloods/gr06/gr06_maps_jpg_p01/GR06_P01_S6_soil.jpg.

Technical report on flood hazard maps for the Attica Water District on behalf of SSW. Available at: http://thyamis.itia.ntua.gr/egyFloods/gr06/report/%CE%99_3_P05_GR06.pdf.

Technical report on the methodology of construction of ombrian curves at country level, SSW. Available at: http://floods.ypeka.gr/index.php/methodologies-ergaleia/omvries-kampyles.

WorldView-4 satellite image of 21 November 2017. Available at: https://www.satimagingcorp.com/satellite-sensors/geoeye-2/.

Conclusions

Kleomenis Kalogeropoulos, Nikolaos Stathopoulos and Andreas Tsatsaris

The use, protection, and management of natural resources is a vital issue, and overtime research and design of best practices are becoming increasingly necessary and essential. Climate change, overpopulation, overexploitation, and other similar phenomena, which are evolving rapidly, are putting strong pressure on the quantitative and qualitative characteristics of groundwater systems, with ever-increasing environmental, economic, and social impacts.

The methods developed in this book have used satellite data of various types and characteristics. In addition to the ready-made background data from satellite programs (e.g., CORINE and ASTER DEM), primary satellite data, SENTINEL-1 and LANDSAT-8, were also utilized and analyzed through appropriate processing steps to produce the desired levels of information required for the development of each method for each chapter of the book. In addition, the development and application of the methods, through the analysis and processing techniques utilized, was mainly carried out in a Geographic Information System (GIS) environment, and commercial and free software was utilized.

RADAR satellite data, such as SENTINEL products, are suitable for studying phenomena such as flooding, as they are not affected by atmospheric conditions, unlike optical satellites (e.g., LANDSAT) which, for example, when there is cloud cover, are unable to produce usable products. The main negative aspect of RADAR satellite data, as shown in the application of the R-M-R methodological framework, is that the satellite acquisitions have a time interval of about 6 days, which makes it difficult to approximate flooding events, which are finite in time and usually quite short in duration. An optimal approximation of such a natural phenomenon would require satellite images of a frequency of hours if not less.

Passive remote sensing satellite data, such as those from LANDSAT-8, have the limitation of atmospheric conditions (e.g., cloud cover) on the quality of the data they receive. LANDSAT-8 is a new-generation optical satellite, whose products are being exploited with very good results in many environmental applications, for example, soil moisture estimation. Important for these applications is the large and satisfactory time series of products available from the LANDSAT program, which partly helps to overcome the problem of unsatisfactory image quality (due to weather conditions). Of course, as with SENTINEL, the frequency of reception (satellite pass from the same point) remains a major limitation that could be overcome with the development of satellite program technology in the coming years.

The satellite data, used by the authors of this book, are emphasized as being able to serve as stand-alone tools for the analysis and investigation of natural processes and hazards. The analysis and mapping of flooded areas from SENTINEL-1 data, for example, can in itself be an integrated tool with a variety of environmental, management, and similar applications.

The methods developed in this book are modern and effective tools for research and management of geo-environmental as well as major natural resources. One of the most important features of these methods, which enhances their potential, is their potential for evolution, which is based on the modern way these methods were developed and structured (modern software, data, techniques, etc.). Consequently, ever-increasing scientific knowledge and technological advances will therefore work beneficially on an ongoing basis in the evolution of the methods applied in this book.

In conclusion, with regard to the SENTINEL and LANDSAT satellite programs, the future challenge is the time densification of data acquisition and production, which requires the launching of new satellites and their integration into the existing constellations. At the same time, the aim should be to improve the spatial resolution of the downloads (and hence the products) of these satellites, which will require a new generation of sensors (transmitter and receiver) in the equipment of the satellites.

Summarizing the above, it is noted that in this book, new geo-environmental methods of research and management of natural processes—risks of soil erosion, flooding, and groundwater resources—have been developed, using modern technologies such as GIS, remote sensing, modeling, and many other new tools and analysis techniques. At the same time, the main effort in developing these methods was the low or no cost for their application, fast and usable results, simplicity in their application by researchers, their use for targeted field research, and, in general, their reliability as research, management, and decision-making tools. In addition, a wealth of free data of different types and characteristics were integrated, such as remote sensing data from passive and RADAR satellites and different spatially resolved DTMs (Digital Terrain Models), from open and global databases of geospatial, meteorological, satellite, and other data.

Index

Note: Page numbers followed by "*f*" and "*t*" refer to figures and tables, respectively.

D

E

F

H

I

J

K

L

M

Q

R

T

W

X

Z

Printed in the United States
by Baker & Taylor Publisher Services